31 Springer Series in Solid-State Sciences

Edited by Peter Fulde

Springer Series in Solid-State Sciences

Editors: M. Cardona P. Fulde H.-J. Queisser

40 **Semiconductor Physics** An Introduction
3rd Edition By K. Seeger

41 **The LMTO Method**
Muffin-Tin Orbitals and Electronic Structure
By H. L. Skriver

42 **Crystal Optics with Spatial Dispersion, and Excitons**
By V. M. Agranovich and V. L. Ginzburg

43 **Resonant Nonlinear Interactions of Light with Matter**
By V. S. Butylkin, A. E. Kaplan, Yu. G. Khronopulo, and E. I. Yakubovich

44 **Elastic Media with Microstructure II**
Three-Dimensional Models
By I. A. Kunin

45 **Electronic Properties of Doped Semiconductors**
By B. I. Shklovskii and A. L. Efros

46 **Topological Disorder in Condensed Matter**
Editors: F. Yonezawa and T. Ninomiya

47 **Statics and Dynamics of Nonlinear Systems**
Editors: G. Benedek, H. Bilz, and R. Zeyher

48 **Magnetic Phase Transitions**
Editors: M. Ausloos and R. J. Elliott

49 **Organic Molecular Aggregates,** Electronic Excitation and Interaction Processes
Editors: P. Reineker, H. Haken, and H. C. Wolf

50 **Multiple Diffraction of X-Rays in Crystals**
By Shih-Lin Chang

51 **Phonon Scattering in Condensed Matter**
Editors: W. Eisenmenger, K. Laßmann, and S. Döttinger

52 **Superconductivity in Magnetic and Exotic Materials**
Editors: T. Matsubara and A. Kotani

53 **Two-Dimensional Systems, Heterostructures, and Superlattices**
Editors: G. Bauer, F. Kuchar, and H. Heinrich

54 **Magnetic Excitations and Fluctuations**
Editors: S. Lovesey, U. Balucani, F. Borsa, and V. Tognetti

55 **The Theory of Magnetism II**
Thermodynamics and Statistical Mechanics
By D. C. Mattis

56 **Spin Fluctuations in Itinerant Electron Magnetism** By T. Moriya

57 **Polycrystalline Semiconductors,**
Physical Properties and Applications
Editor: G. Harbeke

58 **The Recursion Method and Its Applications**
Editors: D. Pettifor and D. Weaire

59 **Dynamical Processes and Ordering on Solid Surfaces**
Editors: A. Yoshimori and M. Tsukada

60 **Excitonic Processes in Solids**
By M. Ueta, H. Kanzaki, K. Kobayashi, Y. Toyozawa, and E. Hanamura

61 **Localization, Interaction, and Transport Phenomena**
Editors: B. Kramer, G. Bergmann, and Y. Bruynseraede

62 **Theory for the Valence of Fluctuating State**
Editors: T. Saso and T. Kasuya

63 **Electronic Properties of Polymers and Related Compounds**
Editors: H. Kuzmany, M. Mehring, and S. Roth

Volume 1–39 are listed on the back inside cover

R. Kubo M. Toda N. Hashitsume

Statistical Physics II

Nonequilibrium Statistical Mechanics

With 28 Figures

Springer-Verlag
Berlin Heidelberg New York Tokyo

Professor Dr. *Ryogo Kubo*
Professor Emeritus, University of Tokyo
Department of Physics, Faculty of Science and Technology, Keio University
3-14-1 Hiyoshi, Kohoku-ku, Yokohama 223, Japan

Professor Dr. *Morikazu Toda*
Professor Emeritus, Tokyo University of Education
5-29-108 Yoyogi, Shibuya-ku, Tokyo, Japan

Professor Dr. *Natsuki Hashitsume*
Department of Physics, Faculty of Science, Ochanomizu University
1-1, Otsuka 2 Chome Bunkyo-ku, Tokyo 112, Japan

Revised translation of the revised original Japanese edition:
Tokei Butsurigaku

Originally published by Iwanami Shoten, Publishers, Tokyo (1978).
English translation by Morikazu Toda, Ryogo Kubo, Nobuhiko Saitô
and Natsuki Hashitsume

ISBN 3-540-11461-0 Springer-Verlag Berlin Heidelberg New York Tokyo
ISBN 0-387-11461-0 Springer-Verlag New York Heidelberg Berlin Tokyo

Library of Congress Cataloging in Publication Data. Toda, Morikazu, 1917 – Statistical physics. (Springer series in solid-state sciences ; 30–) Rev. translation of: Tōkei-butsurigaku. Bibliography: v. 1, p. [239]–240. Includes index. Contents: 1. Equilibrium statistical mechanics – 2. Nonequilibrium statistical mechanics / R. Kubo, M. Toda, N. Hashitsume. 1. Statistical physics. I. Kubo, Ryōgo, 1920– . II. Saitō, N. (Nobuhiko), 1919– . III. Title. IV. Series. QC174.8.T613 1983 530.1′3 82-5865

Printed in Germany

Typesetting, printing and bookbinding: Konrad Triltsch, Graphischer Betrieb, Würzburg
2153/3130-543210

Foreword to Statistical Physics I and II

The search to discover the ultimate structure of matter, from molecules to atoms, from atoms to electrons and nuclei, to nucleons, elementary particles, and now to quarks, has formed the mainstream of modern physics. The ultimate structure is still buried in darkness, but the efforts of mankind from the beginning of this century in probing the fundamentals of the physical world have been extremely successful in revealing the grandeur of order in nature. The glory of this success may hide another great endeavor of modern physics which is equally important in discovering another side of the order in nature. This other endeavor is directed at the synthesis of analyzed elements into organized systems which are encountered in our more common experience. Analysis and synthesis are the two sides of the evolution of science. They are very often so interconnected that they cannot be separated in any simple way. Briefly characterized, statistical physics is the methodology of synthesis. It is the subject of this treatment.

The construction of macroscopic system from microscopic elements on the basis of analyzed elements at the microscopic level is not limited only to physics. Macrosystems are synthesized from microscopic structure and dynamics in biology, the social sciences, psychology and other sciences as well. This activity of synthesizing is undoubtedly one of the most powerful methods of science. However, we may say that it is best organized and best developed in physics. This is, of course, because the objects studied in *physics are simpler and more concrete* than those in other sciences, and theories can be more easily tested by means of experiments. The synthesis of a macroscopic system from microscopic elements is not simply a collection of fragments. The macroscopic system has a unity characteristically different from that existing at the microscopic level. The most typical example is perhaps the second law of thermodynamics. Despite the reversibility of the microscopic law of dynamics, macroscopic phenomena are indeed irreversible, and entropy always increases.

As is well known, the law of increasing entropy is interpreted in terms of probability. To describe a macroscopic system consisting of an enormous number of microscopic elements, the extremely complex motion of the microscopic elements has to be projected, so to speak, onto a much smaller number of macroscopic variables. This projection is necessarily statistical. In this sense, it is statistical physics that synthesizes the microscopic world to the macroscopic world. Statistical physics cover a very large area, from statistical thermodynamics (statistical mechanics in the classical sense which constructs thermodynamics on the basis of statistical physics) to the generalized statistical mechanics of irreversible processes and the kinetic theories which inherit the tradition of the classical kinetic theory of gases. The great breadth of these subjects makes it impossible to treat all subjects in statistical physics within the limited framework of this treatment. Fortunately, two volumes of the *Iwanami Series in Fundamental Physics* are also included in this series[1] and are devoted to the physics of condensed matter with various applications of statistical physics. Therefore the emphasis in this book can be placed upon methodological aspects rather than upon specific applications.

The year 1972, during which the first Japanese edition of this book was prepared, was the hundredth anniversary of the Boltzmann equation first proposed by L. Boltzmann. This equation determines the evolution of the velocity distribution function of molecules in a dilute gas. The stationary solution of this equation gives the Maxwell–Boltzmann distribution law on which the statistical thermodynamics of dilute gases was founded. An even more important aspect of this equation is that it provided a method for calculating the properties of dilute gases in nonequilibrium states. The Boltzmann equation was the prototype of the general kinetic method which treats the temporal evolution of the distribution functions of microscopic elements.

Although the kinetic method is very useful and powerful, generalizing it to apply to denser systems is very difficult. It can hardly be regarded as the general basis of statistical thermodynamics. Here again, Boltzmann made a great contribution and in so doing created statistical mechanics. He recognized that the assumption of equal weights of microscopic states is sufficient to build a general scheme for the statistical mechanics of equilibrium states, namely statistical thermodynamics. The inscription

$$S = k \log W$$

on Boltzmann's gravestone in the central cemetery of Vienna is the essence by means of which Max Planck summarized Boltzmann's somewhat obscure statements. This simple equation gave birth to statistical mechanics. The whole structure was beautifully reconstructed by W. Gibbs some years later. Although there was a difference in viewpoints between Boltzmann and Gibbs, this is no longer so important.

The true mechanics of the microscopic world is quantum mechanics. Thus, statistical mechanics based on classical mechanics was doomed to show inconsistencies when applied to real physical problems. The most well-known case of this was the problem of blackbody radiation which led Planck

1 S. Nakajima, Y. Toyozawa, R. Abe: *The Physics of Elementary Excitations.* T. Matsubara (ed.): *The Structure and Properties of Matter,* Springer Ser. Solid-State Sci., Vols. 12 and 28 (Springer, Berlin, Heidelberg 1980 and 1982)

to the discovery of energy quanta. Moreover, the quantal structure of nature is reflected in the very existence of thermodynamics. For example, the Gibbs paradox for the extensiveness of entropy cannot be resolved by the classical picture. If nature had a different structure, macroscopic thermodynamics would have been totally different from what we know in the physical world. The logical structure of statistical mechanics, particularly that constructed by Gibbs, received the new mechanics as if it had been anticipated. The logic bridging the microscopic and macroscopic worlds does not much depend upon the mechanics governing the former. Here we see the general character of the methods of statistical physics.

At least as far as thermodynamic properties are concerned, quantum-statistical mechanics is the most general scheme for elucidating the properties of a macroscopic system on the basis of its microscopic structure. Developments since 1930 in the modern physics of condensed matter have been theoretically supported by quantum mechanics and quantum-statistical mechanics. Some of the basic problems will be treated in this book, but most of the physical problems had to be left to the two volumes of this series mentioned previously. It should be kept in mind that there is no distinct boundary between statistical physics and condensed matter physics. Indeed, the progress of the former was made through efforts in the latter. Theoretical methods were developed in the treatment of real physical problems. It is only in the last decades that statistical physics has grown to quantum-statistical physics, which includes some general aspects of nonequilibrium theories. In these years this progress was truly remarkable and was made hand in hand with developments in solid-state physics and related fields.

This treatment includes such recent development in the foundations of statistical physics. It is not possible to cover the entirety of these subjects in these few pages. Our intention is to make this an elementary introduction to the subjects on the one hand, and to indicate to the reader the directions of future developments on the other. This may not have been achieved to the extent we wished, but we believe that this treatment has a uniqueness which justifies its addition to the already long list of textbooks on statistical physics.

The treatment is divided into two volumes. *Statistical Physics I*[2] is on equilibrium theories and *Statistical Physics II*[3] is on nonequilibrium theories. The first three chapters of Volume I form an introduction to statistical thermodynamics, specifically to the statistical mechanics of equilibrium states. The reader is expected to be acquainted only with elementary mechanics, such as the Hamiltonian equation of motion in classical mechanics and the concepts of quantum states in quantum mechanics.

2 M. Toda, R. Kubo, N. Saitô: *Statistical Physics I,* Equilibrium Statistical Mechanics, Springer Ser. Solid-State Sci., Vol. 30 (henceforth denoted by I)

3 R. Kubo, M. Toda, N. Hashitsume: *Statistical Physics II,* Nonequilibrium Statistical Mechanics, Springer Ser. Solid-State Sci., Vol. 31 (henceforth denoted by II)

Chapter 1 (I) discusses some elements of mechanics and treats some simple problems which use only the concept of the average. These do not depend upon the precise meaning of the average and are therefore very illuminating. Chapter 2 (I) reviews the skeleton structure of statistical mechanics. As mentioned earlier, statistical mechanics is based on a probabilistic assumption for microscopic states, namely the principle of equal weight. The logical problem of justifying this principle is discussed later in Chap. 5 (I). Here the pinciple is accepted as a postulate, and we will see how the whole structure of statistical thermodynamics is constructed on this basis. This is a standpoint commonly taken in textbooks, so that the construction in this chapter is not much different from that in other books. A newcomer to the field should study this chapter carefully.

Chapter 3 (I) is devoted to a few applications. Their number is limited because of limited space. These problems are basic and will be a good preparation for further study.

Chapter 4 (I) treats the problem of phase change within the limitations of statistical thermodynamics. The dynamic aspects are not treated and remain as future problems. Even within this limitation, the problem is the most difficult and fascinating one in statistical physics. The first successful theory of the so-called order–disorder problem was the Weiss theory of ferromagnets from which followed a number of approximate theories. Except for one- and two-dimensional models, no rigorous theory of phase transition exists. Rigorous treatments are discussed in this chapter for examples of lattice gases and low-dimensional Ising models. Approximations for three dimensions are discussed. Recently, more examples of rigorous solutions have been found and their intrinsic relations elucidated. These solutions are highly mathematical and are not treated here. While our treatment is also somewhat mathematical, a newcomer need not read this through in detail. The problem of singularities associated with a second-order phase transition has been a central topic of statistical mechanics in recent years and is related to many important aspects of statistical physics. This is briefly touched upon along with the problem of critical indices. The reader is referred to other textbooks on the scaling and renormalization group theories.

Chapter 5 (I) is devoted to a fundamental consideration of the mechanical basis of statistical mechanics, namely ergodic problems. We have limited ourselves primarily to classical ergodic problems based on classical mechanics. Quantum-mechanical ergodic theories are only briefly sketched. It is questionable whether classical ergodic theories can really be meaningful as a foundation of statistical mechanics; nevertheless, such theories have their own significance as a branch of physics and have made remarkable progress in recent years. Still, it is indeed a great pity for those engaged in research work in statistical physics that basic principles such as that of equal weight lack rigorous proof. The hope that someone among the readers may someday accomplish this ambitious task is one reason why this chapter was incorporated.

Nonequilibrium processes for which temporal evolution is explicitly to be considered are regarded as stochastic processes. This view is discussed in Chapters 1 and 2 (II). The theory of stochastic processes is an important field of mathematics. We emphasize its physical aspects and introduce the reader to the subject using Brownian motion as an example. Brownian motion is not merely random motion of a very fine particle; in general it is random motion of a physical quantity to be observed in a macrosystem. Such random motion is idealized to ideal Brownian motion in the same way that real gases are idealized to ideal gases in statistical thermodynamics. In this sense, the subject matter of Chap. 1 (II) is basic to the whole framework of statistical physics. In particular, the fluctuation–dissipation theorem is the heart of the theory that provides a stepping-stone to the treatments in Chaps. 3–5 (II).

To reach the macroscopic level of observation starting from the very fundamental microscopic level, we have to climb up successive levels of coarse graining. In going up each level of this staircase, a certain amount of information gets lost and corresponding uncertainty is added to the probabilistic description. This is precisely the fundamental theme of statistical physics. However, it is difficult to formulate the program of coarse graining in a general way. Therefore, in Chap. 2 (II) we treat a few relatively simple examples to show how this program is carried out, and finally we discuss the derivation of the master equation. Boltzmann's equation is somewhat removed from the main theme of this chapter, but we have included some basic matter on this subject. The Boltzmann equation is also important from a historical point of view, as well as from a conceptual and practical one. Difficult problems still remain in its derivation and generalization, but these are not touched upon here.

Chapters 2–5 (II) review the developments in nonequilibrium statistical mechanics which have occurred in the past few decades. Chapter 3 (II) is an introduction to these problems from their phenomenological side. It treats relaxation processes from nonequilibrium to equilibrium states and the response of a system near equilibrium to a weak external disturbance. These are linear irreversible processes belonging to a category of physics easily accessible as an extension of the well-founded statistical mechanics of equilibrium states. Theoretical methods for deriving from microscopic physics the relaxation and response functions for such linear processes are discussed in the linear-response theory in Chap. 4 (II).

Chapter 5 (II) treats new developments in quantum-statistical mechanics on the basis of Chap. 4 (II). These developments are applications of the Green's functions and their perturbative calculations. These are among the most remarkable developments of recent years. The Green's-function method can join the kinetic approach and equilibrium-statistical mechanics, though to a somewhat limited extent. The formalism takes advantage of the fact that microdynamics is quantum mechanics. The content of this chapter is rather condensed. For further study of the subject the reader is referred to a few well-known textbooks.

The number of pages far exceeded the total originally planned. In spite of this, there are many things which the authors regretfully omitted. We hope that the reader will feel inspired enough to go on and study the subjects at more advanced levels.

R. Kubo and *M. Toda*

Preface

This volume of *Statistical Physics* consititutes the second part of Statistical Physics (Springer Series in Solid-State Science, Vols. 30, 31) and is devoted to nonequilibrium theories of statistical mechanics. We start with an introduction to the stochastic treatment of Brownian motion and then proceed to general problems involved in deriving a physical process from an underlying more basic process. Relaxation from nonequilibrium to equilibrium states and the response of a system to an external disturbance form the central problems of nonequilibrium statistical mechanics. These problems are treated both phenomenologically and microscopically along the lines of recent developments.

Emphasis is placed on fundamental concepts and methods rather than on applications which are too numerous to be treated exhaustively within the limited space of this volume. For information on the general aim of this book, the reader is referred to the Foreword. For further reading, the reader should consult the bibliographies, although these are not meant to be exhaustive.

The original title of this book was Tokei-Butsurigaku (Statistical Physics) and it is part of the Iwanami Kiso-Butsurigaku Koza (Iwanami Series on Fundamental Physics). The first edition was published in 1973 and the second in 1978. The English edition has been divided into two volumes at the request of the publisher, the first treating equilibrium, the second nonequilibrium, statistical mechanics. Chapters 1 and 2 were written by R. Kubo and Chaps. 3, 4, and 5 by N. Hashitsume. In this edition, a considerable part of Chap. 4 was rewritten by the two authors and parts of other chapters were also revised.

It is a pleasure to express my gratitude to Professor P. Fulde for his detailed improvements in the manuscript, and to Dr. H. Lotsch of Springer-Verlag for his continued cooperation.

February 1985 *Ryogo Kubo*

Contents

1. Brownian Motion 1
1.1 Brownian Motion as a Stochastic Process 2
1.2 Central Limit Theorem and Brownian Motion 8
1.3 Langevin Equation and Harmonic Analysis 14
1.4 Gaussian Processes 22
1.5 Brownian Motion Modeled by a Gaussian Process 27
1.6 Fluctuation-Dissipation Theorem 31

2. Physical Processes as Stochastic Processes 40
2.1 Random Frequency Modulation 40
2.2 Brownian Motion Revisited 48
2.3 Markovian Processes 55
2.4 Fokker-Planck Equation 62
2.5 Contraction of Information. Projected Processes 69
2.6 Derivation of Master Equations 77
2.7 Brownian Motion of a Quantal System 84
2.8 Boltzmann Equation 90
2.9 Generalized Langevin Equation and the Damping Theory . . 97

3. Relaxation and Resonance Absorption 109
3.1 Linear Irreversible Processes 109
3.1.1 Mechanical and Thermal Forces vs Displacements and Currents 110
3.1.2 Linear Relations 110
3.1.3 Response to a Pulsed Force 112
3.1.4 Relaxation Phenomena 114
3.2 Complex Admittance 115
3.2.1 Harmonic (Fourier) Analysis 116
3.2.2 Energy Dissipation 117
3.3 Debye Relaxation 120
3.3.1 Dielectric Relaxation 120
3.3.2 Response Functions with Exponential Damping 121
3.3.3 Solution of Polar Molecules 123
3.4 Resonance Absorption 125
3.4.1 Van Vleck-Weisskopf-Fröhlich Type Resonance Absorption 126

3.4.2 Nuclear Magnetic Resonance 128
3.4.3 Failure at High Frequencies 131
3.5 Wave Number-Dependent Complex Admittance 131
3.5.1 Non-Markovian Nonlocal Linear Relations 132
3.5.2 Complex Admittance for the Diffusion Phenomenon . . 134
3.6 Dispersion Relations 136
3.6.1 Proof of the Dispersion Relations 136
3.6.2 Dispersion Relations and Causality 138
3.6.3 Analytical Continuation into the Complex Plane . . . 138
3.7 Sum Rules and Interpolation Formulas 139
3.7.1 Moment Sum Rules 140
3.7.2 Non-Markovian Law of Diffusion 143

4. Statistical Mechanics of Linear Response 146
4.1 Static Response to External Force 146
4.1.1 Static Admittance and the Canonical Correlation 147
4.2 Dynamic Response to External Force 150
4.2.1 The Response Function and the Poisson Bracket 150
4.2.2 Kubo Formula 154
4.2.3 Initial Values of the Response Function and Its Derivatives 158
4.3 Symmetry and the Dispersion Relations 158
4.3.1 Spectral Function and Its Symmetry 158
4.3.2 Symmetry in the Current Response 162
4.3.3 Symmetry in the Displacement Response 163
4.3.4 Proof of the Dispersion Relations 164
4.4 Fluctuation and Dissipation Theorem 167
4.4.1 Symmetrized Correlation 167
4.4.2 The Equivalence Between the Symmetrized Correlation Function and the Response or the Relaxation Function . 168
4.4.3 Fluctuation-Dissipation Theorem 170
4.5 Density Response, Conduction and Diffusion 171
4.5.1 Density and Current in Response to the External Field . 171
4.5.2 Relaxation of the Density Response and the Density Fluctuation 174
4.5.3 Shielding of the External Potential 176
4.5.4 Resistivity Formula 178
4.5.5 Dielectric Shielding and Electric Conductivity 179
4.5.6 Kramers-Kronig Relations and the Sum Rules 181
4.6 Response to Thermal Internal Forces 183
4.6.1 Onsager's Postulate 184
4.6.2 Fluctuation of Macrovariables as Brownian Motion . . . 185
4.6.3 A General Formulation of Onsager's Postulate 187
4.6.4 Nonequilibrium Density Matrix 190

4.7 Some Remarks on the Linear-Response Theory 196
4.7.1 The Kinetic Method versus the Linear-Response Theory 196
4.7.2 Van Kampen's Objection 198
4.7.3 Spurious Singularities at the Zero Value of the External Field 199
4.7.4 Singularities at $k = 0$, $\omega = 0$ 200

5. Quantum Field Theoretical Methods in Statistical Mechanics . . . 203
5.1 Double-Time Green's Functions 203
5.1.1 Retarded Green's Functions 204
5.1.2 Advanced Green's Functions 206
5.2 Chain of Equations of Motion and the Decoupling Approximation 207
5.2.1 Chain of Equations of Motion 208
5.2.2 Complex Dielectric Function of a Plasma in a Decoupling Approximation 210
5.3 Relation to the Kinetic Equation 216
5.3.1 Klimontovich Operator 216
5.3.2 Self-Consistent Field Approximation 217
5.3.3 Plasma Oscillation 220
5.4 Single-Particle Green's Function and the Causal Green's Function 222
5.4.1 Single-Particle Green's Functions 222
5.4.2 Single-Particle Green's Functions for Free Particles . . . 224
5.4.3 Causal Green's Functions 225
5.5 Basic Formula for Perturbational Expansion 226
5.5.1 Perturbational Expansion of the Equilibrium Density Matrix 227
5.5.2 Perturbational Expansion of the Thermodynamic Potential 228
5.6 Temperature Green's Function 229
5.6.1 Temperature Green's Functions (Matsubara-Green's Functions) 229
5.6.2 Fourier Analysis of the Temperature Green's Function . 230
5.6.3 Single-Particle Temperature Green's Function for Noninteracting Particles 232
5.6.4 Abrikosov-Gor'kov-Dzyaloshinskii-Fradkin Theorem . 233
5.7 Diagram Technique 235
5.7.1 Bloch-De Dominicis Theorem 235
5.7.2 Perturbational Expansion of $\langle \hat{\mathscr{S}}(\beta) \rangle_0$ 238
5.7.3 Correspondence with Feynman Diagrams 240
5.7.4 Matsubara Formula 242
5.8 Dyson Equation 244
5.8.1 Single-Particle Temperature Green's Function 244

5.8.2 Graphical Summation 246
5.8.3 Feynman Rules 248
5.9 Relationship Between the Thermodynamic Potential and the Temperature Green's Function 253
5.10 Special Case of Two-Particle Green's Function 255
5.10.1 Two-Particle Green's Function of Zeroth-Order for a Plasma 257
5.10.2 Polarization Operator 258
5.10.3 Electric Charge Density Green's Function 260

General Bibliography of Textbooks 265

References 266

Subject Index 271

1. Brownian Motion

In 1827 the botanist *Brown* discovered under his microscope vigorous irregular motion of small particles originating from pollen floating on water [1.1]. He also observed that very fine particles of minerals undergo similar incessant motion as if they were living objects. This discovery must have been a great wonder at that time. The idea of combining such a motion – Brownian motion – with molecular motion became fairly widespread in the latter half of the nineteenth century when atomism had not yet been fully recognized as reality. It was, however, the celebrated work of *Einstein*, which appeared in 1905, that gave the first clear theoretical explanation of such a phenomenon which could be directly verified quantitatively by experiments and thus established the very basic foundation of the atomic theory of matter [1.2]. *Einstein* did not know that Brownian motion had actually been observed many years before when he first came upon this idea to verify the reality of the atomic concept. At any rate, Einstein's theory had a great impact at that time, finally convincing people of the theory of heat as molecular motion, and so paved the way to modern physics of the twentieth century. It also greatly influenced pure mathematics, that is, the theory of stochastic processes.

The theory of stochastic processes, called the Wiener process, was initiated by N. Wiener as a mathematical model of Brownian motion. Some years later this was combined with Feynman's path integral formulation of quantum mechanics. R. P. Feynman did not know of Wiener's work when he devised this method independently. It is very instructive that such unconscious coincidences often arise at very decisive moments in the progress of science in seemingly far separated disciplines.

The theory of Brownian motion was further developed by P. Langevin, M. Smoluchowski, G. E. Uhlenbeck, L. S. Ornstein, and many others. The classical theory was excellently reviewed by *Wang* and *Uhlenbeck* [1.3]. The present treatment owes a great deal to this review, which still remains a standard reference. But our intention is to extend the theory a bit further and to base on it the developments of nonequilibrium statistical mechanics, treated in some detail in the following chapters. Grasping physical phenomena as stochastic processes is one of the very fundamental methods of this approach. Brownian motion is an excellent example of this.

Einstein proved that the diffusion constant D of a Brownian particle is related to its mobility μ by

$$D = \mu k T .$$

This relationship, which is called the *Einstein relation,* provides us with a very good basis of experimental verification that Brownian motion is in fact related to the thermal motion of molecules. It is very important to realize that this has an even deeper meaning. Namely, it is the first example of one of the most general theorems of statistical physics, called the *fluctuation-dissipation theorem* in recent literature [1.4]. This theorem is, as shown in Sect. 1.6, one of the most fundamental cornerstones supporting statistical mechanics of nonequilibrium states.

Fig. 1.1. Brownian motion caused by incessant impacts of gas molecules against a mirror suspended in a dilute gas [1.5]

It turns out that Brownian motion is not only limited to the motion of small particles, but is actually very common. For example, irregular oscillatory motion of a small mirror suspended in a dilute gas is caused by incessant impacts of gas molecules (Fig. 1.1). In electrical circuits, thermal motion of electrons in conductors gives rise to fluctuations of electric currents as well as potential differences between different components. When suitably amplified, such fluctuations can be heard by ear as, so-called, thermal noise. More generally, every physical quantity we observe is accompanied by similar fluctuations due to thermal motion of microscopic degrees of freedom in matter. In a great many cases, such fluctuations are small in comparison with the average values of the quantity under observation and can generally be ignored. However, such fluctuations reflect the microscopic motions in the system under study, so analyzing them provides very important keys for studying the system. In this chapter, the basic concepts of stochastic processes are discussed, taking mainly the Brownian motion of Brownian particles as the simplest example. Even though other examples are not mentioned explicitly, the reader should keep in mind that these concepts and methods are not confined to the simplest model but are general and applicable to a wide class of physical phenomena.

1.1 Brownian Motion as a Stochastic Process

Suppose that we observe a Brownian particle under a microscope over a time interval $0 \leqq t \leqq T$ and obtain a record of its position $x(t)$ as a function of time. For simplicity, we consider in the following only the projection onto the x axis and treat the problem as one-dimensional motion, but the essentials are the same for the three-dimensional motion. The observations are repeated in time to get N readings of the particle position

$$x_1(t), x_2(t), \ldots, x_N(t) . \tag{1.1.1}$$

These readings all differ, namely, the motion of the Brownian particle is not reproducible.

Then we ask, "what can physics predict about Brownian motion?" Obviously, unlike in mechanics, we are not able to make deterministic predictions: we must rather take a probabilistic outlook. The value of the displacement $x(t)$ of the Brownian particle at time t is probabilistic and each of the observed series $x_j(t)$ is a sample from a statistical ensemble. If we repeat the observations a great many times to make N very large, we should be able to find empirically the distribution law obeyed by $x(t)$.

The stochastic variable (random variable) is $\boldsymbol{x}(t)$[1]. This is a series of random variables having t as a parameter. Such a time series of random variables is generally called a *stochastic process*. If a continuous observation is made, a function $x(t)$ with a continuous parameter t is obtained as a sample of the process. If observations are made at discrete times

$$0 < t_1 < t_2 \ldots < t_n < T, \tag{1.1.2}$$

then a set of n real numbers

$$x(t_1), x(t_2), \ldots, x(t_n)$$

is a sample obtained by the observations. If we regard the set as a vector, then an n-dimensional real space R^n is the sample space of the process $\boldsymbol{x}(t)$ for the selected time points (1.1.2). An element of the sample space may also be represented by a zig-zag path (Fig. 1.2). One may consider the limit of very large n and vanishing lengths of time segments to attain a path with a continuous time. This is an intuitive conclusion, not easily made rigorous mathematically. In fact, the introduction of a proper measure in the space consisting of all possible paths $x(t)$ $(0 \leqq t \leqq T)$ requires considerable mathematics, into which we shall not go here since the physical aspects of the problem are emphasized instead.

In understanding Brownian motion as such a stochastic process, how can $\boldsymbol{x}(t)$ be described in terms of probability theory? Firstly, what is the

---

1 A sample of the observed values of random variable $\boldsymbol{x}(t)$ is $x(t)$. This is similar to the relationship between an observable (dynamical quantity) and its observed value in quantum mechanics.

A bold-face italic letter is used for a random variable to distinguish it from its sample value. However, this special lettering will be dispensed with when there is no fear of confusion between these two concepts

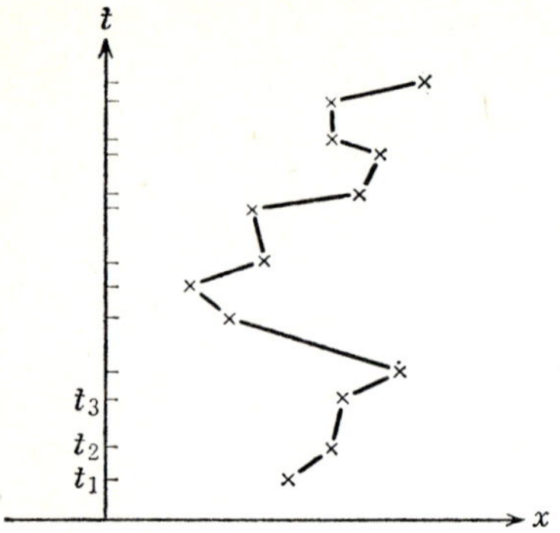

Fig. 1.2. A sample reading of the process $x(t)$ by intermittent observation

probability of finding an observed value $x(t)$ of $\boldsymbol{x}(t)$ at time t in the interval, x and $x+dx$? If its probability density is W_1, then the probability is

$$W_1(x,t)\,dx = \Pr\{x < \boldsymbol{x}(t) \leqq x + dx\}\,. \tag{1.1.3}$$

On the right-hand side, Pr {...} means the probability of the event in the curly bracket occurring. Next, what is the probability that two observed values $x(t_1)$ and $x(t_2)$ at times t_1 and t_2 are found in the intervals (x_1, x_1+dx_1) and (x_2, x_2+dx_2), respectively? For this, the probability density W_2 is defined by

$$\begin{aligned} &W_2(x_1,t_1;x_2,t_2)\,dx_1\,dx_2 \\ &\quad = \Pr\{x_1 < \boldsymbol{x}(t_1) \leqq x_1 + dx_1, x_2 < \boldsymbol{x}(t_2) \leqq x_2 + dx_2\}\,. \end{aligned} \tag{1.1.4}$$

More generally, for a set of observed values $x(t_1), x(t_2), \ldots x(t_n)$ at $t_1, t_2, \ldots$ and t_n

$$\begin{aligned} &W_n(x_1,t_1;x_2,t_2;\ldots;x_n,t_n)\,dx_1\,dx_2\ldots dx_n \\ &\quad = \Pr\{x_j < \boldsymbol{x}(t_j) \leqq x_j + dx_j, j = 1, 2, \ldots, n\}\,. \end{aligned} \tag{1.1.5}$$

This is the joint probability distribution for n random variables, $\boldsymbol{x}(t_1)$, $\boldsymbol{x}(t_2), \ldots$, and $\boldsymbol{x}(t_n)$. The stochastic process $\boldsymbol{x}(t)$ is defined when such probabilities are given for any set of $n\,(n = 1, 2, \ldots \infty)$ time points. In other words, each possible path of the Brownian motion $x(t)$ has a probability assigned to it so that the probability (1.1.5) is defined as the sum of these probabilities for all possible paths going through the gates $dx_1, dx_2, \ldots dx_n$ set at the selected time points. This is the basic idea of constructing a path integral, but we shall not go into this topic here.

Many kinds of probabilities can be derived from definitions (1.1.3–5). Particularly important is the transition probability, which is defined as the probability that the Brownian particle is found at time t_1 between x_1 and

$x_1 + dx_1$ when it was at x_0 at time t_0:

$$P(x_0, t_0 | x_1, t_1)\, dx_1 = \frac{W_2(x_0, t_0; x_1, t_1)\, dx_1}{W_1(x_0, t_0)}. \tag{1.1.6}$$

Transition probabilities for two time points are most commonly used, but a more general definition of transition probabilities is

$$P(x_0, t_0 | x_1, t_1; \ldots; x_n, t_n)\, dx_1 \ldots dx_n = \frac{W_{n+1}(x_0, t_0; x_1, t_1; \ldots; x_n, t_n)\, dx_1 \ldots dx_n}{W_1(x_0, t_0)} \tag{1.1.7}$$

for n observations at n time points when the initial state x_0 is sharply defined at time t_0.

Brownian motions in reality are complex in many respects, so that idealization and abstraction are necessary to formulate them in physical or mathematical terms. There are many different levels of such idealization, each corresponding to a stage of our understanding of their physical nature. The primary purpose of the discussions in this chapter is to make the meaning of these levels as clear as possible. Let us now start from the most simplified model of Brownian motion.

Consider a medium which contains a large number of Brownian particles and define the particle density as $n(x, t)$. Brownian motion of particles makes the distribution of particles tend toward uniformity. This process is called *diffusion.* Corresponding to the gradient of the density distribution, flow is produced

$$j_d = -D \frac{\partial n}{\partial x} \tag{1.1.8}$$

which then induces change of the density according to

$$\frac{\partial n(x, t)}{\partial t} = -\frac{\partial j_d}{\partial x} = D \frac{\partial^2 n}{\partial x^2}. \tag{1.1.9}$$

This is the *diffusion equation.* When a uniform force field such as gravitation exists, a uniform flow is produced with the terminal velocity u_0 determined by the balance of the driving force K and the frictional force from the surrounding fluid acting on a particle. This flow is denoed by j_K and is given by

$$j_K = n u_0 = \frac{nK}{m\gamma}, \tag{1.1.10}$$

where m is the mass of the particle and $m\gamma$ is the friction constant. Therefore the total flow is

$$j = j_K + j_d = \frac{nK}{m\gamma} - D \frac{\partial n}{\partial x}. \tag{1.1.11}$$

In the presence of the external field K, the diffusion equation becomes

$$\frac{\partial n(x,t)}{\partial t} = -\frac{\partial}{\partial x}\left(\frac{nK}{m\gamma}\right) + D\frac{\partial^2 n}{\partial x^2} \tag{1.1.12}$$

instead of (1.1.9).

Starting from any distribution of density, the particles attain a final equilibrium distribution after a sufficiently long time. A uniform distribution of density arises when there is no external field of force, but in a gravitational field, sedimentation equilibrium arises, represented by

$$n(x) = n(x_0)\exp\left(\frac{K(x-x_0)}{kT}\right) \tag{1.1.13}$$

at temperature T of the fluid. This equilibrium is the balance of two flows j_d and j_K, a simple example of *detailed balance*. This means that the distribution (1.1.13) makes the flow j in (1.1.11) vanish. Therefore

$$\frac{D}{kT} = \frac{1}{m\gamma} \quad \text{or}$$

$$D = \mu kT \tag{1.1.14}$$

must hold. Here $\mu = 1/m\gamma$ is the mobility, which is the ratio of u_0 to the force K. Equation (1.1.14) is nothing but the *Einstein relation* given in the introductory section. This relationship is further discussed below; but note here that it is obtained from a very simple idea.

As long as the particle density is not too high, the interactions between Brownian particles can be ignored so that the diffusion described by (1.1.9 or 12) is the result of independent particle motion. Namely, the density $n(x,t)$ at time t and the spatial point x is

$$n(x,t) = \int n(x_0,t_0)\,dx_0\,P(x_0,t_0|x,t)\,, \tag{1.1.15}$$

where $n(x_0,t_0)$ is the density at t_0 and x_0. The transition probability $P(x_0,t_0|x,t)$ satisfies the diffusion equation (we consider Brownian motion in the absence of an external field)

$$\frac{\partial}{\partial t}P(x_0,t_0|x,t) = D\frac{\partial^2}{\partial x^2}P(x_0,t_0|x,t) \tag{1.1.16}$$

because (1.1.9) must be satisfied by $n(x,t)$ given by (1.1.15) for an arbitrary initial condition $n(x_0,t_0)$. Then (1.1.16) simply becomes

$$\frac{\partial}{\partial t}P(x,t) = D\frac{\partial^2}{\partial x^2}P(x,t)\,. \tag{1.1.17}$$

The transition probability $P(x_0, t_0 | x, t)$ is the fundamental solution of (1.1.17) for the initial condition

$$P(x_0, t_0 | x, t) = \delta(x - x_0) \tag{1.1.18}$$

and is given by

$$P(x_0, t_0 | x, t) = \frac{1}{\sqrt{4\pi D(t-t_0)}} \exp\left(-\frac{(x-x_0)^2}{4D(t-t_0)}\right). \tag{1.1.19}$$

If boundaries or sources exist, some appropriate boundary conditions must be imposed.

This is the simplest possible idealization of Brownian motion. The probability of finding a particle at x at time t when it was certainly located at x_0 at t_0 is independent of the knowledge of where the particle was before t_0. Its history previous to time t_0 is summarized, so to say, in the information that the particle was located at x_0 at time t_0, expressed by

$$P(x', t'; x_0, t_0 | x, t) = P(x_0, t_0 | x, t) \quad (t' < t_0) \tag{1.1.20}$$

and hence

$$\begin{aligned} &P(x_0, t_0 | x_1, t_1; x_2, t_2) \\ &\quad = P(x_0, t_0 | x_1, t_1)\, P(x_1, t_1 | x_2, t_2) \quad (t_0 < t_1 < t_2). \end{aligned} \tag{1.1.21}$$

Namely, the evolution of the process in the time interval (t_0, t_2) can be constructed by evolution in the two intervals (t_0, t_1) and (t_1, t_2), where t_1 is an arbitrary time point between t_0 and t_2. Therefore, integrating over all possible values of x_1 at t_1 gives

$$P(x_0, t_0 | x_2, t_2) = \int P(x_0, t_0 | x_1, t_1)\, dx_1\, P(x_1, t_1 | x_2, t_2). \tag{1.1.22}$$

Generally, a stochastic process $\boldsymbol{x}(t)$ is called *Markovian* if it satisfies conditions (1.1.21, 22). That the Brownian motion defined by (1.1.16 or 17) is Markovian is a consequence of the fact that these equations are first order with respect to t. It can be easily proved directly that the transition probability (1.1.19) satisfies (1.1.22).

In the presence of an external force field, (1.1.17) is replaced by

$$\frac{\partial P}{\partial t} = -\frac{\partial}{\partial x}(\mu K P) + D\frac{\partial^2}{\partial x^2} P\,,$$

but everything stated above remains true, with only minor modifications.

1.2 Central Limit Theorem and Brownian Motion

The probability distribution of the displacement $X = x - x_0$ over the time interval $(0, t)$ (for simplicity, $t_0 = 0$) makes a *normal distribution,* or a *Gaussian distribution,* and its variance grows in proportion to time:

$$\langle X^2 \rangle = 2Dt \,. \tag{1.2.1}$$

Now the time interval $(0, t)$ is divided into $n\,(\gg 1)$ segments Δt_i $(i = 1, 2, \ldots, n)$ and displacements in each segment are denoted by ΔX_i. Then naturally

$$X = \sum_{i=1}^{n} \Delta X_i \quad \text{and} \tag{1.2.2}$$

$$\langle \Delta X_i \rangle = 0 \,. \tag{1.2.3}$$

The expectation of the total displacement is zero

$$\langle X \rangle = 0 \,.$$

Further, displacements in different time segments are statistically independent, as implied by (1.1.21):

$$\langle \Delta X_i \Delta X_j \rangle = 0 \qquad (i \neq j) \,. \tag{1.2.4}$$

Therefore it follows from (1.2.2) that

$$\langle X^2 \rangle = \sum_{i=1}^{n} \langle \Delta X_i^2 \rangle \,. \tag{1.2.5}$$

Taking, for simplicity, equal lengths for the time segments, $\langle \Delta X_i^2 \rangle$ are then all equal so that

$$\langle X^2 \rangle = n \langle \Delta X^2 \rangle = t \, \frac{\langle \Delta X^2 \rangle}{\Delta t} \,.$$

Comparing this with (1.2.1) gives for the diffusion constant

$$D = \frac{\langle \Delta X^2 \rangle}{2 \Delta t} \,. \tag{1.2.6}$$

As long as the diffusion model of Sect. 1.1 holds true for the displacement over each time interval Δt, the result (1.2.6) is simply a repetition of (1.2.1). However, the above consideration has a deeper meaning.

The well-known Gaussian law of errors teaches us that an observation error X follows a normal distribution if the error is an accumulation of a large number of small errors. The displacement X of a Brownian particle is also a sum of a large number of successive small displacements ΔX_i. Therefore, we should expect that the distribution law of displacement X over a sufficiently long time interval t is normal even if the diffusion equation (1.1.12) loses its validity for displacement in too a short time interval Δt [in fact, Sect. 1.6 shows that (1.1.17, 19) are not valid in general for short time intervals], and that the variance $\langle X^2 \rangle$ is given by (1.2.1) with the diffusion constant D defined by (1.1.14). Then the transition probability $P(x_0, t_0 \mid x, t)$ has the form (1.1.19) and satisfies the diffusion equation (1.1.16).

The Gaussian law of errors is contained in a very general theorem of probability theory called the *central limit theorem*, which is of fundamental importance in statistical physics. Therefore let us discuss it with some generality. In the same way as for (1.2.2), consider a sum of $n\,(\gg 1)$ independent random variables $\Delta \boldsymbol{X}_1, \Delta \boldsymbol{X}_2, \ldots \Delta \boldsymbol{X}_n$ and set

$$\boldsymbol{X}_n = \Delta \boldsymbol{X}_1 + \Delta \boldsymbol{X}_2 + \ldots + \Delta \boldsymbol{X}_n . \tag{1.2.7}$$

Here $\Delta \boldsymbol{X}_1, \Delta \boldsymbol{X}_2, \ldots$ and $\Delta \boldsymbol{X}_n$ are assumed to have zero expectation values as (1.2.3) and the variances

$$\langle \Delta X_j^2 \rangle = \sigma_j^2 .$$

We set

$$s_n^2 = \sigma_1^2 + \sigma_2^2 + \ldots + \sigma_n^2 . \tag{1.2.8}$$

The central limit theorem now states that if a certain appropriate condition is met by the random variables $\Delta \boldsymbol{X}_1, \Delta \boldsymbol{X}_2, \ldots \Delta \boldsymbol{X}_n$, the probability distribution of the random variable

$$\boldsymbol{Y}_n = \frac{\boldsymbol{X}_n}{s_n} \tag{1.2.9}$$

approaches a normal distribution with variance equal to 1; namely, its distribution density $f_n(Y_n)$ tends to $f(Y)$ as

$$f_n(Y) \to \frac{1}{\sqrt{2\pi}} \exp\left(-\frac{1}{2} Y^2\right) \tag{1.2.10}$$

asymptotically as n increases to infinity. Therefore the probability distribution density $P(X_n)$ of X_n has the property

$$P(X_n) \approx \frac{1}{\sqrt{2\pi s_n}} \exp\left(-\frac{X_n^2}{2 s_n^2}\right) \qquad (n \gg 1) . \tag{1.2.11}$$

This is the Gaussian law of errors already mentioned.

The essential point of the conditions for the validity of the central limit theorem is that n random variables $\Delta \boldsymbol{X}_1, \Delta \boldsymbol{X}_2, \ldots$ and $\Delta \boldsymbol{X}_n$ are all alike and there are no eminent few that dominate the others. When expressed mathematically, this qualitative condition is formulated as various kinds of sufficient conditions. There are different forms of the central limit theorem, such as Lindeberg's or Ljapunov's, for which the reader is referred to mathematical textbooks [1.6]. Here our treatment is greatly simplified by introducing rather strong conditions.

For a general treatment of such problems, it is most convenient to use the *characteristic function.* The characteristic function for a random variable $\boldsymbol{x}$ is defined by

$$\Phi(\xi) = \langle e^{i\xi x} \rangle . \tag{1.2.12}$$

In particular, if the probability distribution density $f(x)$ exists, it is expressed by

$$\Phi(\xi) = \int_{-\infty}^{\infty} e^{i\xi t} f(x)\, dx , \tag{1.2.13}$$

which is nothing but the Fourier transform of $f(x)$. Then $f(x)$ is obtained from $\Phi(\xi)$ as its inverse Fourier transform. But it is not necessary that the density function $f(x)$ exists. General theorems of probability theory state that the characteristic function $\Phi(\xi)$ exists even if the density function f does not, and that the probability distribution of x is uniquely determined from the knowledge of $\Phi(\xi)$ [1.7].

If two random variables $\boldsymbol{x}$ and $\boldsymbol{y}$ are independent, then obviously

$$\langle e^{i\xi(x+y)} \rangle = \langle e^{i\xi x} \rangle \langle e^{i\xi y} \rangle . \tag{1.2.14}$$

More generally, the characteristic function of a sum of an arbitrary number of independent random variables is equal to the product of characteristic functions of the respective random variables. This is one of the basic properties of the characteristic function. The partition function introduced in [Ref. 1.8, Chap. 2] as the fundamental function in equilibrium statistical mechanics is a kind of characteristic function for unnormalized probability distribution of microscopic variables (where a real parameter $-\beta$ was used instead of imaginary $i\xi$).

If the moments

$$\langle x^n \rangle = \int_{-\infty}^{\infty} x^n f(x)\, dx \qquad (n = 0, 1, 2, \dots) \tag{1.2.15}$$

exist for all n's, the characteristic function $\Phi(\xi)$ is analytic in the neighborhood of $\xi = 0$ and is expanded as

$$\Phi(\xi) = \sum_{n=0}^{\infty} \frac{(i\xi)^n}{n!} \langle x^n \rangle . \tag{1.2.16}$$

Conversely, the moment $\langle x^n \rangle$ is then obtained from $\Phi(\xi)$ as

$$\frac{1}{i^n} \left[\left(\frac{d}{d\xi} \right)^n \Phi(\xi) \right]_{\xi=0} = \langle x^n \rangle . \tag{1.2.17}$$

However, it should be remembered that the moments do not necessarily exist. For example, for the Cauchy distribution

$$f(x) = \frac{1}{\pi} \frac{1}{1 + x^2} , \tag{1.2.18}$$

the second and higher moments are all divergent. Correspondingly, the characteristic function is not analytic at $\xi=0$ as is clear from

$$\Phi(\xi)=e^{-|\xi|}. \tag{1.2.19}$$

For such an expansion as (1.2.16) to be possible, it is necessary that the distribution function $f(x)$ tends to zero sufficiently fast as $x \to \pm\infty$.

For example, the characteristic function for the normal distribution

$$f(x)=\frac{1}{\sqrt{2\pi\sigma^2}}\exp\left(-\frac{(x-m)^2}{2\sigma^2}\right) \tag{1.2.20}$$

is calculated as follows:

$$\Phi(\xi)=\frac{1}{\sqrt{2\pi}\,\sigma}\int_{-\infty}^{\infty}\exp\left(-\frac{(x-m)^2}{2\sigma^2}+ix\xi\right)dx$$

$$=\frac{1}{\sqrt{2\pi}\,\sigma}\int_{-\infty}^{\infty}\exp\left[im\xi-\frac{\sigma^2}{2}\xi^2-\frac{1}{2\sigma^2}(x-m-i\sigma^2\xi)^2\right]dx\,.$$

Here note the equality

$$\int_{-\infty}^{\infty}\exp\left[-\frac{1}{2}(y-a)^2\right]dy=\int_{-\infty}^{\infty}\exp\left(-\frac{y^2}{2}\right)dy=\sqrt{2\pi} \tag{1.2.21}$$

obtained by shifting the path of integration as shown in Fig. 1.3 from the path AB on the real axis to CD parallel to AB through point a. Since the function $\exp(-y^2/2)$ is analytic everywhere on the complex plane, the difference of the integrations is due only to integrations along AC and BD, but these vanish as A and B are pushed to $-\infty$ and ∞ respectively. Therefore

$$\Phi(\xi)=\exp\left(im\xi-\frac{\sigma^2}{2}\xi^2\right) \tag{1.2.22}$$

is the characteristic function for the normal distribution. This simple result is worth remembering.

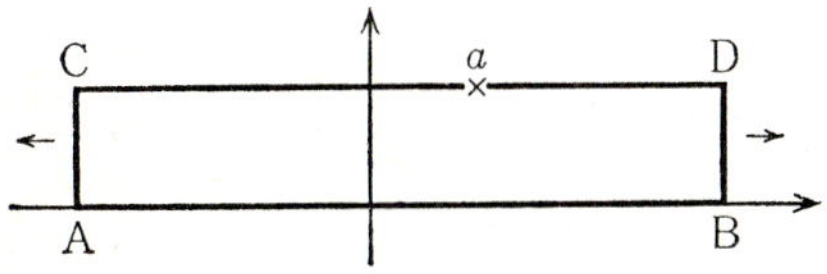

Fig. 1.3. The integration path to prove (1.2.21)

A cumulant function $\Psi(\xi)$ is defined by

$$\Phi(\xi)=e^{\Psi(\xi)}. \tag{1.2.23}$$

This corresponds to the thermodynamic characteristic function (free energy divided by kT) in statistical mechanics. It is written as

$$\Psi(\xi)=\ln\Phi(\xi)\,. \tag{1.2.24}$$

If the expansion (1.2.16) is possible, this may be expanded to

$$\Psi(\xi)=\sum_{n=1}^{\infty}\frac{(\mathrm{i}\,\xi)^n}{n!}\langle x^n\rangle_c\,, \tag{1.2.25}$$

where the expansion coefficient $\langle x^n\rangle_c$ is called the nth cumulant and is related to the moments $\langle x^m\rangle\, m\leqq n$ by (1.2.23 or 24). Explicit relations are

$$\begin{aligned}
\langle x\rangle_c &=\langle x\rangle,\\
\langle x^2\rangle_c &=\langle x^2\rangle-\langle x\rangle^2, \quad \langle x^2\rangle=\langle x^2\rangle_c+\langle x\rangle_c^2\\
\langle x^3\rangle_c &=\langle x^3\rangle-3\langle x^2\rangle\langle x\rangle+2\langle x\rangle^3\\
\langle x^3\rangle &=\langle x^3\rangle_c+3\langle x\rangle_c\langle x^2\rangle_c+\langle x\rangle_c^3
\end{aligned} \tag{1.2.26}$$

for $n\leqq 3$. Generally, the nth cumulant is expressed in terms of moments not higher than the nth. Conversely, the nth moment is expressed in terms of cumulants not higher than the nth. In particular, $\langle x\rangle_c$ is the expectation and $\langle x^2\rangle_c$ is the variance.

For normal distribution (1.2.20), from (1.2.22)

$$\langle x\rangle_c=m\,,\quad \langle x^2\rangle_c=\sigma^2\,,\quad \langle x^n\rangle_c=0 \qquad (n\geqq 3)\,. \tag{1.2.27}$$

A characteristic feature of the normal distribution is that cumulants higher than the third are all zero.

Now the characteristic function of X_n, (1.2.7), is

$$\Phi(\xi)=\langle \mathrm{e}^{\mathrm{i}\xi X_n}\rangle=\prod_{j=1}^{n}\langle \mathrm{e}^{\mathrm{i}\xi\Delta X_1}\rangle\,. \tag{1.2.28}$$

Denoting the cumulant function by $\psi_j(\xi)$ for ΔX_j $(j=1,2,\dots)$ and that by $\Psi_n(\xi)$ for X_n, then from (1.2.28)

$$\Psi_n(\xi)=\sum_{j=1}^{n}\psi_j(\xi)\,.$$

If the expansions

$$\psi_j(\xi)=\mathrm{i}\xi\langle\Delta X_j\rangle_c-\frac{\xi^2}{2}\langle\Delta X_j^2\rangle_c+\frac{(\mathrm{i}\,\xi)^3}{3!}\langle\Delta X_j^3\rangle_c+\dots \tag{1.2.29}$$

are all possible, then

$$\Psi_n(\xi)=-\frac{\xi^2}{2}s_n^2+\frac{(\mathrm{i}\,\xi)^3}{3!}\sum\langle\Delta X_j^3\rangle_c+\dots \tag{1.2.30}$$

from (1.2.7, 8) by assuming $\langle\Delta X_j\rangle=0$. The characteristic function for Y_n, (1.2.9),

$$\langle \mathrm{e}^{\mathrm{i}\eta Y_n}\rangle=\langle \mathrm{e}^{\mathrm{i}\eta X_n/s_n}\rangle\,,$$

is

$$\langle \mathrm{e}^{\mathrm{i}\eta Y_n}\rangle=\exp\left(-\frac{1}{2}\eta^2+\frac{(\mathrm{i}\eta)^3}{3!}\sum_{j=1}^{n}\frac{\langle\Delta X_j^3\rangle_c}{s_n^3}+\dots\right) \tag{1.2.31}$$

obtained by replacing ξ in (1.2.30) by η/s_n. Assume that the mth moments of ΔX_j are all finite and of the same order of magnitude. Then s_n^2 increases, according to (1.2.8), in the order of n with increasing n. Then the mth cumulant in (1.2.31) tends to zero

$$\frac{O(n)}{O(n^{m/2})} \to 0$$

from $m \geqq 3$. Therefore

$$\langle e^{i\eta Y_n} \rangle \to e^{-\eta^2/2}. \qquad (1.2.32)$$

This shows, as noted previously, that Y_n approaches a normal distribution with the variance equal to 1.

In the above we have imposed a very strict condition, namely the existence of moments at all orders, which is not in fact necessary for proving the central limit theorem. However, this is not unreasonable to assume for many physical processes. Whether or not this is allowed in reality, the central limit theorem indicates that the probabilistic motion of a Brownian particle is, for a sufficiently long time, described very well by a diffusion process defined by (1.1.19). For shorter times, there is no reason to expect the diffusion process to be valid for a physical process of particle motion. If a particle moves with a velocity u at a certain time t, we have to wait a finite time before we find different velocities. This time τ_c is the *correlation time* of the velocity. However, when the time segment Δt is much longer than the correlation time τ_c, displacements in each time segment are regarded as independent of each other. So, if t is so long that $n = t/\Delta t$ is much larger than 1, the total displacement X in (1.2.2) must have a normal distribution (1.1.19) with the variance

$$\langle X^2 \rangle = 2Dt. \qquad (1.2.33)$$

This is a consequence of the central limit theorem.

The random walk problem is often considered as a model of Brownian motion. The simplest model is random walk with steps $\pm a$ to the right or to the left randomly at every τ. After n steps the displacement $x = ma$ has the binomial distribution

$$P_n(m) = \frac{n!}{2^n}\left(\frac{n+m}{2}!\,\frac{n-m}{2}!\right)^{-1}. \qquad (1.2.34)$$

When n is large, this is approximated by a normal distribution. From (1.2.6) the diffusion constant is

$$D = \frac{a^2}{2\tau}. \qquad (1.2.35)$$

This is also easily seen by using the Stirling formula valid for (1.2.34). The relationship (1.2.35) is, however, more general and not limited to any particular random walk model. In three dimensions, it becomes

$$D = \frac{l^2}{6\tau}, \tag{1.2.36}$$

where l is the mean free path and τ the mean free time. Equation (1.2.35) for a one-dimensional case is recovered by setting $l^2 = 3a^2$. Depending on the way in which definitions are made, extra factors may be introduced into these equations. Most important is the physical meaning of the diffusion constant implied by these equations.

1.3 Langevin Equation and Harmonic Analysis

So far we have concentrated on the displacement of Brownian particles. However, the physical model should start from the motion itself. The equation of motion of a particle is, of course,

$$m\frac{du}{dt} = F, \tag{1.3.1}$$

where u is the velocity and F the force acting on the particle from molecules of the fluid surrounding the Brownian particle. As before, we treat one-dimensional cases, but three-dimensional cases are the same.

The force F may be divided into two parts. The first part is the *frictional force* and is proportional to the velocity of the particle. If the frictional coefficient is denoted by $m\gamma$ as in (1.1.10), the frictional force is assumed to be

$$F_u = -m\gamma u. \tag{1.3.2}$$

If the Stokes law is assumed for a spherical particle, for example, the frictional coefficient is

$$m\gamma = 6\pi a\eta, \tag{1.3.3}$$

where a is the radius of the particle and η the viscosity of the fluid.

The second part of the force is the remainder of the force, F_u subtracted from F, and is regarded as random, independent of the motion of the particle. This part is called the *random force* and is denoted as $\boldsymbol{R}(t)$ hereafter.

Then (1.3.1) is written as

$$m\frac{d\boldsymbol{u}}{dt} = -m\gamma\boldsymbol{u} + \boldsymbol{R}(t). \tag{1.3.4}$$

as a stochastic equation.

The same consideration can be applied to Brownian motion in the presence of a force field, for example, the gravitational field or an harmonic force binding the particle elastically to the origin. If the potential of the force is denoted by V, the equation of motion becomes

$$\frac{d\boldsymbol{p}}{dt} = -\frac{\partial V}{\partial \boldsymbol{x}} - \gamma \boldsymbol{p} + \boldsymbol{R}(t), \tag{1.3.5}$$

$$\frac{d\boldsymbol{x}}{dt} = \frac{\boldsymbol{p}}{m} \equiv \boldsymbol{u}(t). \tag{1.3.6}$$

A set of equations of motion containing a random force, like (1.3.4) or (1.3.5, 6), is called a Langevin equation.

The random force $\boldsymbol{R}(t)$ is a stochastic process randomly changing in time. Brownian motion, $\boldsymbol{u}(t)$ [or $\boldsymbol{p}(t)$] and $\boldsymbol{x}(t)$ are also stochastic processes driven by [generated by] $\boldsymbol{R}(t)$. They are related to $\boldsymbol{R}(t)$ by (1.3.4 or 5, 6). If we consider the force as causing the motion, then the random force $\boldsymbol{R}(t)$ produces Brownian motion. Thus our problem is to determine the stochastic processes $\boldsymbol{u}(t)$ and $\boldsymbol{x}(t)$ from knowing $\boldsymbol{R}(t)$. This is what is meant by solving stochastic equations like (1.3.4) or (1.3.5, 6).

A standard method of solving a stochastic equation is harmonic analysis. This expresses a motion by a superposition of oscillating functions, which is an orthodox method for linear systems. Since the environment is considered in a stationary condition with constant temperature and pressure, the Brownian motion must also be stationary if the particle is aged (kept for a sufficiently long time) in the environment. In other words, the probability such as (1.1.5) must be invariant with respect to a shift in time:

$$W_n(x_1, t_1; x_2, t_2; \ldots; x_n, t_n) = W_n(x_1, t_1 + \tau; x_2, t_2 + \tau; \ldots; x_n, t_n + \tau). \tag{1.3.7}$$

The processes $\boldsymbol{R}(t)$, $\boldsymbol{u}(t)$ and $\boldsymbol{x}(t)$ are stationary in this sense.

Let us consider in general a stationary process $z(t)$, a sample of which is $z(t)$ observed over a time interval $0 \leqq t \leqq T$. This function $z(t)$ is expanded in a Fourier series as

$$z(t) = \sum_{n=-\infty}^{\infty} a_n \mathrm{e}^{\mathrm{i}\omega_n t}, \tag{1.3.8}$$

where the frequencies are

$$\omega_n = \frac{2\pi n}{T} \quad (n = 0, \pm 1, \pm 2, \ldots), \tag{1.3.9}$$

corresponding to the interval T. The function $z(t)$ is considered real so that the Fourier coefficient a_n has the form

$$a_n = a_n' + \mathrm{i}a_n'', \quad a_{-n} = a_n^* = a_n' - \mathrm{i}a_n''. \tag{1.3.10}$$

The stochastic process $\boldsymbol{z}(t)$ is now expanded as

$$\boldsymbol{z}(t) = \sum_{n=-\infty}^{\infty} \boldsymbol{a}_n \, \mathrm{e}^{\mathrm{i}\omega_n t} \qquad (0 \leqq t \leqq T) . \tag{1.3.11}$$

Then the Fourier coefficient a_n in (1.3.8) is a sample of the random variable $\boldsymbol{a}_n$ defined by

$$\boldsymbol{a}_n = \frac{1}{T} \int_0^T \boldsymbol{z}(t) \, \mathrm{e}^{-\mathrm{i}\omega_n t} \, dt . \tag{1.3.12}$$

Equation (1.3.11) shows that the stochastic process $\boldsymbol{z}(t)$ is expressed by a set of a countably infinite number of random variables $\{\boldsymbol{a}_n\}$. For the process $\boldsymbol{z}(t)$, the probability is defined as explained in Sect. 1.2 and so the probabilities and expectations are also defined for $\{\boldsymbol{a}_n\}$. The expectation value of $\boldsymbol{a}_n$ is given by

$$\langle a_n \rangle = \frac{1}{T} \int_0^T \langle z(t) \rangle \, \mathrm{e}^{-\mathrm{i}\omega_n t} \, dt .$$

It follows that

$$\langle a_n \rangle = 0 \qquad (n \neq 0) , \tag{1.3.13}$$

since $\langle z(t) \rangle$ is a constant as it is stationary. For $n = 0$, obviously

$$\langle a_0 \rangle = \frac{1}{T} \int_0^T \langle z(t) \rangle \, dt = \langle z \rangle . \tag{1.3.14}$$

A sample of $\boldsymbol{a}_0$ is the time average of a sample $z(t)$ of the process $\boldsymbol{z}(t)$ over the interval $(0, T)$:

$$a_0 = \overline{z(t)}^{\,T} \equiv \frac{1}{T} \int_0^T z(t) \, dt . \tag{1.3.15}$$

Generally, this need not be equal to $\langle z \rangle$. A process $\boldsymbol{z}(t)$ is called *ergodic* if

$$\lim_{T \to \infty} \overline{z(t)}^{\,T} = \langle z \rangle \tag{1.3.16}$$

holds.

On the other hand, the expectation $\langle z \rangle$ is the average of $z(t)$ over the set of *all* possible values of z. This set is not necessarily identical with the set of values of $z(t)$ for $0 \leqq t \leqq T$. It may well be that different subsets are reached from a different starting state $z(0)$, in which case the time average becomes different for different $z(0)$. For example, a combination of two distinct processes is obviously nonergodic. A nonergodic process is generally decomposes into ergodic processes. In most cases, a stationary stochastic process is already reduced to such a simple one, so that we may hereafter assume ergodicity. In this sense, $\boldsymbol{a}_0 = \langle z \rangle = \text{const}$ and consider $\boldsymbol{z}(t) - \langle z \rangle$ to assume that

$$\langle a_n \rangle = 0 \quad (n = 0, \pm 1, \ldots, \pm \infty) \tag{1.3.17}$$

without losing generality. The averaged strength of the Fourier component $\boldsymbol{a}_n$ is defined by

$$\langle |a_n|^2 \rangle = \langle |a_n'|^2 \rangle + \langle |a_n''|^2 \rangle . \tag{1.3.18}$$

The right-hand side of this expression is the sum of mean square averages of the real and imaginary parts of the amplitude $\boldsymbol{a}_n$. When a suitable filter is used to select only the frequencies lying in the interval $\Delta\omega$, strengths of these Fourier components are observable. The average intensity $I(\omega)$ is

$$I(\omega)\, \Delta\omega = \sum_{\omega_n \text{ in } \Delta\omega} \langle |a_n|^2 \rangle . \tag{1.3.19}$$

The right-hand side of this expression is a sum over all frequencies contained in the frequency band. The number of such modes is

$$\frac{\Delta\omega}{2\pi/T} = \frac{T}{2\pi} \Delta\omega$$

because the interval $\delta\omega$ of successive frequencies in (1.3.9) is $2\pi/T$. We may suppose $\langle a_n^2 \rangle$ to be continuous in the frequency ω_n and write (1.3.19) as

$$I(\omega) = \lim_{T \to \infty} \frac{T}{2\pi} \langle |a_n|^2 \rangle \tag{1.3.20}$$

to define the intensity spectrum of the process $z(t)$ at frequency ω. In fact, if $z(t)$ is the noise voltage produced between two terminals of an electrical network, $I(\omega)$ is the intensity of the noise heard by filtering frequencies to a narrow bandwidth $\Delta\omega$ around ω. In this sense, $I(\omega)$ defined by (1.3.20) is called the power spectrum of the process $z(t)$.

For the power spectrum, the well-known *Wiener-Khintchine* theorem [1.9] is fundamental. The correlation function for the process $z(t)$ is

$$\phi(t) = \langle z(t_0)\, z(t_0 + t) \rangle . \tag{1.3.21}$$

This represents the correlation between the observed values of $z(t)$ at two time points t_0 and $t_0 + t$ and is independent of t_0 since $z(t)$ is stationary. The Wiener-Khintchine theorem asserts that

$$I(\omega) = \frac{1}{2\pi} \int_{-\infty}^{\infty} \phi(t)\, e^{-i\omega t}\, dt \tag{1.3.22}$$

holds. Conversely, it holds that

$$\phi(t) = \int_{-\infty}^{\infty} I(\omega)\, e^{i\omega t}\, d\omega . \tag{1.3.23}$$

This theorem is proved as follows. From (1.3.12)

$$\langle |a_n|^2 \rangle = \frac{1}{T^2} \int_0^T dt_1 \int_0^T dt_2 \langle z(t_1)\, z(t_2) \rangle \exp\left[-i\omega_n (t_1 - t_2)\right]. \tag{1.3.24}$$

In the integrand, the correlation function is dependent only on the time difference $t_1 - t_2$. The integration over t_1 and t_2 is carried out over the square $0 \leqq t_1 \leqq T$, $0 \leqq t_2 \leqq T$ as shown in Fig. 1.4. The integration is divided into two parts. For $t_1 > t_2$ the integration variables are changed to $t_1 - t_2 = t$ and t_2, for which the Jacobian $\partial(t_1, t_2)/\partial(t, t_2) = 1$. Integration over t_2 is carried out from 0 to $T - t$, resulting in

$$\int_0^T (T - t)\, \phi(t)\, \mathrm{e}^{-\mathrm{i}\omega_n t}\, dt, \tag{1.3.25}$$

since the integrand does not depend on t_2. Similarly, integration for the part $t_1 < t_2$ yields

$$\int_0^T (T - t)\, \phi(-t)\, \mathrm{e}^{\mathrm{i}\omega_n t}\, dt. \tag{1.3.26}$$

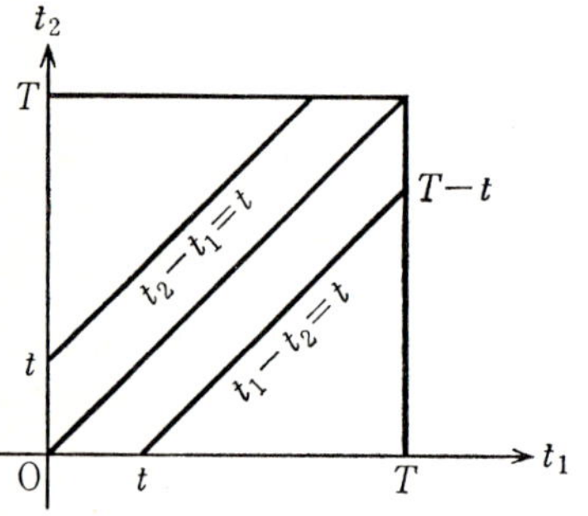

Fig. 1.4. Changing the integration variables in (1.3.24)

These results are inserted into (1.3.24) and T is made infinity. As long as the integrals

$$\int_0^\infty \phi(\pm t)\, \mathrm{e}^{\mp\mathrm{i}\omega_n t}\, dt, \quad \int_0^\infty t\,\phi(\pm t)\, \mathrm{e}^{\mp\mathrm{i}\omega_n t}\, dt \tag{1.3.27}$$

are convergent, then

$$I(\omega) = \frac{1}{2\pi} \left[\int_0^\infty \phi(t)\, e^{-\mathrm{i}\omega t}\, dt + \int_0^\infty \phi(-t)\, \mathrm{e}^{\mathrm{i}\omega t}\, dt \right]. \tag{1.3.28}$$

Converting the second term to an integral over $(-\infty, 0)$ gives (1.3.22).

Instead of Fourier-analyzing over a finite time, we may Fourier-analyze over an infinite time interval and write

$$z(t) = \int_{-\infty}^{\infty} \boldsymbol{a}(\omega)\, e^{i\omega t}\, d\omega \tag{1.3.29}$$

$$\boldsymbol{a}(\omega) = \frac{1}{2\pi} \int_{-\infty}^{\infty} z(t)\, e^{-i\omega t}\, dt \tag{1.3.30}$$

as Fourier integrals. If $z(t)$ is real,

$$\boldsymbol{a}(-\omega) = \boldsymbol{a}^*(\omega)$$

is the complex conjugate of $\boldsymbol{a}(\omega)$. Furthermore, if $z(t)$ is stationary, it holds that

$$\langle a(\omega)\, a(\omega')\rangle = I(\omega)\, \delta(\omega + \omega') \tag{1.3.31}$$

or

$$\langle a(\omega)\, a^*(\omega')\rangle = I(\omega)\, \delta(\omega - \omega'), \tag{1.3.32}$$

where $I(\omega)$ is the power spectrum given by (1.3.20). In fact, the left-hand side of (1.3.31) is calculated as

$$\begin{aligned}\langle a(\omega_1)\, a(\omega_2)\rangle &= \frac{1}{(2\pi)^2} \int_{-\infty}^{\infty} dt_1 \int_{-\infty}^{\infty} dt_2\, \phi(t_1 - t_2)\, \exp(-i\omega_1 t_1 - i\omega_2 t_2) \\ &= \frac{1}{2\pi} \int_{-\infty}^{\infty} \phi(t)\, \exp(i\omega_1 t)\, dt \cdot \frac{1}{2\pi} \int_{-\infty}^{\infty} \exp[-i(\omega_1 + \omega_2)\tau]\, d\tau\,.\end{aligned}$$

With (1.3.22) we obtain (1.3.31). It is obvious that (1.3.23) follows conversely from (1.3.9, 31).

Harmonic analysis is particularly useful for linear stochastic equations like (1.3.4). We Fourier-expand the random force as a stationary stochastic process,

$$\boldsymbol{R}(t) = \sum_{n=-\infty}^{\infty} \boldsymbol{R}_n\, e^{i\omega_n t}. \tag{1.3.33}$$

In the same way the velocity $\boldsymbol{u}(t)$ of a Brownian particle is expanded as

$$\boldsymbol{u}(t) = \sum_{n=-\infty}^{\infty} \boldsymbol{u}_n e^{i\omega_n t}. \tag{1.3.34}$$

Then the stochastic differential equation (1.3.4)

$$m\dot{\boldsymbol{u}}(t) = -m\gamma \boldsymbol{u}(t) + \boldsymbol{R}(t) \tag{1.3.35}$$

is converted into the relation

$$\boldsymbol{u}_n = \frac{1}{i\omega_n + \gamma} \frac{\boldsymbol{R}_n}{m} \tag{1.3.36}$$

between the Fourier components. If we write the power spectra of $\boldsymbol{R}(t)$ and $\boldsymbol{u}(t)$ as I_R and I_u, respectively, we immediately obtain from (1.3.20, 36)

$$I_u(\omega) = \frac{1}{|\mathrm{i}\omega + \gamma|^2} \frac{I_R(\omega)}{m^2} = \frac{1}{\omega^2 + \gamma^2} \frac{I_R(\omega)}{m^2} . \tag{1.3.37}$$

If the Brownian particle is bound to the origin by an elastic force, the Langevin equation is assumed to be

$$\frac{d^2\boldsymbol{x}}{dt} + \gamma \frac{d\boldsymbol{x}}{dt} + \omega_0^2 \boldsymbol{x} = \frac{1}{m} \boldsymbol{R}(t) , \tag{1.3.38}$$

where $\boldsymbol{x}$ is the displacement and ω_0 is the natural frequency of the elastic binding. Torsional oscillation of a small mirror suspended in a dilute gas is an example to which harmonic analysis is also directly applicable. The power spectrum I_x of the displacement is easily seen to be

$$\begin{aligned} I_x(\omega) &= \frac{1}{|\omega_0^2 - \omega^2 + \mathrm{i}\gamma\omega|^2} \frac{I_R(\omega)}{m^2} \\ &= \frac{1}{(\omega_0^2 - \omega^2)^2 + \gamma^2\omega^2} \frac{I_R(\omega)}{m^2} \end{aligned} \tag{1.3.39}$$

in the same way, as we derived (1.3.37).

The Wiener-Khintchine theorem (1.3.22, 23) shows that knowing the power spectrum is equivalent to knowing the correlation function. When $\boldsymbol{R}(t)$ is known, (1.3.37 or 39) converts it into $\boldsymbol{u}(t)$ or $\boldsymbol{x}(t)$, so that solving (1.3.35 or 38) to the same extent. As shown in Sect. 1.2, the solution is complete if $\boldsymbol{R}(t)$ is a Gaussian process.

The nature of the random force $\boldsymbol{R}(t)$ is considered in more detail in the following sections. Here we make the simplest possible assumption that the power spectrum I_R is independent of frequency

$$I_R(\omega) = I_R = \text{const} , \tag{1.3.40}$$

when, the spectrum is said to be *white*. Obviously, from (1.3.23) it follows that the correlation function of a process having a white power spectrum has a vanishingly short correlation time,

$$\phi_R(t_1 - t_2) \equiv \langle R(t_1) R(t_2) \rangle = 2\pi I_R \delta(t_1 - t_2) . \tag{1.3.41}$$

If the random force is such, it follows from (1.3.37) with (1.3.23) that

$$\phi_u(t) = \int_{-\infty}^{\infty} \frac{\mathrm{e}^{\mathrm{i}\omega t}\, d\omega}{\omega^2 + \gamma^2} \frac{I_R}{m^2} , \tag{1.3.42}$$

namely,

$$\langle u(t_1)\, u(t_2)\rangle = \frac{\pi I_R}{m^2 \gamma}\, e^{-\gamma |t_1 - t_2|}. \tag{1.3.43}$$

Thus, the correlation function of velocity of a free Brownian particle decays exponentially in time with the decay constant γ. In particular, for $t_1 = t_2$, (1.3.43) reduces to

$$\langle u^2 \rangle = \frac{\pi I_R}{m^2 \gamma}. \tag{1.3.44}$$

If the Brownian particle has been kept for a sufficiently long time in the fluid at temperature T, the equipartition law

$$m \langle u^2 \rangle = kT \tag{1.3.45}$$

must hold for the energy distribution. For (1.3.44) to be consistent with this,

$$I_R = \frac{m\gamma kT}{\pi} \tag{1.3.46}$$

must hold. In other words, the random force R (if it has a white spectrum) must have the intensity given by (1.3.46) so that the Langevin equation (1.3.35) represents free Brownian motion in thermal equilibrium at temperature T.

For harmonic Brownian motion, (1.3.39) leads to

$$\phi_x(t) = \int_{-\infty}^{\infty} \frac{e^{i\omega t}\, d\omega}{(\omega_0^2 - \omega^2)^2 + \gamma^2 \omega^2} \frac{I_R}{m^2}$$

$$= \frac{\pi I_R}{m^2 \gamma \omega_0^2} \left(\cos \omega_1 t + \frac{\gamma}{2\omega_1} \sin \omega_1 t \right) e^{-\gamma t/2} \quad (t > 0), \tag{1.3.47}$$

if I_R is assumed constant (white). For a derivation of this result, the residues at the poles

$$\omega = \pm \frac{i}{2} \gamma \pm \omega_1, \quad \text{where} \quad \omega_1 = \left(\omega_0^2 - \frac{\gamma^2}{4} \right)^{1/2},$$

of the integrand are calculated. In the limit $t \to 0$, this reduces to

$$\phi_x(0) = \langle x^2 \rangle = \frac{\pi I_R}{m^2 \gamma \omega_0^2} = \frac{kT}{m \omega_0^2}. \tag{1.3.48}$$

Thus (1.3.46) again guarantees the equipartition law of energy.

1.4 Gaussian Processes

A general stochastic process is defined by giving the probabilities (1.1.5) for all possible sets of $t_1, t_2, \ldots, t_n$ ($n = 1, 2, \ldots$). Probabilities of lower hierarchy are derived from those of higher hierarchy, but the latter generally contain new information not contained in the former. The situation becomes simpler for Markovian processes, in which all higher probabilities are determined by the transition probability $P(x_1, t_1 | x_2, t_2)$. This kind of stochastic process is considered below. Here we take up another class of simple processes, namely the Gaussian processes. This is an extension of the normal distribution discussed in Sect. 1.2 to stochastic processes. Just as a normal distribution is defined by its second moment or the variance, a Gaussian process is completely defined by the correlation function (1.3.21).

A stochastic process $z(t)$ is Gaussian if the probability distribution of its observed values $z_1, z_2, \ldots, z_n$ at n time points $t_1, t_2, \ldots, t_n$ is an n-dimensional Gaussian (normal) distribution; namely W_n in (1.1.5) has the form

$$W_n(z_1, t_1; z_2, t_2; \ldots; z_n, t_n) = C \exp\left[-\frac{1}{2}\sum_{j=1}^{n}\sum_{k=1}^{n} a_{jk}(z_j - m_j)(z_k - m_k)\right], \tag{1.4.1}$$

where

$$m_j = \langle z_j \rangle \equiv \langle z(t_j) \rangle \tag{1.4.2}$$

is the expectation value of $z(t)$ at time t_j and the matrix

$$(a_{jk}) \equiv A \tag{1.4.3}$$

is positive definite. The elements of its inverse matrix A^{-1} are the correlation functions of the process $z(t)$

$$\begin{aligned}(A^{-1})_{jk} &= \langle (z_j - m_j)(z_k - m_k) \rangle \\ &= \langle [z(t_j) - \langle z(t_j) \rangle][z(t_k) - \langle z(t_k) \rangle] \rangle .\end{aligned} \tag{1.4.4}$$

In order to see this, we use the characteristic function explained in Sect. 1.2 in a slightly generalized form. We introduce the parameters $\zeta_1, \zeta_2, \ldots, \zeta_n$ corresponding to the n random variables $z_1, z_2, \ldots, z_n$ and write the characteristic function of (1.4.1) as

$$\Phi(\zeta_1, \ldots, \zeta_n) = \int_{-\infty}^{\infty} dz_1 \ldots \int_{-\infty}^{\infty} dz_n\, W_n(z_1, t_1; \ldots; z_n, t_n) \exp\left(\mathrm{i}\sum_{j=1}^{n} \zeta_j z_j\right). \tag{1.4.5}$$

For brevity, we use the vector notations[2]

$$z = (z_1, z_2, \ldots, z_n), \quad \zeta = (\zeta_1, \zeta_2, \ldots, \zeta_n)$$

2 Note that bold-faced letters are not used for these vectors to avoid confusion with random variables

and carry out the integration in the following way. Inserting (1.4.1) into W_n, the exponential function in (1.4.5) is rewritten as

$$\begin{aligned}&\exp[-\tfrac{1}{2}(z-m)A(z-m)+\mathrm{i}\zeta z]\\&\quad=\exp(\mathrm{i}\zeta m-\tfrac{1}{2}yAy+\mathrm{i}\zeta y)\\&\quad=\exp(\mathrm{i}\zeta m-\tfrac{1}{2}uAu-\mathrm{i}uAv+\tfrac{1}{2}vAv+\mathrm{i}\zeta u-\zeta v)\,,\end{aligned}$$

setting

$$m=(m_1,m_2,\dots,m_n),\quad z-m=y=u+\mathrm{i}v\,.$$

Now we choose the vector v by the condition

$$Av=\zeta,\quad \text{namely}\quad v=A^{-1}\zeta\,.$$

Then the first-order term of u vanishes, and the integral becomes

$$\Phi(\zeta)=\exp(\mathrm{i}m\zeta-\tfrac{1}{2}\zeta A^{-1}\zeta)\int_{-\infty}^{\infty}du_1\dots\int_{-\infty}^{\infty}du_n\,C\exp(-\tfrac{1}{2}uAu)\,. \tag{1.4.6}$$

Integration along the real axes of $z_1,z_2,\dots,z_n$ was here transformed to that along the real axes of $u_1,u_2,\dots,u_n$ just as for (1.2.21). The integral can be explicitly calculated by achieving orthogonal transformation to diagonalize the quadratic form, uAu. But this is not necessary, because we should have $\Phi=1$ for $\zeta_1=\zeta_2=\dots=\zeta_n=0$, as is seen by the fact that W_n is normalized by the constant C. Therefore,

$$\Phi(\zeta_1,\zeta_2,\dots,\zeta_n)=\exp\left(\mathrm{i}\sum_{j=1}^{n}m_j\zeta_j-\frac{1}{2}\sum_{j=1}^{n}\sum_{k=1}^{n}(A^{-1})_{jk}\zeta_j\zeta_k\right). \tag{1.4.7}$$

The moment and cumulant definitions introduced by (1.2.15, 16, 25) can easily be generalized to an n-dimensional random variable $(z_1,z_2,\dots z_n)$. Namely, the $(r_1,r_2,\dots,r_n)$th moment is

$$\langle z_1^{r_1}z_2^{r_2}\dots z_n^{r_n}\rangle=\int dz_1\dots\int dz_n\,W(z_1,\dots,z_n)\,z_1^{r_1}\dots z_n^{r_n}\,, \tag{1.4.8}$$

$[W(z_1,z_2,\dots,z_n)$ is the joint distribution of $z_1,z_2,\dots,z_n]$ and the characteristic function (1.4.5) is expanded in a power series

$$\Phi(\zeta)=\sum_{r_1=0}^{\infty}\dots\sum_{r_n=0}^{\infty}\frac{(\mathrm{i}\zeta_1)^{r_1}\dots(\mathrm{i}\zeta_n)^{r_n}}{r_1!\dots r_n!}\langle z_1^{r_1}\dots z_n^{r_n}\rangle\,. \tag{1.4.9}$$

This gives all the moments, provided that such an expansion is possible. The cumulant function $\Psi(\zeta)$ is defined by

$$\Phi(\zeta)=\exp\Psi(\zeta),\quad \Psi(\zeta)=\ln\Phi(\zeta)\,. \tag{1.4.10}$$

The cumulants are defined by

$$\Psi(\zeta)=\sum{}'\frac{(\mathrm{i}\zeta_1)^{r_1}\dots(\mathrm{i}\zeta_n)^{r_n}}{r_1!\dots r_n!}\langle z_1^{r_1}\dots z_n^{r_n}\rangle_{\mathrm{c}} \tag{1.4.11}$$

if the expansion is possible, where $\sum'$ means the omission of the term with $r_1=r_2=\dots=r_n=0$. Cumulants and moments are mutually related by (1.4.10), e.g.,

$$\begin{aligned}\langle z_1 z_2\rangle &=\langle z_1 z_2\rangle_{\mathrm{c}}+\langle z_1\rangle\langle z_2\rangle \\ \langle z_1 z_2 z_3\rangle &=\langle z_1 z_2 z_3\rangle_{\mathrm{c}}+\langle z_1\rangle\langle z_2 z_3\rangle_{\mathrm{c}}+\langle z_2\rangle\langle z_1 z_3\rangle_{\mathrm{c}} \\ &\quad+\langle z_3\rangle\langle z_1 z_2\rangle_{\mathrm{c}}+\langle z_1\rangle\langle z_2\rangle\langle z_3\rangle\end{aligned} \tag{1.4.12}$$

As evident in (1.2.27), all cumulants, $n\geqq 3$, are identically zero for a one-dimensional Gaussian distribution. This statement is generalized to an n-dimensional Gaussian distribution for which all cumulants vanish except the first and second, as is seen in (1.4.7) which contains terms only to second order in ζ. The coefficients of second-order terms are the variance matrix (1.4.4). Its element

$$\begin{aligned}\langle z(t_j)\,z(t_k)\rangle_{\mathrm{c}} &=\langle z(t_j)\,z(t_k)\rangle-\langle z(t_j)\rangle\langle z(t_k)\rangle \\ &=\langle[z(t_j)-\langle z(t_j)\rangle][z(t_k)-\langle z(t_k)\rangle]\rangle\end{aligned} \tag{1.4.13}$$

is the correlation function of $z(t)$. Therefore, (1.4.7) becomes

$$\Phi(\zeta_1,\dots,\zeta_n)=\exp\left(\mathrm{i}\sum_{j=1}^{n}\zeta_j m(t_j)-\frac{1}{2}\sum_{j=1}^{n}\sum_{k=1}^{n}\phi(t_j,t_k)\,\zeta_j\,\zeta_k\right), \tag{1.4.14}$$

where

$$m(t_j)=\langle z(t_j)\rangle,\quad \phi(t_j,t_k)=\langle[z(t_j)-\langle z(t_j)\rangle][z(t_k)-\langle z(t_k)\rangle]\rangle. \tag{1.4.15}$$

Thus, the process $z(t)$ is completely determined by the expectations and the correlation functions since the characteristic function is completely defined by these quantities.

Assume for simplicity that

$$m(t)=0.$$

For an arbitrary set of n time points $(t_1, t_2, \dots, t_n)$,

$$\langle z(t_1)\dots z(t_n)\rangle=\begin{cases}0 & \text{for odd } n,\\ \sum\limits_{\text{pairing}}\prod\limits_{\text{pairs}}\phi(t_j,t_k) & \text{for even } n,\end{cases} \tag{1.4.16}$$

holds. This is easily seen by comparing the power series expansion of (1.4.14) [setting $m(t_j)=0$] in $\zeta_1,\zeta_2,\dots\zeta_n$ and (1.4.9). In (1.4.16), the summation means the following: we divide the set $t_1, t_2, \ldots, t_n$ (with an even n; any of these time points may coincide) into pairs and construct the product of $\phi(t_j, t_k)$ for this pairing and sum up such terms for all possible ways of pairing. For example, thus

$$\langle z(t_1)\, z(t_2)\, z(t_3)\, z(t_4)\rangle$$
$$= \phi(t_1, t_2)\, \phi(t_3, t_4) + \phi(t_1, t_3)\, \phi(t_2, t_4) + \phi(t_1, t_4)\, \phi(t_2, t_3).$$

In the definition of the characteristic function (1.4.5), we set

$$z_j = z(t_j), \quad \zeta_j = \zeta(t_j)\, \Delta t_j \quad (j = 1, 2, \ldots, n)$$

and take the limit of $n \to \infty$ and $\Delta t_j \to 0$ for $t_0 < t_1 < t_2 \ldots < t_n < t$ to attain the limit

$$\sum_{j=1}^{n} \zeta_j z_j = \sum_{j=1}^{n} \zeta(t_j)\, z(t_j)\, \Delta t_j \to \int_{t_0}^{t} \zeta(t')\, z(t')\, dt'.$$

This defines

$$\Phi[\zeta(t)] = \left\langle \exp\left[i \int_{t_0}^{t} \zeta(t')\, z(t')\, dt' \right] \right\rangle. \tag{1.4.17}$$

This is the most general form of the characteristic function for the processes $z(t)$ and is called the *characteristic functional* because it contains an arbitrary function $\zeta(t)$.

In particular, if $z(t)$ is Gaussian, its characteristic functional is

$$\Phi[\zeta(t)] = \exp\left[i \int_{t_0}^{t} \zeta(t')\, m(t')\, dt' - \frac{1}{2} \int_{t_0}^{t} dt_1 \int_{t_0}^{t} dt_2\, \phi(t_1, t_2)\, \zeta(t_1)\, \zeta(t_2) \right] \tag{1.4.18}$$

corresponding to (1.4.14). In other words, the characteristic functional of a Gaussian process is completely defined in terms of the expectation $m(t)$ and the correlation function $\phi(t_1, t_2)$. If it is stationary, $m(t)$ is a constant so that it can be set equal to zero without losing generality. Furthermore, the correlation function $\phi(t_1, t_2)$ is a function of $t_1 - t_2$ only. Thus the characteristic function has the form

$$\Phi[\zeta(t)] = \exp\left[-\frac{1}{2} \int_{t_0}^{t} dt_1 \int_{t_0}^{t} dt_2\, \phi(t_1 - t_2)\, \zeta(t_1)\, \zeta(t_2) \right]. \tag{1.4.19}$$

If the characteristic functional $\Phi[\zeta(t)]$ is known, a situably chosen $\zeta(t)$ gives a desired characteristic function. For example, setting

$$\zeta(t) = \sum_{j=1}^{n} \zeta_j\, \delta(t - t_j), \tag{1.4.20}$$

(1.4.17) is reduced to (1.4.5). By taking a functional derivative of the characteristic functional, we can calculate various sorts of expectations, provided that the derivative is analytic.

Gaussian processes are rather common in the real world and are regarded as standards. The reason for this seems to lie in the central limit theorem just in the same way as the Gaussian law of error. Assume that the random variable $z(t)$ is a sum of independent random variables

$$z(t) = \Delta z_1(t) + \Delta z_2(t) + \ldots + \Delta z_n(t)\,, \tag{1.4.21}$$

where each component is by itself a stochastic process. Then the characteristic functional of $z(t)$ has the form

$$\Phi[\zeta(t)] = \exp\left\{\sum_{j=1}^{n} \psi_j[\zeta(t)]\right\},$$

where $\psi_j[\zeta(t)]$ is the cumulant function for the characteristic functional of $\Delta z_j(t)$ and is assumed to be expanded as

$$\psi_j[\zeta(t)] = \mathrm{i}\int_{t_0}^{t} \zeta(t_1)\,\langle \Delta z_j(t_1)\rangle\, dt_1 - \frac{1}{2}\int_{t_0}^{t} dt_1 \int_{t_0}^{t} dt_2\, \langle \Delta z_j(t_1)\,\Delta z_j(t_2)\rangle_{\mathrm{c}}\, \zeta(t_1)\,\zeta(t_2) + \ldots .$$

If the sums of cumulants are all of the order of n like

$$\sum_{j=1}^{n} \langle \Delta z_j(t)\rangle = O(n), \quad \sum_{j=1}^{n} \langle \Delta z_j(t_1)\,\Delta z_j(t_2)\rangle_{\mathrm{c}} = O(n), \ldots ,$$

we change the variable to

$$y(t) = \frac{z(t)}{\sqrt{n}} \tag{1.4.22}$$

and obtain its characteristic functional as

$$\begin{aligned}\Phi[\eta(t)] &= \left\langle \exp\left[\mathrm{i}\int_{t_0}^{t} \eta(t')\,y(t')\,dt'\right]\right\rangle \\ &= \exp\left\{\sum_{j=1}^{n} \psi_j[n^{-1/2}\,\eta(t)]\right\} \\ &= \exp\left[\mathrm{i}\int_{t_0}^{t} \eta(t_1)\,\langle y(t_1)\rangle\, dt_1 - \frac{1}{2}\int_{t_0}^{t} dt_1 \int_{t_0}^{t} dt_2\, \langle y(t_1)\,y(t_2)\rangle\, \eta(t_1)\,\eta(t_2) \right. \\ &\qquad \left. + O(n^{-1/2}) + O(n^{-1}) + \ldots\right]\end{aligned} \tag{1.4.23}$$

which approaches the form (1.4.18) as $n \to \infty$. Here, $O(n^{-1/2})$ and $O(n^{-1})$ mean that the higher-order terms in the cumulant expansion become asymptotically small at these orders. Like the previous discussion of the central limit theorem in Sect. 1.2, the above argument is not quite satisfactory as a mathematical proof, but it indicates that we may generally expect a Gaussian property if a phyiscal process results from a large number of independent random processes. More generally, even such a strict independence is not necessarily required for the Gaussian property to hold, provided that the interaction satisfies certain appropriate conditions not specified here. Furthermore, there are certainly important cases where the Gaussian properties no longer hold. Physical processes in the neighborhood of a phase change are interesting examples. The non-Gaussian nature of critical fluctuations has been a central topic in recent progress in statistical mechanics.

Gaussian distributions have remarkable stabilities. If the joint distribution of the random variables, $X_1, X_2, \ldots$ and X_n is Gaussian, any linear combination

$$Y_j = \sum_{k=1}^{n} C_{jk} X_k$$

of these variables also has a Gaussian distribution. Likewise, if $z(t)$ is a Gaussian process, any linear transformation of it

$$y(t) = \int_a^b C(t, t')\, z(t')\, dt', \tag{1.4.24}$$

is also Gaussian because it holds that

$$\langle y(t) \rangle = \int_a^b C(t, t') \langle z(t') \rangle\, dt', \tag{1.4.25}$$

$$\langle y(t_1)\, y(t_2) \rangle_c = \int_a^b dt_1' \int_a^b dt_2'\, C(t_1, t_1')\, C(t_2, t_2') \langle z(t_1')\, z(t_2') \rangle_c \tag{1.4.26}$$

and cumulants of $y(t)$ higher than the third vanish together with those of $z(t)$. In particular, the Fourier coefficients of $z(t)$ defined by (1.3.12) have a Gaussian distribution if $z(t)$ is a Gaussian process.

1.5 Brownian Motion Modeled by a Gaussian Process

The Langevin equation (1.3.35) depicts Brownian motion as driven by the random force $\boldsymbol{R}(t)$. As an idealization of Brownian motion, $\boldsymbol{R}(t)$ is assumed to satisfy the following conditions:

i) $\boldsymbol{R}(t)$ is a Gaussian process,
ii) $\boldsymbol{R}(t)$ has a white spectrum, namely (1.3.40) holds.

The process $\boldsymbol{u}(t)$ then represents the random change of velocity of a Brownian particle and is often called the *Ornstein-Uhlenbeck process* after the authors who treated the problem extensively for the first time [1.10].

In fact, these assumptions seem very plausible if Brownian particles are far larger than the molecules of the surrounding fluid. The force $\boldsymbol{R}(t)$ acting on a Brownian particle results from a great many impacts from the fluid molecules, so that the Gaussian property is expected to hold by the central limit theorem. Secondly, the time constant of the motion of fluid molecules will be much shorter than the characteristic time of the Brownian particle if the mass of a Brownian particle is much larger than that of fluid molecules. (Rigorously speaking, this is not quite sufficient. As Sect. 1.6 shows, this idealization of $R(t)$ is legitimate only in the limit of very large mass density of the Brownian particle.) If that is the case, as an idealization the characteristic time of successive impacts from fluid molecules may be considered as infinitely short.

The Langevin equation

$$\dot{\boldsymbol{u}}(t) = -\gamma \boldsymbol{u} + \frac{\boldsymbol{R}(t)}{m} \tag{1.5.1}$$

is easily integrated to give

$$\boldsymbol{u}(t) = \boldsymbol{u}(t_0)\, \mathrm{e}^{-\gamma(t-t_0)} + \int_{t_0}^{t} \mathrm{e}^{-\gamma(t-t')}\, \frac{\boldsymbol{R}(t')}{m}\, dt' . \tag{1.5.2}$$

Obviously this has the form of (1.4.24) so that $\boldsymbol{u}(t)$ must be a Gaussian process if $\boldsymbol{R}(t)$ is Gaussian.

The power spectrum of $\boldsymbol{u}(t)$ has already been obtained in (1.3.37) and its correlation function by (1.3.43) under assumption (ii). Therefore, the process $\boldsymbol{u}(t)$ is completely defined. The transition probability $P(u_0, t_0 | u, t)$ to find the velocity u at time t when the velocity was u_0 at the initial time t_0 is derived from (1.5.2) as follows.

The characteristic functional of $\boldsymbol{u}(t)$ is

$$\begin{aligned}\langle \mathrm{e}^{\mathrm{i}\xi u(t)} \rangle = \exp\Bigg(&\mathrm{i}\,\xi\, u_0 \exp[-\gamma(t-t_0)] - \frac{\xi^2}{2} \int_{t_0}^{t} dt_1 \int_{t_0}^{t} dt_2 \\ &\times \exp[-\gamma(t-t_1) - \gamma(t-t_2)] \frac{\langle R(t_1)\, R(t_2)\rangle}{m^2} \Bigg) .\end{aligned} \tag{1.5.3}$$

This is obtained by replacing $z(t)$ in (1.4.18) by $R(t)$ and $\zeta(t')$ by $\xi \exp[-\gamma(t-t')/m]$. However, this is rather obvious from the derivation of (1.4.18). If (1.3.41) is assumed further for $\boldsymbol{R}(t)$, the second term in the exponent in the integrand of (1.5.3) is calculated as

$$\int_{t_0}^{t} dt_1 \int_{t_0}^{t} dt_2 \exp\left[-\gamma(t-t_1)-\gamma(t-t_2)\right] \frac{2\pi I_R}{m^2}\,\delta(t_1-t_2)$$

$$= \frac{2\pi I_R}{m^2} \int_{t_0}^{t} dt'\, \mathrm{e}^{-2\gamma(t-t')} = \frac{\pi I_R}{m^2}\, \frac{1-\mathrm{e}^{-2\gamma(t-t_0)}}{\gamma}.$$

With (1.3.46), then (1.5.3) becomes

$$\left\langle \mathrm{e}^{\mathrm{i}\xi u(t)} \right\rangle = \exp\left[\mathrm{i}\,\xi\, u_0\, \mathrm{e}^{-\gamma(t-t_0)} - \frac{kT}{2m}\left(1-\mathrm{e}^{-2\gamma(t-t_0)}\right)\xi^2\right]. \tag{1.5.4}$$

This indicates that $u(t)$ has the Gaussian distribution

$$P(u_0, t_0 \,|\, u, t) = \left(\frac{m}{2\pi kT}\right)^{1/2} \frac{1}{\left(1-\mathrm{e}^{-2\gamma(t-t_0)}\right)^{1/2}} \exp\left(-\frac{m}{2kT}\,\frac{\left(u-u_0\,\mathrm{e}^{-\gamma(t-t_0)}\right)^2}{1-\mathrm{e}^{-2\gamma(t-t_0)}}\right) \tag{1.5.5}$$

which is the transition probability for $(u_0, t_0) \rightarrow (u, t)$. The expectation of the velocity decays exponentially as

$$\langle u(t) \rangle = u_0\, \mathrm{e}^{-\gamma(t-t_0)}, \tag{1.5.6}$$

if the initial was u_0 at time t_0. This is, of course, to be expected. The variance around the expectation grows in time as

$$\left\langle \left[u(t) - u_0\, \mathrm{e}^{-\gamma(t-t_0)}\right]^2 \right\rangle = \frac{kT}{m}\left(1-\mathrm{e}^{-2\gamma(t-t_0)}\right) \tag{1.5.7}$$

and approaches the Maxwellian value at $t \rightarrow \infty$. As shown below, the distribution (1.5.5) is the fundamental solution of the diffusion equation in the velocity space

$$\frac{\partial}{\partial t} P = \frac{\partial}{\partial u}\left(\gamma u + D_u \frac{\partial}{\partial u}\right) P \quad \left(D_u = \frac{\gamma kT}{m}\right). \tag{1.5.8}$$

The displacement $\boldsymbol{x}(t)$ in the time interval $(0, t)$ is obtained by integrating (1.5.2) as

$$\boldsymbol{x}(t) = \int_0^t \boldsymbol{u}(t')\, dt = \boldsymbol{u}_0 \frac{1-\mathrm{e}^{-\gamma t}}{\gamma} + \int_0^t dt_1 \int_0^{t_1} dt_2\, \mathrm{e}^{-\gamma(t_1-t_2)} \frac{\boldsymbol{R}(t_2)}{m} \tag{1.5.9}$$

($t_0 = 0$, for brevity). Obviously this has a Gaussian distribution if u_0 and $\boldsymbol{R}(t)$ are Gaussian. The second term is transformed into

$$\int_0^t dt_2 \int_{t_2}^{t} dt_1\, \mathrm{e}^{-\gamma(t_1-t_2)} \frac{\boldsymbol{R}(t_2)}{m} = \int_0^t dt' \frac{1-\mathrm{e}^{-\gamma(t-t')}}{\gamma} \frac{\boldsymbol{R}(t')}{m}.$$

The characteristic functional of $\boldsymbol{x}(t)$,

$$\langle e^{i\xi x(t)}\rangle = \left\langle \exp\left(i\,\xi\,u_0 \frac{1-e^{-\gamma t}}{\gamma}\right)\right\rangle \left\langle \exp\left(i\,\xi \int_0^t dt' \frac{1-e^{-\gamma(t-t')}}{\gamma} \frac{R(t')}{m}\right)\right\rangle$$

is calculated similarly as for (1.5.3). Assuming the distribution of $\boldsymbol{u}_0$ to be Maxwellian and using (1.3.44), then

$$\langle e^{i\xi x(t)}\rangle = \exp\left[-\xi^2 \frac{\langle u^2\rangle}{\gamma}\left(t - \frac{1-e^{-\gamma t}}{\gamma}\right)\right]. \tag{1.5.10}$$

Further we set

$$\langle u^2\rangle = \frac{kT}{m} \quad \text{and} \quad D = \frac{\langle u^2\rangle}{\gamma} = \frac{kT}{m\gamma} \tag{1.5.11}$$

to rewrite (1.5.10) as

$$\langle e^{ix(t)}\rangle = \exp\left[-\frac{\xi^2}{2} 2D\left(t - \frac{1-e^{-\gamma t}}{\gamma}\right)\right]. \tag{1.5.12}$$

This corresponds to the transition probability $P(0,0|x,t)$ of a Brownian particle to arrive around x at time t when it was certainly at $x=0$ at the initial time. The probability is Gaussian and is given by

$$P(0,0|x,t) = \left[4\pi D\left(t - \frac{1-e^{-\gamma t}}{\gamma}\right)\right]^{-1/2} \exp\left\{-x^2\left[4D\left(t - \frac{1-e^{-\gamma t}}{\gamma}\right)\right]^{-1}\right\}. \tag{1.5.13}$$

The mean square average (variance) of the displacement in the time interval $(0,t)$ is

$$\langle x(t)^2\rangle = 2D\left(t - \frac{1-e^{-\gamma t}}{\gamma}\right). \tag{1.5.14}$$

For a short time $t \ll 1/\gamma$ the distribution of $\boldsymbol{x}(t)$ is Gaussian with the variance

$$\langle x^2\rangle \approx \langle u^2\rangle\, t^2 \tag{1.5.15}$$

as is seen from (1.5.11, 14). In such a short time, the Brownian particle still keeps its initial velocity. The result (1.5.15) is simply a reflection of the initial Maxwellian distribution of the velocity.

For longer times $t \gg 1/\gamma$, the Brownian particle repeatedly zig-zags and loses the memory of its initial velocity. It is natural to expect that displacements for such a long time become a diffusion process as discussed in Sect. 1.1. In fact, (1.5.13) is then approximated by

$$P(0,0|x,t) \approx \frac{1}{(4\pi Dt)^{1/2}} \exp\left(-\frac{x^2}{4Dt}\right) \quad (t \gg \gamma^{-1}) \tag{1.5.16}$$

in accordance with (1.1.19). Furthermore, (1.5.11) is identical with (1.1.14). Namely, the Einstein relation is obtained here again. The above reasoning is not quite the same as that discussed in Sect. 1.1, but the essential point lies in the assertion that Brownian motion in a medium in thermal equilibrium also tends to attain thermal equilibrium.

1.6 Fluctuation-Dissipation Theorem

In the Langevin equation (1.3.4), the force on a Brownian particle was divided into the frictional force $-m\gamma \boldsymbol{u}$ and the random force $\boldsymbol{R}(t)$, between which a relationship like (1.3.46) exists, indicating that the power intensity of $\boldsymbol{R}(t)$ is proportional to the friction coefficient and the thermal energy kT. We have also seen that the Einstein relation relates the diffusion constant to the friction coefficient. Both express that such a mechanism of energy dissipation is closely related to fluctuations in thermal equilibrium and they are simple examples of a more general principle called the fluctuation-dissipation theorem [1.5]. Chapter 4 deals with a quantum-statistical derivation of this theorem. Here we consider this from the viewpoint of the Brownian motion theory.

As already mentioned, Brownian motion is not limited to Brownian particles. It is, generally speaking a fluctuating motion of a mode in a macroscopic dynamical system with a very large number of particles or a large number of degrees of freedom. It is particularly simple for a particle much heavier than molecules in the medium or for a mirror in a gas, described well by the simple Langevin equation discussed in Sect. 1.3. However, various modifications are required for the Langevin equation to be applied to more general sorts of Brownian motion. One modification is to abandon the assumption of a white spectrum for the random force $\boldsymbol{R}(t)$. This means, as seen in the following, that retarded friction is accounted for. This is very necessary for applying the theory to more realistic problems for which the idealizations are not legitimate.

In the Langevin equation (1.3.4) the friction is assumed to be determined by the instantaneous velocity of the particle. However, in general, friction will be retarded so that the Langevin equation should be generalized to

$$\frac{d}{dt}\boldsymbol{u}(t) = -\int_{-\infty}^{t} \gamma(t-t')\,\boldsymbol{u}(t')\,dt' + \frac{1}{m}\boldsymbol{R}(t) + \frac{1}{m}K(t)\,, \tag{1.6.1}$$

where $\gamma(t)$ expresses the friction retardation. Equation (1.6.1) is called a generalized Langevin equation. On the right-hand side, $\boldsymbol{R}(t)$ is the random force and $K(t)$ is an external force. The random force is zero on average, satisfying the condition

$$\langle R(t)\rangle = 0\,. \tag{1.6.2}$$

Suppose now that the external force is periodic as

$$K(t) = K_0 \cos \omega t = \mathrm{Re}\,\{K_0\, \mathrm{e}^{\mathrm{i}\omega t}\}\,.$$

Then the average velocity induced by this force is

$$\langle u(t)\rangle = \mathrm{Re}\,\{\mu(\omega)\, K_0\, \mathrm{e}^{\mathrm{i}\omega t}\}\,, \tag{1.6.3}$$

where $\mu(\omega)$ is the complex mobility for the frequency ω and is given by

$$\mu(\omega) = \frac{1}{m}\,\frac{1}{\mathrm{i}\,\omega + \gamma[\omega]}\,, \tag{1.6.4}$$

where

$$\gamma[\omega] = \int_0^\infty \gamma(t)\, \mathrm{e}^{-\mathrm{i}\omega t}\, dt \tag{1.6.5}$$

is the Fourier-Laplace transform[3] of the retardation function of friction. This is obtained from the averaged equation of (1.6.1)

$$\frac{d}{dt}\langle u(t)\rangle = -\int_{-\infty}^{t} \gamma(t-t')\,\langle u(t')\rangle\, dt' + \mathrm{Re}\left\{\frac{1}{m}\, K_0\, \mathrm{e}^{\mathrm{i}\omega t}\right\}.$$

If the particles are charged with e and the particle density is n, the current induced by an electric field E is

$$j(t) = en\,\langle u(t)\rangle = \mathrm{Re}\,\{e^2 n\, \mu(\omega)\, E_0\, \mathrm{e}^{\mathrm{i}\omega t}\}\,,$$

so that the complex conductivity $\sigma(\omega)$ is

$$\sigma(\omega) = e^2\, n\, \mu(\omega) = \frac{e^2\, n}{m}\,\frac{1}{\mathrm{i}\omega + \gamma[\omega]}\,. \tag{1.6.6}$$

In fact, if we write conductivity, or more generally a complex admittance in this form, $\gamma[\omega]$ is usually not constant but depends on the frequency ω. If we treat such a system from the viewpoint of Brownian motion theory, the retardation function $\gamma(t)$ must be introduced, given as the inverse of (1.6.5).

---

3 A Fourier-Laplace transform is defined for the integration range $(0, \infty)$ in contrast to an ordinary Fourier transform with $(-\infty, \infty)$. Conventional Laplace transforms use a complex parameters s instead of $\mathrm{i}\,\omega$

The generalized Langevin equation (1.6.1) is linear so that it can be treated by harmonic analysis, Sect. 1.3. The Brownian motion follows the equation

$$\frac{d}{dt}\boldsymbol{u}(t) = -\int_{-\infty}^{t} \gamma(t-t')\,\boldsymbol{u}(t')\,dt' + \frac{1}{m}\boldsymbol{R}(t), \tag{1.6.7}$$

if the external force K is zero and the motion is driven only by the random force $\boldsymbol{R}(t)$. Now we Fourier-analyze $\boldsymbol{R}(t)$ and $\boldsymbol{u}(t)$ as

$$\boldsymbol{R}(t) = \int_{-\infty}^{\infty} \boldsymbol{R}(\omega)\,\mathrm{e}^{\mathrm{i}\omega t}\,d\omega, \qquad \boldsymbol{u}(t) = \int_{-\infty}^{\infty} \boldsymbol{u}(\omega)\,\mathrm{e}^{\mathrm{i}\omega t}\,d\omega$$

to obtain

$$\boldsymbol{u}(\omega) = \frac{1}{\mathrm{i}\omega + \gamma[\omega]}\,\frac{\boldsymbol{R}(\omega)}{m}$$

from (1.6.7). If $\boldsymbol{R}(t)$ is stationary, $\boldsymbol{u}(t)$ becomes stationary for large enough t. The power spectra of the two processes are related to each other by

$$I_u(\omega) = \frac{1}{m^2}\,\frac{I_R(\omega)}{|\mathrm{i}\omega + \gamma[\omega]|^2}, \tag{1.6.8}$$

as seen from (1.3.20).

When the spectrum $I_R(\omega)$ for the random force $\boldsymbol{R}(t)$ is given, (1.6.8) yields $I_u(\omega)$, from which the correlation function $\langle u(0)\,u(t)\rangle$ is obtained by the Wiener-Khintchine theorem. If it should represent the velocity distribution in thermal equilibrium, the spectrum $I_R(\omega)$ is required to fulfill a certain condition. The condition is a generalization of (1.3.46) and is given by

$$I_R(\omega) = \frac{mkT}{\pi}\,\mathrm{Re}\,\{\gamma[\omega]\} \quad \text{or} \tag{1.6.9}$$

$$\langle R(\omega)\,R^*(\omega')\rangle = \frac{mkT}{\pi}\,\mathrm{Re}\,\{\gamma[\omega]\,\delta(\omega-\omega')\}\,. \tag{1.6.10}$$

This means

$$\langle R(t_1)\,R(t_2)\rangle = mkT\gamma(t_1-t_2) \tag{1.6.11}$$

for the correlation function of $\boldsymbol{R}(t)$ as is seen in the following way. The function $\gamma(t)$ in (1.6.5) is defined only for $t > 0$ but is extended to $t < 0$ by assuming $\gamma(t) = \gamma(-t)$. Then

$$\mathrm{Re}\,\{\gamma[\omega]\} = \frac{1}{2}\int_{-\infty}^{\infty} \gamma(t)\,\mathrm{e}^{-\mathrm{i}\omega t}\,dt\,,$$

from (1.6.5). Equation (1.6.11) follows from (1.6.9, 1.3.23). In (1.6.9) it must be assumed that

$$\mathrm{Re}\,\{\gamma[\omega]\} \geqq 0 \quad (\text{for real } \omega), \tag{1.6.12}$$

since the power spectrum should never become negative.

If $\gamma^*[\omega]$ denotes the complex conjugate of $\gamma[\omega]$, from (1.6.8, 9)

$$\langle u(t_0)\,u(t_0+t)\rangle = \frac{kT}{2\pi m}\int_{-\infty}^{\infty}\left(\frac{1}{\mathrm{i}\omega+\gamma[\omega]}+\frac{1}{-\mathrm{i}\omega+\gamma^*[\omega]}\right)e^{\mathrm{i}\omega t}\,d\omega\,. \tag{1.6.13}$$

We now show the contribution from the second term of the integrand vanishes for $t>0$. To see this, observe that the function $\gamma(\omega)$ defined by (1.6.5) is analytic in the lower half-plane $\mathrm{Im}\,\{\omega\}<0$. For such a function the dispersion relation discussed in Sect. 3.6 holds. This gives

$$\gamma[\omega] = \frac{\mathrm{i}}{\pi}\int_{-\infty}^{\infty}\frac{\gamma'(\nu)}{\nu-\omega}\,d\nu\,.$$

Setting $\omega=\omega'-\mathrm{i}\,\omega''$, then

$$\mathrm{Re}\,\{\gamma[\omega]\} = \frac{1}{\pi}\int_{-\infty}^{\infty}d\nu\,\gamma'(\nu)\,\frac{\omega''}{(\nu-\omega')^2+\omega''^2}\,.$$

It follows from (1.6.12) that

$$\mathrm{Re}\,\{\gamma[\omega]\} > 0 \quad (\mathrm{Im}\,\{\omega\}<0)$$

and

$$\mathrm{Re}\,\{\mathrm{i}\omega+\gamma[\omega]\} > 0 \quad (\mathrm{Im}\,\{\omega\}<0)\,.$$

Therefore, on the right-hand side of (1.6.13) the first term in the bracket is analytic for $\mathrm{Im}\,\{\omega\}<0$. Correspondingly, the second term is analytic for $\mathrm{Im}\,\{\omega\}>0$ and henceforth the integral containing this vanishes in (1.6.13), because the integral can be supplemented by a large semicircle on the half-plane $\mathrm{Im}\,\{\omega\}>0$ on which $\exp(\mathrm{i}\omega t)$ tends to zero for $t>0$ and the integrand is analytic inside the closed path of integration. Thus (1.6.13) is simplified to

$$\langle u(t_0)\,u(t_0+t)\rangle = \frac{kT}{2\pi m}\int_{-\infty-\mathrm{i}\varepsilon}^{\infty-\mathrm{i}\varepsilon}\frac{e^{\mathrm{i}\omega t}}{\mathrm{i}\omega+\gamma[\omega]}\,d\omega\,. \tag{1.6.14}$$

Here the integration path lies just below the real axis of ω. If there is no branch cut for the function $\gamma[\omega]$, the path can be made a closed contour by supplementing with a large semicircle as shown in Fig. 1.5. In the limit $t\to 0+$, the sum of the residues of $(\mathrm{i}\omega+\gamma[\omega])^{-1}$ is equal to the residue around the infinity $\omega = \infty$. If the condition

$$\lim_{|\omega| \to \infty} \gamma[\omega] = \text{finite}$$

is assumed, namely if the resistance remains finite for $\omega \to \infty$, the residue is simply equal to one. Therefore the equipartition law results:

$$\lim_{t \to 0+} \langle u(t_0)\, u(t_0 + t) \rangle \equiv \langle u^2 \rangle = \frac{kT}{m}.$$

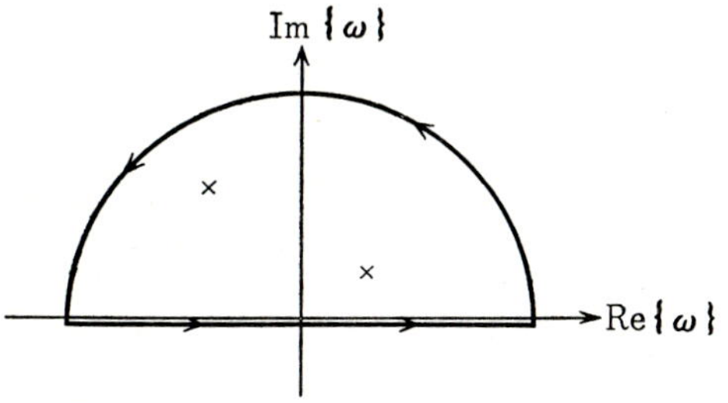

Fig. 1.5. Integration path for (1.6.14). The crosses indicate poles of the integrand

It is interesting to note that the Einstein-Ornstein-Uhlenbeck theory of Brownian motion as formulated by (1.3.1–4) has to be modified because of retardation in viscous resistance due to the hydrodynamic backflow effect. By a hydrodynamic calculation [1.11], it is shown that (1.5.1) is replaced by

$$m^* \frac{d\boldsymbol{u}}{dt} + \beta \boldsymbol{u} + \alpha \int_{-\infty}^{t} (t - t')^{-1/2}\, \dot{\boldsymbol{u}}(t')\, dt' = \boldsymbol{R}(t), \tag{1.6.15}$$

where

$$m^* = m + \tfrac{2}{3}\pi \varrho a^3,$$
$$\alpha = 6\pi \varrho a^2 (\nu/\pi)^{1/2},$$
$$\beta = 6\pi \nu \varrho a.$$

Here ϱ is the density of the fluid surrounding the particle, ϱ_0 is the average density of the matter composing the particle and ν is the kinetic viscosity equal to η/ϱ. The effective mass m^* contains additional inertia due to the dragging motion of the fluid. Retardation of viscous resistance is caused by the back flow of fluid. Then

$$\gamma[\omega] = \{\beta + (\pi \mathrm{i}\omega)^{1/2}\alpha\}/m^*.$$

This function has a branch point at $\omega = 0$ and does not satisfy the condition (1.6.12). Accordingly, the previous argument does not apply. However, the integral (1.6.14) is easily calculated by transforming the path to the contour

$(\mathrm{i}\,\infty - \varepsilon, 0, \mathrm{i}\,\infty + \varepsilon, \varepsilon > 0)$, giving

$$\langle u(t_0)\, u(t_0+t)\rangle = \frac{kT}{m^*}\,\phi(t)\,,$$

where

$$\phi(t) = \frac{1}{\pi}\int_{-\infty}^{\infty} \frac{\exp(-z^2\tau)\,\sigma\, z^2\, dz}{(z^2-1)^2+\sigma^2 z^2}$$

$$= \frac{1}{\pi}\,\sigma\tau^{-3/2}\int_{-\infty}^{\infty} \frac{\exp(-\zeta^2)\,\zeta^2\, d\zeta}{(\zeta^2/\tau - 1)^2+\sigma^2\,\zeta^2/\tau}$$

with

$$\sigma = \left(\frac{9}{2}\,\frac{\varrho}{\varrho_0+\varrho/2}\right)^{1/2}, \quad \tau = \beta t/m^*\,.$$

In the ideal limit of an extremely heavy particle ($\sigma \to 0$) $\phi(t)$ is reduced to $\exp(-\tau)$ and the retardation effect disappears. For a finite value of σ, the retardation effect shows up in a slow decay of the correlation function

$$\phi(t) \sim \frac{\sigma}{2\sqrt{\pi}}\,\tau^{-3/2}\,. \tag{1.6.16}$$

The presence of such a long-time tail was first observed by *Adler* and *Wainwright* in a computer simulation of velocity correlation functions [1.12].

Equation (1.6.14) means

$$\mu(\omega) = \frac{1}{m}\,\frac{1}{\mathrm{i}\omega+\gamma[\omega]} = \frac{1}{kT}\int_0^{\infty}\langle u(t_0)\, u(t_0+t)\rangle\, \mathrm{e}^{-\mathrm{i}\omega t}\, dt \tag{1.6.17}$$

because it is nothing but the inverse transformation of the above expression. Equations (1.6.9, 10) can now be written as

$$m\gamma[\omega] = \frac{1}{kT}\int_0^{\infty}\langle R(t_0)\, R(t_0+t)\rangle\, \mathrm{e}^{-\mathrm{i}\omega t}\, dt \tag{1.6.18}$$

in analogy to (1.6.17), here yielding two fundamental expressions of the fluctuation-dissipation theorem (F-D theorem).

The first expression gives the complex mobility (complex admittance in general) in terms of the Fourier-Laplace transform of the correlation function of velocity (flow) and is a generalization of the Einstein relation (1.5.11). The second gives the complex resistance (complex impedance in general) in terms of the Fourier-Laplace transform of the correlation function of the random force. Formula (1.6.10), which is equivalent to this, was first obtained by *Nyquist* as the power spectrum of noise voltage caused by thermal fluctuations in a resistance [1.13]. These two expressions imply that the response of a system to an external disturbance is related to thermal fluctuations spontaneously produced in the system in the absence of external forces. The relation of the dissipative part of the response to fluctuations was first recognized by *Nyquist* (Nyquist theorem) and obtained the name fluctuation-dissipation theorem. However, it is important to recognize that the theorm is .true both for the dissipative as well as the dispersive (nondissipative) parts.

To distinguish between the two expressions (1.6.17, 18), we call the first the F-D *theorem of the first kind* and the second the F-D *theorem of the second kind.* As shown in Chap. 4, the first theorem can be derived from the linear response theory in a general way [1.14]. The correlation function on the right-hand side of the equation can be analyzed from microscopic theories by statistical mechanics. On the other hand, the random force appearing in the F-D theorem of the second kind is not simple, because the separation of the force into frictional and random forces is itself a complex problem of statistical mechanics. In this sense, the F-D theorem of the first kind should be regarded as basic and the second as a corollary to the first. In the context of this chapter, the two theorems are related to each other through the Langevin equation (1.6.1).

The mean square average of the displacement of a Brownian particle in a time interval $(0, t)$ is given by

$$\langle x(t)^2\rangle = \int_0^t dt_1 \int_0^t dt_2 \langle u(t_1)\, u(t_2)\rangle .$$

This is transformed into

$$\lim_{t\to\infty} \frac{\langle x(t)^2\rangle}{2t} = \int_0^\infty \langle u(t_0)\, u(t_0+t')\rangle\, dt' \tag{1.6.19}$$

as for (1.3.24), since the correlation function $\langle u(t_1)\, u(t_2)\rangle$ is dependent on $t_1 - t_2$ only as a stationary process. Through (1.2.33, 6.17) we obtain

$$D = \mu(0)\, kT .$$

This shows that the Einstein relation (1.5.11) is a special case of the F-D theorem of the first kind.

In (1.6.1, 7) the retarded friction is expressed as an integral from the infinite past to the present. It is possible to modify this to

$$\frac{d}{dt}\boldsymbol{u}(t) = -\int_{t_0}^{t} \gamma(t-t')\, \boldsymbol{u}(t')\, dt' + \frac{1}{m}\bar{\boldsymbol{R}}(t), \quad (t > t_0) . \tag{1.6.20}$$

For stationary Brownian motion, the initial time t_0 in this equation can be chosen arbitrarily. In that sense the correlation function obtained from this equation does not depend on the choice of t_0.

The random force $\bar{\boldsymbol{R}}(t)$ is not the same as $\boldsymbol{R}(t)$ in (1.6.7) and is

$$\bar{\boldsymbol{R}}(t) = \boldsymbol{R}(t) - \int_{-\infty}^{t_0} \gamma(t-t')\, m\boldsymbol{u}(t')\, dt' . \tag{1.6.21}$$

For (1.6.20) to represent the same Brownian motion as represented by (1.6.7), $\bar{\boldsymbol{R}}(t)$ must fulfill the conditions

$$\langle u(t_0)\, \bar{R}(t)\rangle = 0 \quad \text{for} \quad t > t_0 \quad \text{and} \tag{1.6.22}$$

$$\langle \bar{R}(t_0)\, \bar{R}(t_0+t)\rangle = \langle R(t_1)\, R(t_1+t)\rangle = mkT\gamma(t) . \tag{1.6.23}$$

This can be seen as follows. From (1.6.17) follow

$$\int_0^\infty \langle \dot{u}(t_0)\, u(t_0+t)\rangle\, \mathrm{e}^{-\mathrm{i}\omega t}\, dt = -\int_0^\infty \langle u(t_0)\, \dot{u}(t_0+t)\rangle\, \mathrm{e}^{-\mathrm{i}\omega t}\, dt = \frac{\langle u^2\rangle \gamma[\omega]}{\mathrm{i}\omega + \gamma[\omega]} , \tag{1.6.24}$$

$$\int_0^\infty \langle \dot{u}(t_0)\, \dot{u}(t_0+t)\rangle\, \mathrm{e}^{-\mathrm{i}\omega t}\, dt = \frac{\langle u^2\rangle\, \mathrm{i}\omega\gamma[\omega]}{\mathrm{i}\omega + \gamma[\omega]} , \tag{1.6.25}$$

and from (1.6.20)

$$\langle u(t_0)\, \dot{u}(t_0+t)\rangle = -\int_0^t \gamma(t-t')\langle u(t_0)\, u(t_0+t')\rangle\, dt' + \frac{\langle u(t_0)\, \bar{R}(t_0+t)\rangle}{m} .$$

Condition (1.6.22) is necessary for the Laplace transform of the above equation to be satisfied by (1.6.17, 24). Furthermore,

$$\langle \bar{R}(t_0)\, \bar{R}(t_0+t)\rangle = m^2 \left\langle \dot{u}(t_0) \left[\dot{u}(t_0+t) + \int_0^t \gamma(t-t')\, u(t_0+t')\, dt' \right] \right\rangle ,$$

which yields

$$\int_0^\infty \langle \bar{R}(t_0)\, \bar{R}(t_0+t)\rangle\, \mathrm{e}^{-\mathrm{i}\omega t}\, dt = m^2 \langle u^2\rangle\, \gamma[\omega]$$

by (1.6.17, 24, 25). Therefore (1.6.23) should hold. We can show by direct calculation [1.15] using (1.6.21) that for an arbitrary $t_1 > t_0$

$$\langle \bar{R}(t_0)\, \bar{R}(t_0+t)\rangle = \langle \bar{R}(t_1)\, \bar{R}(t_1+t)\rangle ,$$

although $\bar{\boldsymbol{R}}(t)$ is by itself nonstationary since it depends on the arbitrarily chosen initial time t_0. Despite such a somewhat unnatural artifice, the form of the Langevin equation (1.6.20) is sometimes more convenient than that of (1.6.7). As shown in Sect. 2.9, there is a way of transforming the equation of motion into this form which gives a basis to formulate a statistical theory of Brownian motion.

Condition (1.6.22) means that the random force $\bar{\boldsymbol{R}}(t)$ is uncorrelated (in a weak sense) with $u(t_0)$. This does not, however, mean causality, because $\bar{\boldsymbol{R}}(t)$ is generally correlated with $\boldsymbol{u}(t)$ in the future as well as in the past. From (1.6.21) we can show that

$$\langle u(t)\,\bar{R}(t+\tau)\rangle = m\int_0^{t-t_0} \gamma(\tau+t')\,\langle u(0)\,u(t')\rangle\,dt' \tag{1.6.26}$$

and from (1.6.7)

$$\langle u(t)\,\bar{R}(t+\tau)\rangle = m\int_0^{\infty} \gamma(\tau+t')\,\langle u(0)\,u(t')\rangle\,dt'\,. \tag{1.6.27}$$

Both expressions tend to zero with increasing τ to ∞. But they are not equal to zero except for no retardation, where $\gamma(t)$ is a delta function and both correlation functions are zero for $\tau > 0$ and equal to $2m\gamma\langle u^2\rangle\exp(\gamma\tau)$ for $\tau < 0$. In general cases of retarded friction, the random force must be correlated with the velocity in the past, which is not surprising.

2. Physical Processes as Stochastic Processes

The previous chapter treated Brownian motion as the most typical stochastic process in physics. This chapter goes further to discuss the basic ideas of how statistical problems are treated as stochastic processes. In particular, the concept of Markovian processes plays a very important role in physics and so it is treated in detail, including the conditions for their validity. A fundamental problem is how a physical process is treated when it is no longer regarded as Markovian. It is not possible to discuss this thoroughly but we shall touch on it also.

2.1 Random Frequency Modulation

As an introduction to the problems to be treated in this chapter we first consider the following problem. Suppose that an oscillator is randomly perturbed so that its proper frequency is modulated in a random fashion. Namely the frequency ω changes in time as

$$\omega(t) = \omega_0 + \omega_1(t) , \qquad (2.1.1)$$

where the variation $\omega_1(t)$ is a stochastic process[1]. For such an oscillator, a broadened frequency spectrum is observed, whereby the question is how the broadening is related to the stochastic nature of $\omega_1(t)$.

This kind of problem is often encountered when trying to find microscopic information from observing spectral line shapes of resonance absorption by atoms, molecules, or electronic and nuclear spins in matter [2.1–4]. We cannot treat this exhaustively and so shall give a rather simple treatment. The equation of motion of the oscillator is

$$\dot{x}(t) = \mathrm{i}\,\omega(t)\,x(t) , \qquad (2.1.2)$$

where the real part of x corresponds to the coordinate of the oscillator and the imaginary part to the momentum. If the initial value of x is denoted by

---

1 Hereafter we shall not use the special lettering for random variables

x_0, the solution of (2.1.2) is

$$x(t) = x_0 \exp\left[\mathrm{i} \int_0^t \omega(t')\,dt'\right]. \tag{2.1.3}$$

If $\omega_1(t)$ is stochastic, then $x(t)$ is a stochastic process defined by this equation. Its correlation function is

$$\langle x^*(0)\,x(t)\rangle = \left\langle x_0^* x_0 \exp\left[\mathrm{i} \int_0^t \omega(t')\,dt'\right]\right\rangle .$$

The initial value x_0 is assumed to be independent of the process $\omega(t)$. Then the correlation function is essentially

$$\langle x^*(0)\,x(t)\rangle = \left\langle \exp\left[\mathrm{i} \int_0^t \omega(t')\,dt'\right]\right\rangle .$$

By the assumption (2.1.1), ω_0 is simply a constant so that the above expression becomes

$$\begin{aligned} \langle x^*(0)\,x(t)\rangle &= \mathrm{e}^{\mathrm{i}\omega_0 t}\,\phi(t)\,, \\ \phi(t) &= \left\langle \exp\left[\mathrm{i} \int_0^t \omega_1(t')\,dt'\right]\right\rangle . \end{aligned} \tag{2.1.4}$$

The power spectrum is given by the Wiener-Khintchine theorem (Sect. 1.3)

$$I(\omega') = \frac{1}{2\pi} \int_{-\infty}^{\infty} \phi(t)\,\mathrm{e}^{-\mathrm{i}\omega' t}\,dt\,, \tag{2.1.5}$$

where

$$\omega' = \omega - \omega_0 \tag{2.1.6}$$

is the frequency difference measured from the standard frequency ω_0.

The process $\omega_1(t)$ is assumed to be stationary with its average equal to zero

$$\langle \omega_1(t)\rangle = 0\,.$$

Its correlation function is written as

$$\langle \omega_1(t_0)\,\omega_1(t_0+t)\rangle = \langle \omega_1^2\rangle\,\psi(t)\,. \tag{2.1.7}$$

If the process $\omega_1(t)$ is *Gaussian*, every average value concerning $\omega_1(t)$ can be expressed in terms of this correlation function. In particular, (2.1.4) turns out to be

$$\begin{aligned}\phi(t) &= \exp\left[-\frac{1}{2}\int_0^t dt_1 \int_0^t dt_2 \langle \omega_1(t_1)\,\omega_1(t_2)\rangle\right] \\ &= \exp\left[-\int_0^t dt_1 \int_0^{t_1} dt_2 \langle \omega_1(t_1)\,\omega_1(t_2)\rangle\right] \\ &= \exp\left[-\langle \omega_1^2\rangle \int_0^t (t-\tau)\,\psi(\tau)\,d\tau\right]. \end{aligned} \tag{2.1.8}$$

This is obtained easily as follows. In (1.4.18) for the characteristic functional of a Gaussian process $\zeta(t)$ is taken simply as unity. The integral over t_1 and t_2 on a square is changed into an integral over a triangle, half of the square, and is transformed similarly as in deriving (1.3.25) from (1.3.24). Equation (2.1.5) now becomes

$$I(\omega) = \frac{1}{2\pi}\int_{-\infty}^{\infty} dt \exp\left[-\mathrm{i}\,\omega t - \Delta^2 \int_0^t (t-\tau)\,\psi(\tau)\,d\tau\right] \tag{2.1.9}$$

where the frequency is now measured from the standard ω_0, and

$$\Delta^2 = \langle \omega_1^2 \rangle \tag{2.1.10}$$

is a measure of the magnitude of the random frequency modulation. Another quantity characterizing the modulation is the rate of modulation or the correlation time, represented by

$$\tau_c = \int_0^\infty \psi(\tau)\,d\tau = \frac{1}{\langle \omega_1^2\rangle}\int_0^\infty \langle \omega_1(t_0)\,\omega_1(t_0+t)\rangle\,dt\,. \tag{2.1.11}$$

If the correlation function of $\omega_1(t)$ decays simply as

$$\psi(t) = \mathrm{e}^{-t/\tau_c}, \tag{2.1.12}$$

the integral (2.1.11) is just equal to the decay time τ_c. Generally, the decrease of $\psi(t)$ is not as simple as (2.1.12), but τ_c defined by (2.1.11) is regarded as a good parameter for the rate. The spectral form $I(\omega)$ calculated from (2.1.9) depends characteristically on the value of the parameter

$$\tau_c \Delta = \alpha\,. \tag{2.1.13}$$

In particular, when (2.1.12) is assumed, (2.1.8) becomes

$$\begin{aligned}\phi(t) &= \exp\{-\Delta^2\tau_c[t-\tau_c(1-\mathrm{e}^{-t/\tau_c})]\} \\ &= \exp\left[-\alpha^2\left(\frac{t}{\tau_c}-1+\mathrm{e}^{-t/\tau_c}\right)\right]. \end{aligned} \tag{2.1.14}$$

This function is shown in Fig. 2.1 and its Fourier transform

$$I(\omega)=\frac{1}{2\pi}\int_{-\infty}^{\infty}\phi(t)\,\mathrm{e}^{-\mathrm{i}\omega t}\,dt$$

in Fig. 2.2.

In the limit $\alpha\to\infty$ or $\tau_c\to\infty$, (2.1.14) is approximated by

$$\phi(t)=\exp\left(-\frac{\Delta^2}{2}t^2\right). \tag{2.1.15}$$

Correspondingly, the spectrum is Gaussian:

$$I(\omega)=\frac{1}{\sqrt{2\pi}\Delta}\exp\left(-\frac{\omega^2}{2\Delta^2}\right). \tag{2.1.16}$$

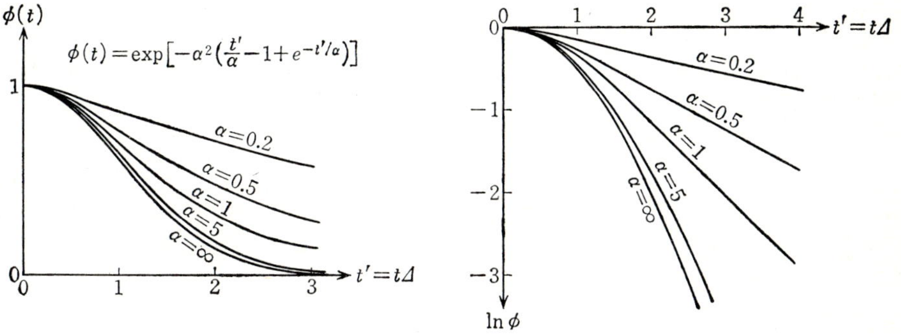

Fig. 2.1. The function $\phi(t)$ for various values of α

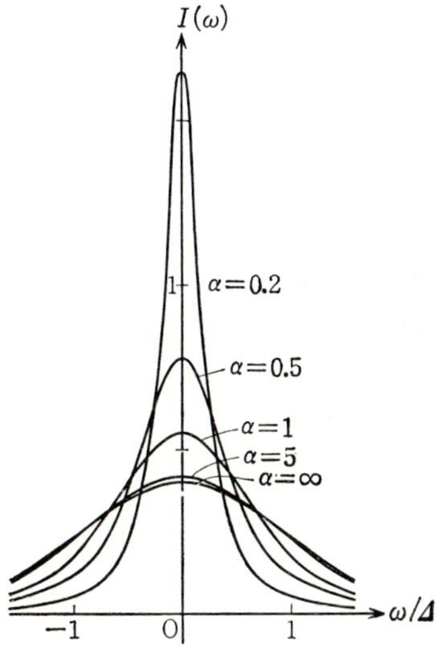

Fig. 2.2. Fourier transform $I(\omega)$ of ϕ(t)

When parameter α decreases, the spectrum becomes narrower and narrower. This narrowing is a common phenomenon, not only limited to this simple example, and is called *motional narrowing* [2.5]. The motion of $x(t)$ expressed by (2.1.3) is written as

$$x(t) = x_0\, e^{i\omega_0 t} \exp\left(i \int_0^t \omega_1(t')\, dt'\right), \tag{2.1.17}$$

which means that it is Brownian motion on a unit circle in the complex plane. A temporary shift of the frequency is meaningful as a frequency change only if it keeps the value for a duration

$$T \approx \frac{2\pi}{\omega_1'}.$$

If the duration time is shorter than T, the modulation is averaged out and cannot be observed. This is a simple interpretation of the above-mentioned phenomenon of motional narrowing.

Let us consider this problem in more detail using (2.1.8, 9). Assume that the correlation function of the frequency modulation decays within the correlation time defined by (2.1.11).

For a very short time $t \ll \tau_c$, the correlation function still remains undamped, so

$$\psi(t) \approx \text{const} = 1 \qquad (t \ll \tau_c).$$

Hence the *short-time approximation* of $\phi(t)$, (2.1.8), is

$$\phi(t) = \exp\left(-\frac{1}{2}\Delta^2 t^2\right), \tag{2.1.18}$$

which is simply the average of $\exp(i\,\omega_1 t)$ over all possible distributions of the modulation ω_1: namely, it is the average

$$\phi(t) = \int \exp(i\,\omega_1 t) \exp(-\omega_1^2/2\Delta^2) \frac{d\omega_1}{\sqrt{2\pi}\,\Delta} \tag{2.1.19}$$

over the ensemble of oscillators with different frequencies ω_1. Each oscillator has coherent motion during this short time. In this sense, the *dynamical coherence* is dominant for the short-time approximation.

On the other hand, for a long time

$$t \gg \tau_c$$

longer than the correlation time τ_c, the function $\phi(t)$ in (2.1.8) behaves as

$$\phi(t) = \exp(-\gamma|t| + \delta), \tag{2.1.20}$$

where

$$\gamma = \Delta^2 \cdot \tau_c = \Delta \cdot \alpha , \tag{2.1.21}$$

$$\delta = \Delta^2 \int_0^\infty \tau \psi(\tau)\, d\tau . \tag{2.1.22}$$

This is called the *long-time approximation,* which is not valid for short times. If it were valid for all t's, then the corresponding spectrum $I(\omega)$ would become

$$I(\omega) = \frac{e^\delta}{\pi} \frac{\gamma}{\omega^2 + \gamma^2} . \tag{2.1.23}$$

This is called the *Cauchy* distribution in probability theory and the *Lorentzian* distribution in physics. With the center frequency ω_0 explicitly included in this expression, it becomes

$$I(\omega) \propto \frac{1}{\pi} \frac{\gamma}{(\omega - \omega_0)^2 + \gamma^2} . \tag{2.1.24}$$

The spectral curves in Fig. 2.2 are in fact nearly *Gaussian* if $\alpha \equiv \tau_c \Delta > 1$ and Lorentzian if $\alpha < 1$. This fact can be understood in the following way. Figure 2.1 shows the change of decaying behavior of the function $\phi(t)$ when the parameter α is changed. In every case, the short-time approximation and the long-time approximation should be valid in appropriate ranges of t. But, depending on α, the decaying behavior as a whole may be well approximated by either of the two approximations. Namely, if $\alpha > 1$, the short-time approximation is suitable and if $\alpha < 1$, the long-time approximation is suitable.

If $\alpha \gg 1$, the short-time approximation gives a very small value $\phi(\tau_c) \ll 1$ at $t = \tau_c$. In other words, the function $\phi(t)$ is already very small at the time where the short-time approximation ceases to be valid. Then the short-time approximation is a good approximation as a whole and the corresponding spectral shape is really observed.

On the other hand, if $\alpha \ll 1$, the approximation (2.1.18) gives $\phi(\tau_c) \approx 1$ so that the decaying behavior of $\phi(t)$ can no longer be described by the short-time approximation. The approximation (2.1.20) gives $\phi(\tau_c) \approx 1$ with $\delta = O(\alpha^2)$, so that most of the decay is described by the long-time approximation. Correspondingly, the spectrum becomes Lorentzian.

The correlation function $\phi(t)$ and the spectrum $I(\omega')$ are mutually Fourier transforms as seen by (2.1.5). By a general property of a Fourier transform, the behavior of $I(\omega')$ in the neighborhood of $\omega' = 0$ is governed by the asymptotic behavior of $\phi(t)$ for large t, and that for large ω' by that of $\phi(t)$ for small t. Accordingly, the shape of the spectrum at the center is close to a Lorentzian form which corresponds to the long-time approximation. However the wings are close to a Gaussian form, corresponding to the short-time approximation. The question as to which is a good approximation for the spectrum as a whole is answered by the condition $\alpha > 1$ or $\alpha < 1$. Transition between the two is observed around the condition $\alpha \approx 1$.

From the inverse formula

$$\phi(t) = \int_{-\infty}^{\infty} I(\omega')\, e^{i\omega' t}\, d\omega'$$

follow the moment formulas

$$1 = \phi(0) = \int_{-\infty}^{\infty} I(\omega')\, d\omega', \tag{2.1.25}$$

$$\left[-\frac{d^2}{dt^2}\phi(t)\right]_{t=0} = \int_{-\infty}^{\infty} I(\omega')\, \omega'^2\, dw' \equiv m_2\,, \tag{2.1.26}$$

$$\left[\frac{d^4}{dt^4}\phi(t)\right]_{t=0} = \int_{-\infty}^{\infty} I(\omega')\, \omega'^4\, d\omega' \equiv m_4\,, \tag{2.1.27}$$

provided that the derivatives exist at $t = 0$. These are simply (1.2.17) because the function $\phi(t)$ is the characteristic function of $I(\omega')$. To calculate these moments, we expand the function $\psi(t)$, (2.1.7), in powers of t as

$$\psi(t) = 1 - \frac{1}{2}\frac{\langle\dot{\omega}_1^2\rangle}{\langle\omega_1^2\rangle}\, t^2 + O(t^4)\,. \tag{2.1.28}$$

This is obtained as follows. Note that the process $\omega_1(t)$ is stationary and that

$$0 = \frac{d}{dt_0}\langle\omega_1(t_0)\,\omega_1(t_0)\rangle = \langle\dot{\omega}_1\,\omega_1\rangle + \langle\omega_1\,\dot{\omega}_1\rangle = 2\langle\omega_1\,\dot{\omega}_1\rangle\,,$$

$$0 = \frac{d^2}{dt_0^2}\langle\omega_1(t_0)\,\omega_1(t_0)\rangle = \langle\ddot{\omega}_1\,\omega_1\rangle + 2\langle\dot{\omega}_1\,\dot{\omega}_1\rangle + \langle\omega_1\,\ddot{\omega}_1\rangle, \quad \text{or}$$

$$\langle\omega_1\,\dot{\omega}_1\rangle = 0\,, \quad \langle\omega_1\,\ddot{\omega}_1\rangle = -\langle\dot{\omega}_1^2\rangle\,.$$

In the expansion of $\langle\omega_1(t_0)\,\omega_1(t_0+t)\rangle$ in t, the term in t vanishes and the coefficient of the term t^2 becomes equal to $-\langle\dot{\omega}_1^2\rangle$. Inserting (2.1.28) into (2.1.8) and taking the terms up to t^4 yields

$$m_2 = \langle\omega_1^2\rangle \equiv \Delta^2\,, \tag{2.1.29}$$

$$m_4 = 3\Delta^4 + \langle\dot{\omega}_1^2\rangle\,. \tag{2.1.30}$$

In other words, the random frequency modulation does not alter the second moment but increases the fourth moment m_4 by $\langle\dot{\omega}_1^2\rangle$ in excess of the

Gaussian value $3\Delta^4$. The excess is larger when the modulation rate is faster. This is another way of expressing the effect of motional narrowing.

When the modulation is no longer Gaussian, the short-time and long-time approximations of $\phi(t)$ can be considered as in (2.1.18, 20). A typical example is when the modulation $\omega_1(t)$ comprises random pulses. The duration time of each pulse is assumed to be very short but the phase shift

$$\int_{\text{pulse}} \omega_1(t)\, dt = \delta \qquad (2.1.31)$$

is finite. Further, the occurrence of pulses is probabilistic and is assumed to be a Poisson process. If the average occurrence is v pulses per unit time, there are

$$\bar{n} = vt \qquad (2.1.32)$$

pulses on average during a time interval t. More precisely, the probability for the occurrence of n pulses is given by the Poisson distribution

$$p(n) = \frac{\bar{n}^n}{n!} \exp(-\bar{n}) \,. \qquad (2.1.33)$$

Therefore the long-time approximation of $\phi(t)$ is

$$\begin{aligned} \phi(t) &= \langle e^{i\Sigma\delta} \rangle = \sum_{n=0}^{\infty} p(n) \langle e^{i\delta} \rangle^n \\ &= \exp[\bar{n}(\langle e^{i\delta}\rangle - 1)] = \exp(-\gamma^* t) \end{aligned} \qquad (2.1.34)$$

where

$$\gamma^* \equiv \gamma' - i\gamma'' = v(1 - \langle e^{i\delta}\rangle) = v(1 - \langle \cos\delta \rangle) - iv\langle \sin\delta \rangle \,. \qquad (2.1.35)$$

It is worth noting that the average $\langle \exp(i\delta)\rangle$ for a pulse contributes in this form to the damping and oscillation of $\phi(t)$. This is quite general. In a more complex many-body system, the contribution of each single scattering process is represented similarly. The imaginary part γ'' of γ^* obviously means a shift of the frequency. The spectral intensity becomes Lorentzian

$$I(\omega) = \frac{1}{\pi} \frac{\gamma'}{(\omega - \omega_0 - \gamma'')^2 + \gamma'^2} \,, \qquad (2.1.36)$$

corresponding to (2.1.34).

It is important that the long-time behavior is an exponential decay for the Gaussian or the Poisson modulation, or even for more general sorts of random modulation. This exponential decay corresponds to the condition for a physical process to be approximated as a Markovian process.

In an external magnetic field, an electronic or a nuclear spin makes a Larmor precession. If the environment interacts with the spin, it produces an additional effective field, which modulates the Larmor frequency according to (2.1.1). Here $\omega_1(t)$ represents the random shift of the frequency caused by the random motion of the environment. The line shape of the observed spectrum reflects the nature of this motion. This is a good example of the random frequency modulation discussed above.

However, this is not why we have considered this problem. This problem is very basic and deeply related to our problem of describing physical processes as stochastic processes.

2.2 Brownian Motion Revisited

The evolution of displacement of a Brownian particle is governed by

$$\dot{x}(t) = u(t)\,, \tag{2.2.1}$$

where $u(t)$ is the velocity. We have discussed this problem already in (1.5.9), but let us come back to this from a slightly different viewpoint. If $u(t)$ is a stochastic process, then $x(t)$ is also a stochastic process to be determined from $u(t)$.

We consider an ensemble of Brownian particles and denote the distribution function by $f(x,t)$; namely $f(x,t)\,dx$ is the probability of finding a particle arbitrarily picked up from the ensemble in dx around x. For a given sample $u(t)$ of the process $u(t)$, the distribution function evolves in time according to

$$\frac{\partial}{\partial t} f(x,t) = -\frac{\partial}{\partial x}[u(t)\,f(x,t)] \tag{2.2.2}$$

because the conservation of particle number (or probability) is

$$f(x,t+\delta t) = f(x-u\,\delta t, t)\,,$$

which results in (2.2.2) in the limit of $\delta t \to 0$. Now we Fourier-analyze $f(x,t)$ as

$$f(x,t) = \frac{1}{2\pi}\int_{-\infty}^{\infty} \mathrm{e}^{\mathrm{i}kx}\, g(k,t)\, dk \tag{2.2.3}$$

and rewrite (2.2.2) as

$$\frac{d}{dt} g(k,t) = -\,\mathrm{i}k\,u(t)\,g(k,t) \tag{2.2.4}$$

for the amplitude $g(k,t)$. Since $u(t)$ is stochastic, this is exactly the same as the equation of an oscillator with random frequency modulation, so the previous treatment can be applied here.

The solution of (2.2.4) for a given sample $u(t)$ is

$$g(k,t) = g(k,0)\exp\left[-ik\int_0^t u(t')\,dt'\right], \tag{2.2.5}$$

where

$$g(k,0) = \int_{-\infty}^{\infty} f(x,0)\,\mathrm{e}^{-ikx}\,dx \tag{2.2.6}$$

corresponds to the initial distribution. Considering all the possibilities of $u(t)$, the probability distribution of Brownian particles at time t is obtained by averaging (2.2.5) over the process $u(t)$.

In particular, if $u(t)$ is Gaussian, then

$$\begin{aligned}\left\langle \exp\left[-ik\int_0^t u(t')\,dt'\right]\right\rangle &= \exp\left[-\frac{k^2}{2}\int_0^t dt_1\int_0^t dt_2\langle u(t_1)\,u(t_2)\rangle\right]\\ &= \exp\left[-k^2\langle u^2\rangle\int_0^t (t-\tau)\,\psi(\tau)\,d\tau\right] \equiv \phi(k,t)\,,\end{aligned} \tag{2.2.7}$$

where

$$\psi(\tau) = \frac{\langle u(t_0)\,u(t_0+\tau)\rangle}{\langle u^2\rangle} \tag{2.2.8}$$

is the normalized correlation function of velocity. Its correlation time is defined by the same equation as (2.1.11). For short times $t \ll \tau_c$ and for long times $t \gg \tau_c$, the short-time and the long-time approximations

$$\phi(t) \approx \exp\left(-\frac{k^2}{2}\langle u^2\rangle t^2\right) \qquad (t \ll \tau_c)\,, \tag{2.2.9}$$

$$\phi(t) \approx \exp(-k^2\langle u^2\rangle\tau_c t + \delta) \qquad (t \gg \tau_c) \tag{2.2.10}$$

may be used, respectively.

According to the discussion in the previous section, the approximation which should be used depends on whether the parameter

$$\alpha = |k|\sqrt{\langle u^2\rangle}\,\tau_c \equiv |k|\,l \tag{2.2.11}$$

is larger or smaller than unity. Here, l is the mean free path so that the meaning of this condition is obvious. Namely, when variations of the probability occur over a distance smaller than the mean free path l, then

$|k|\,l = \alpha > 1$ and so we have to use the short-time approximation. This corresponds to the dynamical coherence which means that particles keep their memory of velocity. But when variations of the probability take place over a distance much larger than l, then $\alpha = |k|\,l \ll 1$ and so the long-time approximation is appropriate. Then

$$P(x,t) \equiv \langle f(x,t) \rangle = \frac{1}{2\pi} \int_{-\infty}^{\infty} e^{ikx} g(k,0) \exp(-k^2 \langle u^2 \rangle \tau_c t + \delta)\, dk \tag{2.2.12}$$

is equivalent to the diffusion equation

$$\frac{\partial P(x,t)}{\partial t} = D \frac{\partial^2 P(x,t)}{\partial x^2}, \qquad D = \langle u^2 \rangle \tau_c . \tag{2.2.13}$$

When the initial condition

$$P(x,0) = \delta(x - x_0), \qquad g(k,0) = e^{-ikx_0}$$

is imposed, (2.2.12) obviously becomes the fundamental solution

$$P(x,t) = \frac{1}{\sqrt{4\pi D t}} \exp\left(- \frac{(x - x_0)^2}{4Dt} \right) \tag{2.2.14}$$

of (2.2.13).

If the correlation function (2.2.8) of velocity has the form (1.5.6), (2.2.7) is found to be

$$\phi(k,t) = \exp\{-k^2 D[t - \tau_c(1 - e^{-t/\tau_c})]\}, \tag{2.2.15}$$

using (2.1.14). As expected, (1.5.13) is obtained by inserting this into (2.2.3) to calculate $\langle f(x,t) \rangle$. The above consideration is instructive in the following way. The transition probability (1.5.13) seemingly coincides with (2.2.14) for values of t much larger than $\tau_c = 1/\gamma$. This is a *coarse graining of time.* But this is not sufficient for a Brownian motion to be approximated by a diffusion process which is typically Markovian. For that, it is necessary to allow also for *coarse graining in space* by limiting the precision of spatial measurements to a scale larger than the mean free path. We repeatedly emphasize the fact that the process $x(t)$ derived from an underlying process $u(t)$ is not Markovian in general, but acquires Markovian character only after some information is lost by achieving both spatial and temporal coarse graining, that is, by limiting the precision of spatial and temporal measurements.

If the aim of statistical physics is to bridge the microscopic and macroscopic worlds, the problems we encounter are always of the same kind. That is, we have to depict a physical process at a cruder stage of description than a more basic physical process allows for, just as in the example of Brownian motion. The process derived can be desired to be as simple as Markovian only if a part of the information is sacrificed by some coarse-graining procedure. We shall come back to this below in Sect. 2.5.

The above consideration leads to the following idealization. In the example of (2.2.1), the velocity correlation function is idealized to

$$\langle u(t_0)\, u(t_0+t)\rangle = 2\langle u^2\rangle \tau_c\, \delta(t) = 2D\, \delta(t)\,. \tag{2.2.16}$$

Although the correlation time τ_c of velocity is finite, it is very short in comparison with the time Δt determined by the precision of spatial measurement Δx:

$$\Delta t = \frac{(\Delta x)^2}{D} \gg \tau_c\,. \tag{2.2.17}$$

In addition, of course, it must be short compared with the precision of temporal measurement. Idealization of the correlation function to a delta function implies these two conditions. Condition (2.2.17) is nothing but

$$\Delta x \gg \sqrt{\langle u^2\rangle}\, \tau_c = l\,.$$

Once an idealization such as (2.2.16) is made, the derivation of (2.2.13) from (2.2.2) is much simpler. The problem can be explained better in a somewhat more general form. Equation (2.2.2) is generalized to

$$\frac{\partial}{\partial t} f(x,t) = \Omega(x,t)\, f(x,t) \tag{2.2.18}$$

for $f(x,t)$, where $\Omega(x,t)$ is a differential operator

$$\Omega(x,t) = -\frac{\partial}{\partial x}\, u(t)\,,$$

in the example of (2.2.2) and is a stochastic operator since $u(t)$ is a stochastic process. For a given sample of $\Omega(x,t)$, (2.2.18) is the equation of motion for determining the evolution of $f(x,t)$ and corresponds to a Liouville equation in phase space of a dynamical system. If $\Omega(x,t)$ is considered as a stochastic process, this is a stochastic equation for a random variable $f(x,t)$. This may be called a *stochastic Liouville equation* [2.6]. Note that a Langevin equation is a stochastic version of an equation of motion whereas a stochastic Liouville equation is a stochastic version of a Liouville equation.

For some given $\Omega(x,t)$ and for the initial condition

$$f(x,t_0) = P(x,t_0)\,,$$

(2.2.18) can be solved in the form

$$f(x,t) = f(x,t;\Omega(x,t), P(x,t_0)) .$$

This is averaged over all possibilities of $\Omega(x,t)$ to result in

$$P(x,t) = \langle f(x,t;\Omega(x,t), P(x,t_0))\rangle . \tag{2.2.19}$$

Then the evolution

$$P(x,t_0) \rightarrow P(x,t) \tag{2.2.20}$$

is the process derived from the stochastic Liouville equation. The above is a general reformulation of what was said about (2.2.1, 2). The Langevin equation

$$\dot{u} = -\gamma u + \frac{1}{m} R(t) \tag{2.2.21}$$

corresponds to the Liouville equation

$$\frac{\partial}{\partial t} f(u,t) = \frac{\partial}{\partial u}\left[\gamma u - \frac{1}{m} R(t)\right] f(u,t) \tag{2.2.22}$$

in the velocity space. By averaging over all possibilities of the process $R(t)$, we should be able to deduce the stochastic evolution of $u(t)$.

Now we integrate (2.2.18) for a time interval $(t, t+\Delta t)$ to obtain

$$f(x,t+\Delta t) = f(x,t) + \int_t^{t+\Delta t} dt_1\, \Omega(x,t_1) f(x,t_1) \tag{2.2.23}$$

$$= \left[1 + \int_t^{t+\Delta t} dt_1\, \Omega(x,t_1) + \int_t^{t+\Delta t} dt_1 \int_t^{t_1} dt_2\, \Omega(x,t_1)\, \Omega(x,t_2) + \dots\right] f(x,t) .$$

The average evolution of $f(x,t)$ is obtained for this time interval by averaging over the possibilities of $\Omega(x,t)$. This is rather complicated in a general case but can be greatly simplified if the correlation time of the process $\Omega(x,t)$ is very short in comparison with the characteristic time of the derived process of $x(t)$. Then we may choose Δt long enough that the correlation of $x(t)$ and $\Omega(x,t)$ for $t < t' < t+\Delta t$ can be ignored and still so short that Δt is considered as infinitesimally small. An average like

$$\langle \Omega(x,t_1)\, \Omega(x,t_2) \dots \Omega(x,t_n)\rangle \tag{2.2.24}$$

can be split into products of cumulant averages in a general term of the expansion (2.2.23). Each cumulant contributes to the time integral in the order of Δt. Therefore only the highest cumulant is relevant in the limit $\Delta t \to 0$. Thus

$$\int_t^{t+\Delta t} dt_1 \int_t^{t_1} dt_2 \ldots \int_t^{t_{n-1}} dt_n \langle \Omega(x,t_1) \ldots \Omega(x,t_n) \rangle \propto \Delta t \tag{2.2.25}$$

(or higher order). If this assumption is made, (2.2.23) becomes

$$P(x,t+\Delta t) - P(x,t) = [\Gamma(x,t)\,\Delta t + o(\Delta t)]\,P(x,t)$$

when averaged in the sense of (2.2.19). Taking the limit of $\Delta t \to 0$ thus gives the evolution equation

$$\frac{\partial}{\partial t} P(x,t) = \Gamma(x,t)\,P(x,t)\,, \tag{2.2.26}$$

where $\Gamma(x,t)$ is an operator determined by the averages (2.2.24). This is independent of t if the process $\Omega(x,t)$ is stationary:

$$\frac{\partial}{\partial t} P(x,t) = \Gamma(x)\,P(x,t)\,. \tag{2.2.27}$$

Equations (2.2.26, 27) are generalizations of the diffusion equation (2.2.13) and represent *Markovian* processes defined by (1.1.22) because they are first-order partial differential equations with respect to t. It is easily proved that the solution of (2.2.26 or 27) with the initial condition

$$P(x,t_0) = \delta(x - x_0) \tag{2.2.28}$$

satisfies (1.1.22).

Let us apply the above consideration to (2.2.2). Assume $u(t)$ to be a Gaussian process with the correlation function (2.2.16). Then

$$\langle \Omega(x,t) \rangle = -\langle u(t) \rangle \frac{\partial}{\partial x} = 0$$

$$\begin{aligned} \int_t^{t+\Delta t} dt_1 \int_t^{t_1} dt_2 \langle \Omega(x,t_1)\,\Omega(x,t_2) \rangle &= \int_t^{t+\Delta t} dt_1 \int_t^{t_1} dt_2 \langle u(t_1)\,u(t_2) \rangle \frac{\partial^2}{\partial x^2} \\ &= \int_t^{t+\Delta t} dt_1 \int_t^{t_1} dt_2\, 2D\,\delta(t_1 - t_2) \frac{\partial^2}{\partial x^2} \\ &= \Delta t\, D \frac{\partial^2}{\partial x^2}\,. \end{aligned}$$

Any term containing as odd number of Ω's vanishes on average, while a term with an even number of Ω's is calculated as

$$\begin{aligned}
&\int_t^{t+\Delta t} dt_1 \int_t^{t_1} dt_2 \dots \int_t^{t_{2n-1}} dt_{2n} \langle u(t_1) \dots u(t_{2n})\rangle \\
&= \frac{1}{(2n)!} \int_t^{t+\Delta t} dt_1 \dots \int_t^{t+\Delta t} dt_{2n} \langle u(t_1) \dots u(t_{2n})\rangle \\
&= \frac{1}{(2n)!} \int_t^{t+\Delta t} dt_1 \dots \int_t^{t+\Delta t} dt_{2n} \sum \prod \langle u(t_j)\, u(t_l)\rangle \propto (\Delta t)^n \,,
\end{aligned} \tag{2.2.29}$$

which shows that terms containing more than two u's do not contribute to Γ in (2.2.27). Thus, (2.2.27) reduces to (2.2.13).

Similarly, if the random force $R(t)$ is Gaussian and has a correlation function (1.3.41)

$$\langle R(t_1)\, R(t_2)\rangle = 2m\gamma kT\, \delta(t_1 - t_2)\,, \tag{2.2.30}$$

it is easy to derive (1.5.8)

$$\frac{\partial}{\partial t} P(u,t) = \frac{\partial}{\partial u} \gamma \left(u + \frac{kT}{m} \frac{\partial}{\partial u}\right) P(u,t) \tag{2.2.31}$$

for (2.2.27). To this end, we set

$$\Omega(t) = \frac{\partial}{\partial u} \left[\gamma u - \frac{1}{m} R(t)\right]$$

and calculate

$$\langle \Omega(t)\rangle = \frac{\partial}{\partial u} \gamma u \qquad \text{and}$$

$$\begin{aligned}
&\int_t^{t+\Delta t} dt_1 \int_t^{t_1} dt_2 \langle \Omega(t_1)\, \Omega(t_2)\rangle \\
&= \frac{1}{m^2} \int_t^{t+\Delta t} dt_1 \int_t^{t_1} dt_2 \langle R(t_1)\, R(t_2)\rangle \frac{\partial^2}{\partial u^2} + O(\Delta t^2) \\
&= \Delta t \frac{\gamma kT}{m} \frac{\partial^2}{\partial u^2} + O(\Delta t^2)\,.
\end{aligned}$$

By the Gaussian assumption for $R(t)$, terms containing more than two Ω's are averaged to higher orders in Δt so that only the above two terms remain in the expression for Γ in (2.2.27), all other terms vanishing in the limit of

$\Delta t \to 0$. Thus we obtain (2.2.31). The stationary solution is determined from

$$\frac{\partial}{\partial t} P(u,t) = 0 \qquad \text{or}$$

$$\gamma\left(u + \frac{kT}{m}\frac{d}{du}\right) P_{\text{eq}}(u) = \text{const} = 0\,. \tag{2.2.32}$$

The constant is zero because $P_{\text{eq}}(u)$ and $P'_{\text{eq}}(u)$ are both zero at $|u| \to \infty$. From (2.2.32), the Maxwellian distribution obviously follows:

$$P_{\text{eq}}(u) = C \exp\left(-\frac{m}{2kT}u^2\right). \tag{2.2.33}$$

From the Langevin equations (1.3.5, 6), a generalization of (2.2.31) is derived with the assumption of Gaussian white random force, giving the *Kramers equation* [2.7]

$$\frac{\partial}{\partial t} P(x,p,t) = \left[-\frac{\partial}{\partial x}\frac{p}{m} + \frac{\partial}{\partial p}\frac{\partial V}{\partial x} + \frac{\partial}{\partial p}\gamma\left(p + mkT\frac{\partial}{\partial p}\right)\right] P(x,p,t)\,. \tag{2.2.34}$$

The stationary solution of this is

$$P_{\text{eq}}(x,p) = C \exp\left[-\frac{1}{kT}\left(\frac{p^2}{2m} + V\right)\right], \tag{2.2.35}$$

as required by equilibrium statistical mechanics.

2.3 Markovian Processes

Brownian motion can be regarded as stochastic because detailed information is lost for molecules in the medium. Similarly, if a large number of degrees of freedom of a many-particle system is masked from observation, the motion of the remaining degrees of freedom is described as a stochastic process. As explained in Sect. 2.2, this becomes a Markovian process if temporal coarse graining is provided for together with masking of partial degress of freedom. Hence it is described by an evolution equation of the form (2.2.26):

$$\frac{\partial}{\partial t} P(q,t) = \Gamma(q,t)\, P(q,t)\,. \tag{2.3.1}$$

Here, q or a set of variables $(q_1, q_2, \ldots, q_n)$ defines the state of the system with regard to the degrees of freedom under observation, $\Gamma(q,t)$ is an operator generally represented by an integral operator such as

$$\frac{\partial}{\partial t} P(q, t) = \int P(q', t)\, dq' (q' | \Gamma(t) | q) \tag{2.3.2}$$

(this can be a differential operator), where $(q' | \Gamma(t) | q)$ is the integral kernel. We use the Dirac notation familiar from quantum mechanics.

Equation (2.3.1) means that once the distribution function is known at a time t_1, then it is determined uniquely from this information. If the evolution $P(q', t_1) \to P(q, t)$ is regarded as a transformation, it is linear because (2.3.1) is linear. Hence

$$P(q, t) = \int P(q', t_1)\, dq'\, P(q', t_1 | q, t), \tag{2.3.3}$$

where the integral kernel $P(q', t_1 | q, t)$ is the transition probability from a state q' to another q in a time interval (t_1, t) and is the solution of (2.3.1) with the inital condition

$$P(q, t_1) = \delta(q - q').$$

If we divide the interval (t_1, t) into (t_1, t_2) and (t_2, t) then the evolution is split into two parts

$$\begin{aligned} P(q, t) &= \int P(q'', t_2)\, dq''\, P(q'', t_2 | q, t) \\ &= \iint P(q', t_1)\, dq'\, P(q', t_1 | q'', t_2)\, dq''\, P(q'', t_2 | q, t). \end{aligned}$$

Therefore the evolution of the process following (2.3.1) satisfies the Markovian condition

$$P(q', t_1 | q, t) = \int P(q', t_1 | q'', t_2)\, dq''\, P(q'', t_2 | q, t). \tag{2.3.4}$$

Conversely, the evolution equation (2.3.1) is derived from the condition (2.3.4). Assume that the transition probability satisfies the condition that

$$P(q', t | q, t + \Delta t) = [1 - \gamma(q')\, \Delta t]\, \delta(q' - q) + (q' | w | q)\, \Delta t + o(\Delta t). \tag{2.3.5}$$

Since the transition probability is normalized by definition, i.e.,

$$\int P(q', t' | q, t)\, dq = 1,$$

it holds for (2.3.5) that

$$\gamma(q') = \int (q' | w | q)\, dq, \tag{2.3.6}$$

which shows that the rate of decay of the probability of remaining in state q' is equal to the sum of the transition probability rates from q' to all other

q's. Using (2.3.5), (2.3.4) is transformed as follows;

$$\begin{aligned} P(q', t_1 | q, t) &= \int P(q', t_1 | q'', t - \Delta t)\, dq''\, P(q'', t - \Delta t | q, t) \\ &= \int P(q', t_1 | q'', t - \Delta t)\, dq'' \,\{[1 - \gamma(q'')\, \Delta t]\, \delta(q'' - q) \\ &\quad + (q'' | w | q)\, \Delta t + o(\Delta t)\} \\ &= P(q', t_1 | q, t - \Delta t)\, [1 - \gamma(q)\, \Delta t] \\ &\quad + \int P(q', t_1 | q'', t)\, dq''\, (q'' | w | q)\, \Delta t + o(\Delta t)\,. \end{aligned}$$

Taking the limit of $\Delta t \to 0$ yields

$$\frac{\partial}{\partial t} P(q', t_1 | q, t) = -P(q', t_1 | q, t)\, \gamma(q) + \int P(q', t_1 | q'', t)\, dq''\, (q'' | w | q) \tag{2.3.7}$$

for the transition rate. Starting from any initial condition, the probability $P(q, t)$ satisfies the same evolution equation

$$\frac{\partial}{\partial t} P(q, t) = -P(q, t)\, \gamma(q) + \int P(q', t)\, dq'\, (q' | w | q)\,, \tag{2.3.8}$$

often called the *Chapman-Kolmogorov* equation. In the form of (2.3.2), the integral kernel is

$$(q' | \Gamma | q) = -\gamma(q')\, \delta(q' - q) + (q' | w | q)\,. \tag{2.3.9}$$

For the probability distribution function $P(q, t)$ to be kept normalized, it is essential that

$$\int (q' | \Gamma | q)\, dq = 0\,, \tag{2.3.10}$$

which is satisfied by (2.3.9). Furthermore it must hold that

$$(q' | w | q) \geqq 0 \tag{2.3.11}$$

because this is the rate of transition from the state q' to q. Equation (2.3.2) can represent a stochastic process if and only if the integral kernel satisfies the conditions of (2.3.9–11). An operator which satisfies these conditions is called a *stochastic operator.* If the variable q is not continuous but is discrete, then integrals in the above argument should be replaced by sums over possible states and the operator Γ is regarded as a matrix.

A stochastic process defined by (2.3.8) is Markovian. Several fundamental theorems can be obtained from the properties of a stochastic operator noted above. However, the reader is referred to [2.8] for the details. Here we note only a few important points for a stationary Markovian process.

For a stationary Markovian process the matrix $(q'|\Gamma|q)$ in (2.3.9) is independent of time t. We now consider the following eigenvalue problems

$$\int \varphi_\alpha(q')\,dq'\,(q'|\Gamma|q) = -\lambda_\alpha\,\varphi_\alpha(q)\,, \tag{2.3.12}$$

$$\int (q|\Gamma|q')\,dq'\,\psi_\alpha(q') = -\lambda_\alpha\,\psi_\alpha(q)\,. \tag{2.3.13}$$

These equations are adjoint to each other and have the same eigenvalues, the corresponding eigenfunctions satisfying the orthogonality condition

$$\int \varphi_\alpha(q)\,dq\,\psi_\beta(q) = \delta_{\alpha\beta}\,. \tag{2.3.14}$$

If an eigenvalue is degenerate the above condition can be satisfied by a suitable choice of linear combinations of eigenfunctions. A solution of (2.3.8) can be expanded to

$$P(q,t) = \sum_\alpha \mathrm{e}^{-\lambda_\alpha t}\,\varphi_\alpha(q)\int \psi_\alpha(q')\,dq'\,P(q',0)\,, \tag{2.3.15}$$

if the eigenfunctions are assumed complete. As shown in the following, in ordinary situations

$$\lambda_0 = 0\,, \quad \mathrm{Re}\{\lambda_\alpha\} > 0 \;\; (\alpha \neq 0)\,, \tag{2.3.16}$$

and the solution approaches a unique final distribution

$$P(q,t) \to \varphi_0(q) \equiv P_{\mathrm{eq}}(q) \tag{2.3.17}$$

from any initial distribution.

It is easily seen that $\lambda_0 = 0$ is one of the eigenvalues because (2.3.10) is just an eigenvalue equation for

$$\psi_0(q) = \mathrm{const} = 1\,.$$

Equation (2.3.12) for $\lambda_0 = 0$ corresponds to the stationary condition

$$-P_{\mathrm{eq}}(q)\,\gamma(q) + \int P_{\mathrm{eq}}(q')\,dq'\,(q'|w|q) = 0\,. \tag{2.3.18}$$

To prove the second inequality of (2.3.16), we take the operator $\exp(\Gamma t)$ instead of Γ and consider

$$\int (q|\mathrm{e}^{\Gamma t}|q')\,dq'\,\psi_\alpha(q') = \mathrm{e}^{-\lambda_\alpha t}\,\psi_\alpha(q) \quad (t > 0) \tag{2.3.19}$$

instead of (2.3.13). From (2.3.2, 7), it is obvious that

$$(q|\mathrm{e}^{\Gamma t}|q') = P(q,0|q'\,t) \tag{2.3.20}$$

is the transition probability from a state q to another q' in a time interval t. This is positive (nonnegative) for any q and q' and satisfies the normalization

$$\int (q|e^{\Gamma t}|q')\,dq' = \int P(q,0|q',t)\,dq' = 1\,. \tag{2.3.21}$$

Now let us denote q_m if q is such that

$$|\psi_\alpha(q)| = \max\,.$$

Then from (2.3.19)

$$\begin{aligned}|e^{-\lambda_\alpha t}|\,|\psi_\alpha(q_m)| &= \int (q_m|e^{\Gamma t}|q')\,dq'\,|\psi_\alpha(q')| \\ &\leqq \int (q_m|e^{\Gamma t}|q')\,dq'\,|\psi_\alpha(q_m)| = |\psi_\alpha(q_m)|\end{aligned}$$

so therefore

$$|e^{-\lambda_\alpha t}| \leqq 1\,. \tag{2.3.22}$$

Generally speaking, there can be pure imaginary eigenvalues for (2.3.12, 13). If this is so, the solution should behave periodically in t and the equilibrium is not uniquely determined. If there is no such periodicity and the solution is unique, the Markovian process is said to be *ergodic.*

Particularly simple are the cases where the transition rate matrix is symmetric, i.e.,

$$(q'|w|q) = (q|w|q') \tag{2.3.23}$$

or where the detailed balance condition

$$P_{\mathrm{eq}}(q')(q'|w|q) = P_{\mathrm{eq}}(q)(q|w|q') \tag{2.3.24}$$

is fulfilled.

When (2.3.23) holds, then (2.3.12, 13) are identical and the equilibrium distribution is

$$\varphi_0(q) = \mathrm{const} = \frac{1}{\Omega} \quad (\text{where } \Omega = \int dq)\,. \tag{2.3.25}$$

For this equilibrium to be unique, all states corresponding to various values of q must be connected by the transition rate matrix w. If the matrix is decomposed into more than two parts and the states are grouped into more than two unconnected groups, the equilibrium is not unique because it can be established separately for each different group.

The H theorem is often referred to in order to show the uniqueness of the approach to equilibrium. The theorem is particularly simple if w is symmetric[2]. We choose the H function

$$H(t) = \int dq\,P(q,t)\ln P(q,t)\,. \tag{2.3.26}$$

---

2 Neither the symmetry nor the detailed balance condition is needed to establish general H theorems. For a general proof, see [2.9]

The time variation of this function is calculated with (2.3.8, 6) in the following way;

$$\begin{aligned}\frac{dH(t)}{dt} &= \int dq \left[\frac{\partial P(q,t)}{\partial t} \ln P(q,t) + \frac{\partial P(q,t)}{\partial t}\right] \\ &= \int dq \frac{\partial P(q,t)}{\partial t} \ln P(q,t) \\ &= \int dq \int dq' \left[-P(q,t)(q|w|q') + P(q',t)(q'|w|q)\right] \ln P(q,t) \\ &= \iint dq\, dq' (q|w|q') \left[P(q',t) - P(q,t)\right] \ln P(q,t) \\ &= \iint dq\, dq' (q'|w|q) \left[P(q,t) - P(q',t)\right] \ln P(q',t) \\ &= \tfrac{1}{2} \iint dq\, dq' (q|w|q') \left[P(q',t) - P(q,t)\right]\left[\ln P(q,t) - \ln P(q',t)\right] \\ &\leqq 0\,. \end{aligned} \tag{2.3.27}$$

Here we have used the symmetry of w repeatedly. The last expression was obtained by averaging the two previous ones. The inequality is due to

$$(a-b)(\ln a - \ln b) \geqq 0$$

which holds for any two postive numbers a and b.

The H function monotonously decreases in time with the evolution of the Markovian process and is stationary when the condition

$$P(q,t) \to P_{\mathrm{eq}}(q)\,, \quad P_{\mathrm{eq}}(q) = P_{\mathrm{eq}}(q')$$

is fulfilled for every pair of states q and q' connected by a non zero matrix element $(q|w|q')$. If the set of states q is indecomposable, namely, if it cannot be divided into more than two independent subsets, the equilibrium is uniquely determined by (2.3.25).

Condition (2.3.24) is more general than (2.3.23). If this holds, we choose the H function

$$H(t) = \int dq\, P(q,t) \ln \frac{P(q,t)}{P_{\mathrm{eq}}(q)} \tag{2.3.28}$$

instead of (2.3.26) and prove that

$$\frac{dH(t)}{dt} \leqq 0\,. \tag{2.3.29}$$

If we define a new matrix $\bar{w}$ by

$$P_{\mathrm{eq}}(q')(q'|w|q) = (q'|\bar{w}|q)\,, \tag{2.3.30}$$

it is symmetric, so that (2.3.8) is written as

$$\frac{\partial}{\partial t} P(q, t) = \int dq' \left(\frac{P(q', t)}{P_{\text{eq}}(q')} - \frac{P(q, t)}{P_{\text{eq}}(q)} \right) (q \,|\, \bar{w} \,|\, q') . \tag{2.3.31}$$

As before, we easily show that

$$\begin{aligned} \frac{dH(t)}{dt} &= \frac{1}{2} \iint dq \, dq' \, (q \,|\, \bar{w} \,|\, q') \left(\frac{P(q', t)}{P_{\text{eq}}(q')} - \frac{P(q, t)}{P_{\text{eq}}(q)} \right) \\ &\quad \times \left(\ln \frac{P(q, t)}{P_{\text{eq}}(q)} - \ln \frac{P(q', t)}{P_{\text{eq}}(q')} \right) \leqq 0 . \end{aligned} \tag{2.3.32}$$

If the system is ergodic, the distribution function approaches equilibrium, i.e.,

$$P(q, t) \to P_{\text{eq}}(q) ,$$

after some elapse of time and the H function attains its minimum value.

When we consider a physical system in terms of a stochastic model, there are various stages of modeling depending on the incompleteness of the description. A stochastic equation for a basic model from which other models for a more incomplete description can be derived is often called a *master equation*[3]. For example, for a assembly of gaseous molecules, the states are incompletely specified by their set of momenta $(p_1, p_2, \ldots, p_N)$, the position variables being suppressed. Molecular collisions cause transitions among the states, say, from $(p_1, \ldots, p_j, \ldots, p_k, \ldots, p_N)$ to $(p_1, \ldots, p_j', \ldots, p_k', \ldots, p_N)$ by a collision between the jth and kth molecules. If the process is assumed to be Markovian, the temporal evolution of the distribution function

$$f(\boldsymbol{p}_1, \ldots, \boldsymbol{p}_N, t)$$

will be described by a master equation from which statistical properties of the gas can be derived. In particular, if we wish, we should be able to derive an equation of the form of a Boltzmann equation for the velocity distribution function of a single molecule. The H function (2.3.26) defined for an N-Particle momentum distribution function is different from the H function for a one-particle distribution function originally introduced by Boltzmann (Sect. 2.8). However, like Boltzmann's H function, it decreases in time until the whole system reaches equilibrium. The generalized entropy is defined by

$$S = -kH .$$

3 In the literature "master equation" often means a Markovian equation modeling a physical system. The usage of the word seems rather ambiguous

This entropy increases with time to a maximum corresponding to the equilibrium.

When a system is in contact with a heat bath, the dynamical variables of the heat bath may be eliminated from the basic dynamical equation, thus leaving a stochastic equation to describe the evolution of the system under observation. Under some general conditions, the stochastic evolution is shown to be Markovian. The stationary solution of the Markovian equation will be a canonical distribution

$$P_{\mathrm{eq}}(q) = C \exp[-\beta E(q)] \quad \left(\beta = \frac{1}{kT}\right) \tag{2.3.33}$$

if the heat bath is at temperature T. Here $E(q)$ is the energy of the system with its dynamical variables q. In such a case, the H function (2.3.28) corresponds to the free energy

$$F(t) \equiv \langle E \rangle - TS = \frac{H(t)}{\beta} \tag{2.3.34}$$

generalized to nonequilibrium states. During contact with the heat bath, the generalized free energy decreases and reaches its minimum when equilibrium is attained.

2.4 Fokker-Planck Equation

The diffusion equations (2.2.13, 31) are parabolic partial differential equations. A generalization of the diffusion equation is called the *Fokker-Planck equation* [2.10], which is a particular case of a Markovian equation but is frequently used in physical problems. We now derive this from the general formula (2.3.8) of Markovian processes. In the transition rate w defined by (2.3.5), a jump from q' to q is

$$r = q - q'$$

and the transition rate is

$$(q' | w | q) \equiv w(q', r). \tag{2.4.1}$$

Then (2.3.8) can be written as

$$\frac{\partial}{\partial t} P(q, t) = -\int w(q, r)\, dr\, P(q, t) + \int w(q - r, r)\, dr\, P(q - r, t) \tag{2.4.2}$$

using (2.3.6). Note that

$$\mathrm{e}^{-r\partial/\partial q} f(q) = \sum_{n=0}^{\infty} \frac{(-r)^n}{n!} \left(\frac{\partial}{\partial q}\right)^n f(q) = f(q - r),$$

for which necessary conditions of convergence are all assumed to be satisfied, and rewrite (2.4.2) as

$$\frac{\partial}{\partial t} P(q, t) = - \int w(q, r)\, dr\, P(q, t) + \int dr \sum_{n=0}^{\infty} \frac{(-)^n}{n!} r^n \left(\frac{\partial}{\partial q}\right)^n w(q, r)\, P(q, t) = \sum_{n=1}^{\infty} \frac{(-)^n}{n!} \left(\frac{\partial}{\partial q}\right)^n \int dr\, r^n\, w(q, r)\, P(q, t)\,. \tag{2.4.3}$$

We define

$$\alpha_n(q) = \int dr\, r^n\, w(q, r) \tag{2.4.4}$$

which following (2.3.5) is

$$\alpha_n(q) = \lim_{\Delta t \to 0} \frac{1}{\Delta t} \int P(q, t \,|\, q + r, t + \Delta t)\, r^n\, dr \equiv \lim_{\Delta t \to 0} \frac{1}{\Delta t} \langle r^n \rangle_q\,, \tag{2.4.5}$$

where $\langle r^n \rangle_q$ is the nth moment of jump r which the system makes from the state q in the time interval Δt after t. This moment is expanded in Δt to

$$\langle r^n \rangle_q = \alpha_n(q)\, \Delta t + O(\Delta t^2)\,. \tag{2.4.6}$$

Then the coefficient of the first-order term is equal to α_n. With these moments of transition rate, (2.4.3) can be expressed as

$$\frac{\partial}{\partial t} P(q, t) = \sum_{n=1}^{\infty} \frac{(-)^n}{n!} \left(\frac{\partial}{\partial q}\right)^n \alpha_n(q)\, P(q, t)\,. \tag{2.4.7}$$

This is a partial differential equation of inifinite order and is a formal expression of the integrodifferential equation (2.3.8). If the condition

$$\alpha_n(q) = 0 \quad (n \geqq 3) \tag{2.4.8}$$

is fulfilled, the right-hand side is terminated at second-order and the equation becomes

$$\frac{\partial}{\partial t} P(q, t) = \left[-\frac{\partial}{\partial q} \alpha_1(q) + \frac{1}{2} \frac{\partial^2}{\partial q^2} \alpha_2(q) \right] P(q, t)\,, \tag{2.4.9}$$

which is the *Fokker-Planck equation.*

If there are many variables, we need to consider only the variable q in the above, consisting of $q_1, q_2, \ldots, q_m$. Then the explicit expression of (2.4.9) is

$$\frac{\partial}{\partial t} P(q_1, \ldots, q_m, t) = \left[-\sum_{j=1}^{m} \frac{\partial}{\partial q_j} \alpha_{1j}(q) + \frac{1}{2} \sum_{j=1}^{m} \sum_{k=1}^{m} \frac{\partial^2}{\partial q_j \, \partial q_k} \alpha_{2jk}(q) \right] \times P(q_1, \ldots, q_m, t) . \tag{2.4.10}$$

It should be noted here that the above derivation also applies when the transition rate w is dependent on t. Even if the α's in (2.4.9, 10) depend on time, the Fokker-Planck equation remains valid as long as condition (2.4.8) is satisfied: namely, it can also be applied to nonstationary Markovian processes.

We have derived this Fokker-Planck equation from the fundamental equation of a Markovian process, but it is more usual to write down a Fokker-Planck equation directly by calculating the quantities (2.4.6), since the transition probability (2.4.5) is what is desired rather than what is given. For example, the Langevin equation (1.3.4) is integrated as

$$\Delta u = -\gamma u \, \Delta t + \frac{1}{m} \int_{t}^{t+\Delta t} R(t') \, dt' + o(\Delta t) . \tag{2.4.11}$$

If the random process $R(t)$ is assumed to be a Gaussian white process, then

$$\lim_{\Delta t \to 0} \frac{\langle \Delta u \rangle}{\Delta t} = -\gamma u, \tag{2.4.12}$$

$$\lim_{\Delta t \to 0} \frac{\langle \Delta u^2 \rangle}{\Delta t} = \frac{1}{m^2} \lim \frac{1}{\Delta t} \int_{t}^{t+\Delta t} \int_{t}^{t+\Delta t} \langle R(t_1) \, R(t_2) \rangle \, dt_1 \, dt_2 = \frac{2\pi I_R}{m^2} = \frac{2\gamma k T}{m},$$

$$\lim_{\Delta t \to 0} \frac{\langle \Delta u^n \rangle}{\Delta t} = 0, \quad (n \geqq 3) . \tag{2.4.13}$$

Equation (2.4.9) becomes (2.2.31) when these expressions are inserted. The process $u(t)$ is known to be Markovian by the assumption that the correlation time of $R(t)$ is infinitely short (white spectrum); $u(t)$ is described by the Fokker-Planck equation. The latter requires the Gaussian assumption of $R(t)$ by which all moments higher than the third vanish, leaving only the first and the second moments, as seen in (2.4.12, 13). The fundamental solution of the Fokker-Planck equation is found to be (1.5.5).

The Fokker-Planck equation can be applied to nonlinear problems. As an example of such an application, we consider here a generalization of the Langevin equation

$$\frac{du}{dt} = -\gamma(u) \, u + \frac{1}{m} R(u, t) , \tag{2.4.14}$$

where the friction coefficient and, correspondingly, the random force R are assumed to depend on the velocity. The random force $R(u, t)$ is a stochastic process which includes u as a parameter as well and is assumed to be Gaussian and white. Its correlation function can be written as

$$\frac{1}{m^2}\langle R(u_1, t_1)\, R(u_2, t_2)\rangle = B(u_1, u_2)\, \delta(t_1 - t_2)\,. \tag{2.4.15}$$

The spectral intensity in (1.3.41) is written here as $m^2 B$.

The nonlinearity of (2.4.14) makes interpretation complex. There are two standard interpretations one by *Ito* and *McKean* [2.11] and the other by *Stratonovich* [2.12]. The *Ito* and *McKean* interpretation is more lucid mathematically. On the other hand, *Strantonovich's* interpretation is more natural from a physical point of view. Here we take the latter.

We integrate (2.4.14) for a short-time interval $(t, t+\Delta t)$ as for (2.4.11). Notice that the u dependence of R complicates matters in

$$\begin{aligned}\Delta u &= u(t+\Delta t) - u(t)\\ &= -\gamma(u)\, u\, \Delta t + \frac{1}{m}\int_t^{t+\Delta t} R(u(t_1), t_1)\, dt_1 + o(\Delta t)\,.\end{aligned} \tag{2.4.16}$$

The second term is calculated as

$$\begin{aligned}&\frac{1}{m}\int_t^{t+\Delta t} dt_1 \left[R(u, t_1) + \frac{\partial R(u, t_1)}{\partial u}\int_t^{t_1} \dot{u}(t_2)\, dt_2 + o(\Delta t)\right]\\ &= \frac{1}{m}\int_t^{t+\Delta t} dt_1 \left[R(u, t_1) + \frac{1}{m}\left(\frac{\partial}{\partial u} R(u, t_1)\right)\int_t^{t_1} R(u, t_2)\, dt_2\right] + o(\Delta t)\,.\end{aligned} \tag{2.4.17}$$

Thus

$$V(u) \equiv \lim_{\Delta t\to 0} \frac{\langle \Delta u\rangle}{\Delta t} = -\gamma(u)\, u + \frac{1}{2}\left[\frac{\partial}{\partial u_1} B(u_1, u)\right]_{u_1=u}, \tag{2.4.18}$$

$$2D(u) \equiv \lim_{\Delta t\to 0} \frac{\langle (\Delta u)^2\rangle}{\Delta t} = B(u, u)\,. \tag{2.4.19}$$

The last term of (2.4.18) is not present if we adopt the *Ito* and *McKean* interpretation of stochastic differentials for the Langevin equation (2.4.14), which replaces $R(u(t_1), t_1)$ in (2.4.16) by $R(u(t), t_1)$. Therefore the resulting Fokker-Planck equation is different from that obtained by the Stratonovich interpretation. Thus, the same Langevin equation has two corresponding different stochastic processes depending on the interpretation. On the other hand, a Fokker-Planck equation defines a stochastic process unambiguously. If the process is modeled by a Langevin equation, its form differs according to the choice of interpretation [2.13].

By (2.4.18, 19) we have the Fokker-Planck equation

$$\frac{\partial P(u,t)}{\partial t} = \left[-\frac{\partial}{\partial u} V(u) + \frac{\partial^2}{\partial u^2} D(u)\right] P(u,t)\,. \tag{2.4.20}$$

We show now that the same result is derived by the method described in Sect. 2.2. Equation (2.2.18) is written as

$$\frac{\partial}{\partial t} f(u,t) = -\frac{\partial}{\partial u}\left[-\gamma(u)\,u + \frac{1}{m} R(u,t)\right] f \equiv \Omega f$$

for the Langevin equation (2.4.14). The solution of the form of (2.2.23) is averaged over all possibilities of the random force R, assuming the Gaussian property and the delta-function type correlation. In the limit $\Delta t \to 0$, only second-order terms of Ω remain. The average of the first-order term is simply

$$\lim_{\Delta t\to 0} \frac{1}{\Delta t} \int_t^{t+\Delta t} dt_1 \,\langle \Omega(t_1)\rangle\, P(u,t) = \frac{\partial}{\partial u} \gamma(u)\, u\, P(u,t)$$

and that of the second-order term is

$$\begin{aligned}
&\lim_{\Delta t\to 0} \frac{1}{\Delta t} \int_t^{t+\Delta t} dt_1 \int_t^{t_1} dt_2 \,\langle \Omega(t_1)\,\Omega(t_2)\rangle\, P(u,t)\\
&= \frac{1}{m^2}\int_0^\infty d\tau \left\langle \frac{\partial}{\partial u} R(u,t+\tau) \frac{\partial}{\partial u} R(u,t)\right\rangle P(u,t)\\
&= \frac{1}{m^2}\int_0^\infty d\tau \left[\frac{\partial}{\partial u}\left\langle R(u,t+\tau)\left(\frac{\partial}{\partial u} R(u,t)\right)\right\rangle\right.\\
&\quad \left. + \frac{\partial}{\partial u}\langle R(u,t+\tau)\, R(u,t)\rangle \frac{\partial}{\partial u}\right] P(u,t)\\
&= \frac{\partial}{\partial u}\left\{\frac{1}{2}\left[\frac{\partial}{\partial u_2} B(u,u_2)\right]_{u_2=u} + \frac{1}{2} B(u,u) \frac{\partial}{\partial u}\right\} P(u,t)\,.
\end{aligned}$$

Therefore

$$\frac{\partial}{\partial t} P(u,t) = \frac{\partial}{\partial u}\left\{\gamma(u)\, u + \frac{1}{2}\left[\frac{\partial}{\partial u_2} B(u,u_2)\right]_{u_2=u} + \frac{1}{2} B(u,u)\frac{\partial}{\partial u}\right\} P(u,t) \tag{2.4.21}$$

which corresponds to (2.2.31). This is identical to (2.4.20) because

$$\frac{d}{du} B(u,u) = \left[\frac{\partial}{\partial u_1} B(u_1,u)\right]_{u_1=u} + \left[\frac{\partial}{\partial u_2} B(u,u_2)\right]_{u_2=u}$$

or

$$-V(u)+\frac{d}{du}D(u)=\gamma(u)\,u+\frac{1}{2}\left[\frac{\partial}{\partial u_2}B(u,u_2)\right]_{u_2=u}.$$

We have seen that a Fokker-Planck equation is generally obtained if the basic random force R is *Gaussian and white*. The driven process u may be linear or nonlinear in R. *The white noise character of R results in the Markovian property of u and the Gaussian assumption leads to the Fokker-Planck description.* In the Fokker-Planck equation, the random force is eliminated from the equation, leaving only its spectral function of intensity. The Fokker-Planck equation can be applied to a number of stochastic processes in physical, biological and sociological phenomena. Its wide applicability is based on the fact that many such processes are generated by an underlying process which is well approximated by a Gaussian process, a manifestation of the central limit theorem.

Section 1.3 described how harmonic analysis is very useful for treating Gaussian processes. The usefulness of the method does not depend on the spectral form of the random force, since it can be used for non-Markovian cases as well. But its usefulness is severely limited if the basic Langevin equation is not linear. On the other hand, the Fokker-Planck equation can be used even for nonlinear problems as well as nonstationary problems, but it is limited to Markovian cases where the basic driving process is white and Gaussian. This should be kept in mind when these methods are applied to physical, biological and sociological models.

Equation (2.4.20 or 21) reduces to

$$\frac{\partial}{\partial t}P(u,t)=\frac{\partial}{\partial u}\left\{D(u)\left[-\left(\frac{d}{du}\ln P_{\mathrm{eq}}(u)\right)+\frac{\partial}{\partial u}\right]\right\}P(u,t) \tag{2.4.22}$$

for a stationary process if the equilibrium distribution denoted here by $P_{\mathrm{eq}}(u)$ is assumed unique. If u is the velocity of a Brownian particle, a Maxwellian distribution

$$P_{\mathrm{eq}}(u)=C\exp\left(-\frac{m}{2kT}u^2\right)$$

may be assumed, so that the Fokker-Planck equation becomes

$$\frac{\partial}{\partial t}P(u,t)=\frac{\partial}{\partial u}\left\{D(u)\left[\frac{m}{kT}u+\frac{\partial}{\partial u}\right]\right\}P(u,t), \tag{2.4.23}$$

which determines the evolution of the velocity distribution function.

For a Brownian particle diffusing in a force field U, the Fokker-Planck equation can be simplified to

$$\frac{\partial}{\partial t}P(x,t)=\frac{\partial}{\partial x}\left\{D(x)\left[\frac{1}{kT}\frac{\partial U}{\partial x}+\frac{\partial}{\partial x}\right]\right\}P(x,t), \tag{2.4.24}$$

allowing spatial dependence of the diffusion constant. This guarantees the equilibrium distribution

$$P_{\text{eq}}(x) = C \exp\left(-\frac{U}{kT}\right). \tag{2.4.25}$$

Before concluding this section we note an important theorem from *Doob* [2.14], whereby if a stationary process $q(t)$ is Gaussian and Markovian, its correlation function ought to decay in time exponentially:

$$\langle q(t)\, q(0)\rangle = \langle q^2\rangle\, \mathrm{e}^{-\gamma t}. \tag{2.4.26}$$

If q is a set of m random variables, it is a vector and its correlation matrix is an $m \times m$ matrix,

$$\langle q_j(t)\, q_k(0)\rangle = \sum_l (\mathrm{e}^{-Mt})_{jl} \langle q_l(0)\, q_k(0)\rangle \tag{2.4.27}$$

or

$$\frac{d}{dt}\langle q_j(t)\, q_k(0)\rangle = -\sum_l M_{jl} \langle q_l(t)\, q_k(0)\rangle. \tag{2.4.28}$$

This theorem can be proved from (1.4.18) and (2.3.4), but the proof is left as an exercise for the reader. We note here only that the content of the theorem is easily understood from what has been explained in this section. If the processes $[q_1(t), q_2(t), \dots, q_m(t)]$ are generated by a set of random forces having white spectra, generally a Fokker-Planck equation of the form (2.4.10) arises. If $q_1(t), \dots, q_m(t)$ are linear in R, i.e., if the Langevin equation has the form

$$\dot{q}_j(t) = -\sum_l M_{jl}\, q_l(t) + R_j(t) \tag{2.4.29}$$

with a constant matrix M, the processes $q_1(t), \dots, q_m(t)$ are Gaussian because the R's are Gaussian. Equation (2.4.28) follows from (2.4.29) or from the Fokker-Planck equation (2.4.10), in which

$$\alpha_{1j}(q) = -\sum_l M_{jl}\, q_l \tag{2.4.30}$$

and α_{2jk} is independent of q. Equation (2.2.31) is the standard form of a one-dimensional Gaussian-Markovian process.

Since (2.4.14) is a nonlinear Langevin equation, the process $u(t)$ generated by $R(t)$ is no longer Gaussian. In (2.2.34), the process $[x(t), p(t)]$ is not Gaussian in general. It is Gaussian only in the case of a harmonic oscillator, because $\partial V/\partial x$ is linear in x.

When (2.4.29) is valid, the expectation of the random variables $q_1(t), q_2(t), \dots$ decreases exponentially in time following

$$\frac{d}{dt}\langle q_j(t)\rangle = -\sum_l M_{jl} \langle q_l(t)\rangle. \tag{2.4.31}$$

This is called *linear relaxation,* which is commonly observed in physics and also in many other cases. Why there are many Gaussian stochastic processes in the real world and the conditions under which a Markovian process is generated have already been touched upon. It is worth remembering that the *Gaussian-Markovian property is related to linear relaxation.*

2.5 Contraction of Information. Projected Processes

We have emphasized repeatedly that the fundamental logical structure of statistical physics is the derivation of the laws of physics by introducing different stages of coarse graining from microscopic laws. By proceeding a step further, our description is made even cruder than the previous one by partial contraction of information. This contraction is a projection of the object onto a certain cross section of our understanding. Here the problem is to find the appropriate way of describing of the projected process.

For example, the stochastic process called Brownian motion is the projection of microscopic motion of a colloid particle and all the molecules of the surrounding liquid onto the dimension of the motion of the colloid particle only. Furthermore, if the motion is projected onto the displacement then it becomes a diffusion process of the Brownian particle.

The complete dynamical description of N gaseous molecules confined in a box is made in terms of their position and momentum variables given as functions of time t. If the molecule positions are not observable, then the projected motion is described by the stochastic evolution of the distribution function $f(p_1, p_2, \ldots, p_N, t)$. The equation governing this evolution is called a master equation. It should furthermore be possible to discover the probability $f_1(r, p, t)$ of finding a molecule with momentum p and position r at time t. This is largely contracted information and useful in describing the gas only extremely crudely.

When a colloid particle is much larger and heavier than molecules of the surrounding liquid, the projected process becomes an ideal Brownian motion which is Gaussian and Markovian. In general, however, there is friction retardation as discussed in Sect. 1.6, so that a Markovian description is not always possible. Contracted distribution functions for gaseous molecule do not necessarily obey simple laws of evolution, such as a Markovian equation like (2.3.1). In general cases, future evolution of such a probability distribution function is not determined only by the information at that instant, but is also governed by the whole history. It is important to recognize that the contraction of information results as a rule in a *non-Markovian* property of the projected process. The main theme of this chapter is a mathematical formulation of *contraction* and *projection* in order to show how such a non-Markovian property comes about and to see how this is turned into a Markovian under a proper coarse-graining procedure.

Assume that a state of a given system is represented by the set of variables $q=(q_1, q_2, \ldots, q_n)$ and the evolution is described by a Markovian process. Then the probability distribution function $f(q_1, q_2, \ldots, q_n, t)$ follows

$$\frac{\partial}{\partial t} f(q_1, q_2, \ldots, q_n, t) = \Gamma f, \tag{2.5.1}$$

where Γ is a linear operator[4]. The q variables are said to be complete if the size of the set is necessary and sufficient to make the process Markovian. Suppose now that our observation is not quite so detailed and only a part of the set

$$q' = (q_1, q_2, \ldots, q_m) \tag{2.5.2}$$

is observed. Then a set of other variables

$$q'' = (q_{m+1}, q_{m+2}, \ldots, q_n) \tag{2.5.3}$$

becomes *hidden variables.*

Our problem now is to investigate how the process is described in this crude observation as a process of these explicit variables q'. This is a projection of the distribution function $f(q_1, q_2 \ldots, q_n, t)$, denoted by the function

$$g = \mathscr{P} f, \tag{2.5.4}$$

dependent only on the variables q'. More explicitly, we assume this to be of the form

$$g(q', q'', t) = \bar{g}(q', t)\, \varphi_0(q''), \tag{2.5.5}$$

where $\varphi_0(q'')$ is a *given* function of the hidden variables q'' so that the change of g is solely due to that of $\bar{g}(q', t)$. The projection $\mathscr{P}$ in (2.5.4) is defined by

$$g(q', q'', t) = \varphi_0(q'') \int \psi_0(q'')\, dq''\, f(q', q'', t), \tag{2.5.6}$$

where $\psi_0(q'')$ is also a given function of the variables q'' and is assumed to satisfy

$$\int \psi_0(q'')\, \varphi_0(q'')\, dq'' = 1\,. \tag{2.5.7}$$

From these definitions

$$\bar{g}(q', t) = \int \psi_0(q'')\, dq''\, f(q', q'', t) \tag{2.5.8}$$

4 Although it need not be a Markovian operator, since it can be classical or a quantal Liouville operator. The following formalism applies to dynamical as well as to stochastic evolution as long as the evolution of f after t is determined by $f(t)$, namely if the evolution is Markovian in a wider sense of the word

and from (2.5.4, 5)

$$\begin{aligned}\mathscr{P}f &= \bar{g}(q',t)\,\varphi_0(q'')\\ \mathscr{P}^2 f &= \mathscr{P}\mathscr{P}f = \bar{g}(q',t)\,\varphi_0(q'')\int \psi_0(q'')\,\varphi_0(q'')\,dq''\\ &= \mathscr{P}f\,,\end{aligned}$$

namely

$$\mathscr{P}^2 = \mathscr{P}\,. \tag{2.5.9}$$

This shows that the operation $\mathscr{P}$ defined by (2.5.6) is, in fact, a projection.

For a given projection operator $\mathscr{P}$, another projection operator

$$\mathscr{P}' = 1 - \mathscr{P} \tag{2.5.10}$$

is defined by

$$\mathscr{P}'f = f - \mathscr{P}f\,.$$

The operator $\mathscr{P}'$ satisfies

$$\mathscr{P}'^2 = \mathscr{P}' \tag{2.5.11}$$

as it should, and so

$$\mathscr{P}\mathscr{P}' = \mathscr{P}'\mathscr{P} = 0 \tag{2.5.12}$$

holds. These relations are obvious from (2.5.11, 9).

Now the distribution function $f(q',q'',t)$ is divided into two parts:

$$f = (\mathscr{P} + \mathscr{P}')\,f = \mathscr{P}f + \mathscr{P}'f\,, \tag{2.5.13}$$

i.e., the projection $g = \mathscr{P}f$ and the remainder $\mathscr{P}'f$. Correspondingly, (2.5.1) is divided into two equations:

$$\begin{aligned}\frac{\partial}{\partial t}\mathscr{P}f &= \mathscr{P}\Gamma\mathscr{P}f + \mathscr{P}\Gamma\mathscr{P}'f\,,\\ \frac{\partial}{\partial t}\mathscr{P}'f &= \mathscr{P}'\Gamma\mathscr{P}'f + \mathscr{P}'\Gamma\mathscr{P}f\,.\end{aligned} \tag{2.5.14}$$

The initial distribution at $t = t_0$ is now assumed to be

$$f(q,t_0) \equiv f_0(q) = \mathscr{P}f_0 + \mathscr{P}'f_0\,. \tag{2.5.15}$$

Then the second equation of (2.5.14) is integrated to

$$\mathscr{P}'f(q,t) = \int_{t_0}^{t} \exp[(t-\tau)\mathscr{P}'\Gamma]\,\mathscr{P}'\Gamma\mathscr{P}f(q,\tau)\,d\tau + \exp[(t-t_0)\mathscr{P}'\Gamma]\,\mathscr{P}'f_0\,, \tag{2.5.16}$$

inserted into the first equation, resulting in

$$\frac{\partial}{\partial t}\mathscr{P}f(q,t) = \mathscr{P}\Gamma\mathscr{P}f(q,t) + \mathscr{P}\Gamma\int_{t_0}^{t} d\tau \exp[(t-\tau)\mathscr{P}'\Gamma]\mathscr{P}'\Gamma\mathscr{P}f(q,\tau)$$
$$+\mathscr{P}\Gamma\exp[(t-t_0)\mathscr{P}'\Gamma]\mathscr{P}'f_0\,. \tag{2.5.17}$$

The equation for $\bar{g}(q',t)$ is obtained when (2.5.5) is inserted and both sides are divided by $\varphi_0(q'')$.

The resulting equation is more lucid if the Laplace transform

$$F(q,s) = \int_{t_0}^{\infty} \mathrm{e}^{-s(t-t_0)} f(q,t)\,dt \tag{2.5.18}$$

is used instead of $f(q,t)$. Then (2.5.17) is expressed as

$$\left(s - \mathscr{P}\Gamma - \mathscr{P}\Gamma\frac{1}{s-\mathscr{P}'\Gamma}\mathscr{P}'\Gamma\right)\mathscr{P}F = \mathscr{P}f_0 + \mathscr{P}\Gamma\frac{1}{s-\mathscr{P}'\Gamma}\mathscr{P}'f_0 \tag{2.5.19}$$

Equation (2.5.19) is nothing but the formalism called the *damping theory* in quantum mechanics [2.15]. This result is easily obtained from (2.5.14) by Laplace-transforming it and eliminating $\mathscr{P}'F$. The derivation is left for the reader as an exercise. In particular, if the initial condition satisfies

$$\mathscr{P}'f_0 = 0\,, \tag{2.5.20}$$

(2.5.17, 19) are simplified, the last terms on the right-hand sides being dropped.

The second term on the right-hand side of (2.5.17) is an integral from the initial time t_0 to the present t and depends on the history of evolution of the function $g = \mathscr{P}f$ over this time interval. This may then be called a *memory term.* The presence of such a memory term governing the evolution of g or $\bar{g}$ means that the process projected onto an incomplete set of variables is, in fact, non-Markovian. To what extent it is non-Markovian depends on the persistence of the memory function in the memory term. If the duration time τ_c of the memory is short enough in comparison to the rate of change of g, it is legitimate to replace the memory term by

$$\left[\mathscr{P}\Gamma\int_{-\infty}^{t} d\tau \exp[(t-\tau)\mathscr{P}'\Gamma]\mathscr{P}'\Gamma\right]\mathscr{P}f(q,t) \tag{2.5.21}$$

after a sufficiently long time elapse from the initial time t_0. Together with this approximation, the last term of (2.5.17) representing the memory of the initial distribution may also be ignored. The inside of the bracket in (2.5.21) is an operator independent of t, so (2.5.17) becomes an equation for $\bar{g}$ with the form

$$\frac{\partial}{\partial t}\bar{g} = \bar{\Gamma}\bar{g}\,. \tag{2.5.22}$$

Under such circumstances, the projected process recovers the Markovian property. The condition that enables this is a proper coarse graining, Sect. 2.2.

Let us consider the Brownian motion (2.2.34) again in order to illustrate the general consideration. The Brownian motion is originally described here as a Markovian process of the random variables x and u. Suppose now we are interested only in observation of the displacement x ignoring the velocity. What will the projected process be? Let us write (2.2.34) as

$$\frac{\partial}{\partial t} f(x,u,t) = (\Gamma_0 + \Gamma_1) f ,$$

$$\Gamma_0 = \gamma \frac{\partial}{\partial u}\left(u + \frac{1}{m\beta}\frac{\partial}{\partial u}\right) \quad (\beta = 1/kT) , \tag{2.5.23}$$

$$\Gamma_1 = - u \frac{\partial}{\partial x} + \frac{1}{m}\frac{\partial V}{\partial x}\frac{\partial}{\partial u} .$$

The projection $\mathscr{P}$ is introduced by

$$\mathscr{P} h(x,u,t) = \varphi_0(u) \int_{-\infty}^{\infty} du\, h(x,u,t) \tag{2.5.24}$$

for an arbitrary function $h(x,u,t)$, where

$$\varphi_0(u) = \left(\frac{2\pi}{m\beta}\right)^{-1/2} \exp\left(-\frac{\beta m u^2}{2}\right) \tag{2.5.25}$$

is a normalized Maxwellian distribution function. Through this definition

$$\Gamma_0 \mathscr{P} = 0 \tag{2.5.26}$$

$$\mathscr{P} \Gamma_0 = 0 . \tag{2.5.27}$$

Equation (2.5.26) is obvious from

$$\Gamma_0 \varphi_0(u) = 0 . \tag{2.5.28}$$

Equation (2.5.27) is evident from

$$\begin{aligned}\mathscr{P}\Gamma_0 h &= \varphi_0(u) \int_{-\infty}^{\infty} du\, \gamma \frac{\partial}{\partial u}\left(u + \frac{1}{m\beta}\frac{\partial}{\partial u}\right) h(x,u,t) \\ &= \varphi_0(u) \left[\gamma\left(u + \frac{1}{m\beta}\frac{\partial}{\partial u}\right) h(x,u,t)\right]_{-\infty}^{\infty} \\ &= 0 ,\end{aligned}$$

where the function $h(x,u,t)$ is assumed to approach zero sufficiently fast as $u \to \pm\infty$. This condition is natural for our distribution functions. The projection of $f(x,u,t)$,

$$g(x,u,t) = \varphi_0(u) \int_{-\infty}^{\infty} du\, f(x,u,t)$$
$$= \varphi_0(u)\, \bar{g}(x,t) \qquad (2.5.29)$$

is governed by (2.5.17) in which the first term on the right-hand side becomes

$$\mathscr{P}\Gamma\mathscr{P}f = \mathscr{P}\Gamma_1\mathscr{P}f$$
$$= \varphi_0(u) \int du' \left(-u' \frac{\partial}{\partial x} + \frac{1}{m}\frac{\partial V}{\partial x}\frac{\partial}{\partial u'}\right) \varphi_0(u') \int du''\, f(x,u'',t)\,,$$

where the integral over u' vanishes when (2.5.25) is inserted. For the spatial distribution function $\bar{g}(x,t)$ [defined by (2.5.29)] (2.5.17) gives

$$\frac{\partial}{\partial t}\bar{g}(x,t) = \int_{-\infty}^{\infty} du\, \Gamma_1 \int_{t_0}^{t} d\tau \exp[(t-\tau)\mathscr{P}'\Gamma]\mathscr{P}'\Gamma_1\, \varphi_0(u)\, \bar{g}(x,\tau)$$
$$+ \int_{-\infty}^{\infty} du\, \Gamma_1 \exp[(t-t_0)\mathscr{P}'\Gamma]\mathscr{P}' f_0(x,u)\,. \qquad (2.5.30)$$

The first term on the right-hand side corresponds to the diffusion of particles, but it is not Markovian as yet because it depends on the history of the spatial distribution. The second term is zero if the velocity is distributed initially as Maxwellian, namely if

$$f_0(x,u) = \varphi_0(u)\, \bar{g}_0(x) \qquad (2.5.31)$$

because $\mathscr{P}f_0 = f_0$. If it is not, the initial non-Maxwellian distribution gives rise to an effect over the evolution of $g(x,t)$, which vanishes after some time. This Brownian motion is rather complex when it is projected onto the spatial dimension.

The projected process becomes a diffusion process and (2.5.30) is reduced to (2.4.24), i.e.,

$$\frac{\partial}{\partial t}\bar{g}(x,t) = \frac{\partial}{\partial x}\left[D\left(\frac{\partial}{\partial x} + \beta\frac{\partial V}{\partial x}\right)\right]\bar{g}(x,t) \qquad (2.5.32)$$

if the distribution function $\bar{g}(x,t)$ and the potential function $V(x,t)$ satisfy the conditions

$$l\left|\frac{1}{\bar{g}}\frac{\partial \bar{g}}{\partial x}\right| \ll 1\,, \qquad l\left|\frac{1}{V}\frac{\partial V}{\partial x}\right| \ll 1\,, \qquad (2.5.33)$$

where

$$l = \frac{\langle u^2 \rangle^{1/2}}{\gamma}$$

is the mean free path of the Brownian particle. This is the same as for (2.2.11). This condition means that the parameter γ is so large that the contribution of the operator Γ_0 is far larger than that of the operator Γ_1, meaning that Γ_1 is treated as a perturbation in comparison with Γ_0. Then the first term of the right-hand side of (2.5.30) is approximated as

$$\exp[(t-\tau)\mathcal{P}'\Gamma] \rightarrow \exp[(t-\tau)\mathcal{P}'\Gamma_0] = \exp[(t-\tau)\,\Gamma_0] \tag{2.5.34}$$

where $\mathcal{P}\Gamma_0 = 0$ is used. In

$$\mathcal{P}'\Gamma_1\,\varphi_0(u)\,\bar{g}\,(x,\tau) = (1-\mathcal{P})\,\Gamma_1\,\varphi_0(u)\,\bar{g}\,(x,\tau)$$

the projection $\mathcal{P}'$ can be omitted because

$$\mathcal{P}\Gamma_1\,\varphi_0(u)\,\bar{g} = \varphi_0(u)\int_{-\infty}^{\infty} du\left[-u\frac{\partial}{\partial x} + \frac{1}{m}\frac{\partial V}{\partial x}\frac{\partial}{\partial u}\right]\varphi_0(u)\,\bar{g} = 0\,.$$

Therefore, the operator (2.5.34) operates on

$$\begin{aligned}\Gamma_1\,\varphi_0(u)\,\bar{g} &= \left[-u\frac{\partial}{\partial x} + \frac{1}{m}\frac{\partial V}{\partial x}\frac{\partial}{\partial u}\right]\varphi_0(u)\,\bar{g}\\ &= -u\,\varphi_0(u)\left[\frac{\partial}{\partial x} + \beta\frac{\partial V}{\partial x}\right]\bar{g}\,(x,\tau)\,.\end{aligned}$$

Note that

$$\Gamma_0\,u\,\varphi_0(u) = \gamma\frac{\partial}{\partial u}\left(u + \frac{1}{m\beta}\frac{\partial}{\partial u}\right)u\,\varphi_0(u) = -\,\gamma\,u\,\varphi_0(u)\,, \tag{2.5.35}$$

i.e., that $u\,\varphi_0(u)$ is an eigenfunction of the operator Γ_0 with the eigenvalue $-\gamma$. Then the term in question becomes

$$-\int_{-\infty}^{\infty} du\left(-u\frac{\partial}{\partial x} + \frac{1}{m}\frac{\partial V}{\partial x}\frac{\partial}{\partial u}\right)u\,\varphi_0(u)\int_{t_0}^{t} d\tau\,\mathrm{e}^{-\gamma(t-\tau)}\left(\frac{\partial}{\partial x} + \beta\frac{\partial V}{\partial x}\right)\bar{g}\,(x,\tau)\,. \tag{2.5.36}$$

When the integration is carried out over u, the term with $\partial/\partial u$ in front vanishes. The neighborhood of $\tau \approx t$ makes the dominant contribution to the integral over τ because $\exp[-\gamma(t-\tau)]$ decays fast for

$$\Delta t = t - \tau > \frac{1}{\gamma} \equiv \tau_{\mathrm{c}}\,.$$

Thus $\bar{g}(x, \tau)$ in (2.5.36) can be replaced by $\bar{g}(x, t)$ and the time integral gives only the factor $1/\gamma$ for $t - t_0 \gg 1/\gamma$. Also note that $\int_{-\infty}^{\infty} du\, \partial/\partial u \ldots = 0$.

Then (2.5.36) can be rewritten as

$$\frac{\partial}{\partial x}\left(\frac{\langle u^2\rangle}{\gamma}\right)\left(\frac{\partial}{\partial x} + \beta\frac{\partial V}{\partial x}\right)\bar{g}(x,t)\,,$$

where

$$D = \frac{\langle u^2\rangle}{\gamma}$$

is the diffusion constant. Equation (2.5.32) is thus obtained. The second term on the right-hand side of (2.5.30) becomes

$$\mathrm{e}^{(t-t_0)\Gamma_0}\mathscr{P}' f_0(x,u) \approx O(\mathrm{e}^{-\gamma(t-t_0)}) \tag{2.5.37}$$

by using (2.5.34) and decreases fast for $t - t_0 \gg \tau_c$ so that the effect of the initial distribution can be ignored for such a time range. The operator Γ_0 has the eigenfunctions $H_n(u)\,\varphi_0(u)$ and eigenvalues $-n\gamma$ $(n=0,1,2,\ldots)$

$$\Gamma_0 H_n(u)\,\varphi_0(u) = -n\gamma H_n(u)\,\varphi_0(u)\,, \tag{2.5.38}$$

where $H_n(u)$ is the nth Hermitian polynomial. The projection $\mathscr{P}'$ implies elimination of the eigenfunction $n=0$, so that the estimation (2.5.37) is correct.

Thus the process of spatial Brownian particle is clearly Markovian under condition (2.5.33), and is described in terms of the diffusion equation (2.5.32). This description is valid for a time in the range

$$t \approx \frac{(\Delta x)^2}{D} = \frac{(\Delta x)^2}{l^2}\,\tau_c\,, \tag{2.5.39}$$

where Δx is the spatial scale

$$\Delta x \approx \left|\frac{1}{\bar{g}}\frac{\partial \bar{g}}{\partial x}\right|^{-1}.$$

Note that the proper time scale of the velocity change of the Brownian particle is $\tau_c = 1/\gamma$. Condition (2.5.39) may be written as

$$\alpha^2\frac{t}{\tau_c} \approx 1 \quad \left(\text{with } \alpha = \frac{l}{\Delta x}\right). \tag{2.5.40}$$

The perturbative calculation deriving (2.5.32) from (2.5.30) is based on the smallness of the parameter (2.5.33), i.e., on

$$\alpha \ll 1 \,. \tag{2.5.41}$$

The validity of the Markovian description thus attained is guaranteed in the time range

$$\frac{t}{\tau_c} = O(\alpha^{-2}) \,. \tag{2.5.42}$$

This situation was already clear for (2.2.10). We have considered the problem from a viewpoint somewhat different from that in Sect. 2.2. It is important to recognize that the present consideration is general and not limited to the particular example of Brownian motion.

2.6 Derivation of Master Equations

In classical statistical mechanics, a statistical ensemble is represented by a distribution function in the phase space of canonical variables describing the system. In quantum mechanics, such a distribution function is replaced by a density matrix, denoted by ϱ hereafter. The equation of motion for the density matrix is given by

$$\frac{\partial \varrho}{\partial t} = \frac{1}{\mathrm{i}\hbar} [\mathscr{H}, \varrho] \equiv \mathrm{i} \mathscr{L} \varrho \,, \tag{2.6.1}$$

where $\mathscr{H}$ is the Hamiltonian of the system. The right-hand side of this equation is a commutator of $\mathscr{H}$ and ϱ and defines a linear operator denoted by $\mathrm{i}\mathscr{L}$ operating on ϱ. Equation (2.6.1) corresponds to the Liouville equation in classical mechanics.

Now suppose that the Hamiltonian $\mathscr{H}$ is divided into an unperturbed part and a perturbation

$$\mathscr{H} = \mathscr{H}_0 + \mathscr{H}_1 , \tag{2.6.2}$$

where the perturbation is proportional to a parameter λ, which is a measure of the ratio of the magnitudes of $\mathscr{H}_1$ and $\mathscr{H}_0$ so that

$$\lambda < 1 \tag{2.6.3}$$

means that the perturbation is weak. For the unperturbed system we assume

$$\mathscr{H}_0 \, \varphi_l = E_l \varphi_l \quad (l = 1, 2, \ldots) \,. \tag{2.6.4}$$

In the representation based on these eigenvectors, (2.6.1) is written as

$$\begin{aligned}\frac{\partial}{\partial t}\langle l|\,\varrho(t)\,|m\rangle &= \frac{1}{\mathrm{i}\hbar}(E_l - E_m)\langle l|\,\varrho\,|m\rangle \\ &+ \frac{1}{\mathrm{i}\hbar}\sum_k (\langle l|\mathcal{H}_1|k\rangle\langle k|\,\varrho\,|m\rangle - \langle l|\,\varrho\,|k\rangle\langle k|\mathcal{H}_1|m\rangle)\,.\end{aligned} \tag{2.6.5}$$

For example, an N-particle system interacting via pair potentials is described by the sum of the unperturbed Hamiltonian

$$\mathcal{H}_0 = \sum_j \frac{1}{2m}\boldsymbol{p}_j^2 \tag{2.6.6}$$

which is the kinetic energy, and the perturbation Hamiltonian

$$\mathcal{H}_1 = \sum_{\langle jk\rangle} v\,(\boldsymbol{r}_j - \boldsymbol{r}_k) \tag{2.6.7}$$

which is the interaction energy. The set of momenta of particles

$$\{\boldsymbol{p}\} = (\boldsymbol{p}_1, \boldsymbol{p}_2, \dots, \boldsymbol{p}_N) \tag{2.6.8}$$

specifies the eigenstates of $\mathcal{H}_0$ so that the quantum number l in (2.6.4) corresponds to a set $\{\boldsymbol{p}\}$.

Equation (2.6.1 or 5) is a quantum-mechanical equation of motion and by itself is not a stochastic equation. However it defines a temporal evolution in terms of a linear operator, which makes it possible to apply the method discussed in the foregoing section by introducing a suitable projection $\mathcal{P}$.

Let a projection operator $\mathcal{P}$ be defined by

$$\langle l|\mathcal{P}A|m\rangle = \langle l|A|l\rangle\,\delta_{lm}\,, \tag{2.6.9}$$

$$\langle l|\mathcal{P}'A|m\rangle = \langle l|A|m\rangle\,(1-\delta_{lm})\,, \tag{2.6.10}$$

indicating that $\mathcal{P}A$ is the projection of an operator A to the diagonal part in the representation diagonalizing $\mathcal{H}_0$, and $\mathcal{P}'$ is the projection which eliminates the diagonal part. In particular, for the density matrix we use the notation

$$\mathcal{P}\varrho = \varrho_{\mathrm{d}}\,. \tag{2.6.11}$$

It is obvious that the operations $\mathcal{P}$ and $\mathcal{P}'$ are in fact projections satisfying (2.5.9–12). Equation (2.5.17) now gives the equation of motion

$$\begin{aligned}\frac{\partial}{\partial t}\mathcal{P}\varrho(t) &= \mathcal{P}\mathrm{i}\mathcal{L}\mathcal{P}\varrho(t) + \mathcal{P}\mathrm{i}\mathcal{L}\int_0^t d\tau\,\exp[(t-\tau)\mathcal{P}'\mathrm{i}\mathcal{L}]\,\mathcal{P}'\mathrm{i}\mathcal{L}\mathcal{P}\varrho(\tau) \\ &+ \mathcal{P}\mathrm{i}\mathcal{L}\,\exp(t\mathcal{P}'\mathrm{i}\mathcal{L})\,\mathcal{P}'\varrho_0\end{aligned} \tag{2.6.12}$$

for $\mathcal{P}\varrho = \varrho_{\mathrm{d}}$.

Equation (2.6.1) is written as

$$\frac{\partial \varrho}{\partial t} = \mathrm{i}(\mathcal{L}_0 + \mathcal{L}_1)\varrho\,, \qquad \text{where}$$

$$\mathrm{i}\mathcal{L}_0\varrho = \frac{1}{\mathrm{i}\hbar}[\mathcal{H}_0, \varrho]\,, \tag{2.6.13}$$

$$\mathrm{i}\mathcal{L}_1\varrho = \frac{1}{\mathrm{i}\hbar}[\mathcal{H}_1, \varrho] \tag{2.6.14}$$

corresponding to the division (2.6.2). Note that

$$\mathrm{i}\mathcal{L}_0\mathcal{P} = 0\,, \tag{2.6.15}$$

$$\mathcal{P}\mathrm{i}\mathcal{L}_0 = 0\,, \tag{2.6.16}$$

$$\mathcal{P}\mathrm{i}\mathcal{L}_1\mathcal{P} = 0\,, \qquad \mathcal{P}\mathrm{i}\mathcal{L}\mathcal{P} = 0\,. \tag{2.6.17}$$

Equation (2.6.15) holds because two diagonal matrices are commutable. Equations (2.6.16, 17) hold because diagonal components of a commutator are all zero if one of the matrices in the commutator is diagonal.

The first term on the right-hand side of (2.6.12) vanishes following (2.6.17). In the second term, the operators $\mathcal{P}\mathrm{i}\mathcal{L}$ and $\mathrm{i}\mathcal{L}\mathcal{P}$ are reduced to $\mathcal{P}\mathrm{i}\mathcal{L}_1$ and $\mathrm{i}\mathcal{L}_1\mathcal{P}$, respectively, so that the equation becomes

$$\frac{\partial}{\partial t}\mathcal{P}\varrho = \mathcal{P}\mathrm{i}\mathcal{L}_1\int_0^t d\tau\,\exp[(t-\tau)\mathcal{P}'\mathrm{i}\mathcal{L}]\,\mathrm{i}\mathcal{L}_1\mathcal{P}\varrho\,, \tag{2.6.18}$$

where the initial condition

$$\mathcal{P}'\varrho_0 = 0 \tag{2.6.19}$$

is assumed. Now we make an approximation to (2.6.18), namely

$$\exp[(t-\tau)\mathcal{P}'\mathrm{i}\mathcal{L}] \;\rightarrow\; \exp[(t-\tau)(1-\mathcal{P})\mathrm{i}\mathcal{L}_0] = \exp[(t-\tau)\mathrm{i}\mathcal{L}_0]\,, \tag{2.6.20}$$

assuming that $\mathcal{H}_1$ is a weak perturbation to $\mathcal{H}_0$. Hence

$$\frac{\partial}{\partial t}\varrho_{\mathrm{d}} = \mathcal{P}\mathrm{i}\mathcal{L}_1\int_0^t d\tau\,\exp[(t-\tau)\mathrm{i}\mathcal{L}_0]\,\mathrm{i}\mathcal{L}_1\varrho_{\mathrm{d}}(\tau) \tag{2.6.21}$$

in the lowest approximation. The right-hand side of this equation is second order in the perturbation $\mathcal{H}_1$ and corresponds to a perturbational calculation at the lowest order familiar in quantum mechanics. This is seen as follows.

Note first the formula

$$A(t) \equiv \exp(\mathrm{i}\mathcal{L}_0 t) A = \exp(t\mathcal{H}_0/\mathrm{i}\hbar) A \exp(-t\mathcal{H}_0/\mathrm{i}\hbar) . \tag{2.6.22}$$

In fact,

$$\frac{\partial}{dt} A(t) = \mathrm{i}\mathcal{L}_0 A(t) = \frac{1}{\mathrm{i}\hbar}[\mathcal{H}_0, A(t)]$$

by differentiating both sides of (2.6.22), which is the solution of this equation of motion with the initial condition $A(0) = A$. Therefore the right-hand side of (2.6.21) is written as

$$\begin{aligned} &\mathscr{P}\frac{1}{\mathrm{i}\hbar}\left[\mathcal{H}_1, \int_0^t d\tau \exp\{(t-\tau)\mathcal{H}_0/\mathrm{i}\hbar\}\frac{1}{\mathrm{i}\hbar}[\mathcal{H}_1, \varrho_\mathrm{d}(\tau)] \exp\{-(t-\tau)\mathcal{H}_0/\mathrm{i}\hbar\}\right] \\ &\quad = -\frac{1}{\hbar^2}\mathscr{P}\left[\mathcal{H}_1, \int_0^t d\tau \{\mathcal{H}_1(\tau - t), \varrho_\mathrm{d}(\tau)\}\right] , \end{aligned} \tag{2.6.23}$$

where

$$\mathcal{H}_1(t) = \exp(-t\mathcal{H}_0/\mathrm{i}\hbar)\mathcal{H}_1 \exp(t\mathcal{H}_0/\mathrm{i}\hbar) \tag{2.6.24}$$

is the so-called interaction representation of $\mathcal{H}_1$. We have also used

$$\mathrm{e}^A[B,C]\,\mathrm{e}^{-A} = [\mathrm{e}^A B\,\mathrm{e}^{-A}, \mathrm{e}^A C\,\mathrm{e}^{-A}] .$$

Disentangling the double commutator and setting $\tau - t = s$, the right-hand side of (2.6.23) becomes

$$\begin{aligned} &-\frac{1}{\hbar^2}\int_{-t}^{0} ds\, \mathscr{P}[\mathcal{H}_1(0)\mathcal{H}_1(s)\varrho_\mathrm{d}(t+s) + \varrho_\mathrm{d}(t+s)\mathcal{H}_1(s)\mathcal{H}_1(0) \\ &\qquad - \mathcal{H}_1(0)\varrho_\mathrm{d}(t+s)\mathcal{H}_1(s) - \mathcal{H}_1(s)\varrho_\mathrm{d}(t+s)\mathcal{H}_1(0)] . \end{aligned} \tag{2.6.25}$$

If the rate of change of $\varrho_\mathrm{d}(t)$ is far slower than that of $\mathcal{H}_1(s)$, the replacement

$$\varrho_\mathrm{d}(t+s) \rightarrow \varrho_\mathrm{d}(t) \tag{2.6.26}$$

is allowed in the integral. We shall examine this assumption later. Equation (2.6.21) can thus be simplified. In the matrix representation it becomes

$$\begin{aligned} \frac{\partial}{\partial t}\langle l|\varrho(t)|l\rangle = &-\sum_m \frac{2}{\hbar(E_m - E_l)} \sin\frac{(E_m - E_l)t}{\hbar} \cdot |\langle l|\mathcal{H}_1|m\rangle|^2 \\ &\times [\langle l|\varrho(t)|l\rangle - \langle m|\varrho(t)|m\rangle] + O(\lambda^3) , \end{aligned} \tag{2.6.27}$$

where $O(\lambda^3)$ means that the remainder is small in the third order of the perturbational parameter λ.

For the example of interacting particles given by (2.6.6, 7), the unperturbed eigenstates are specified by (2.6.8) and their energies are

$$E\{\boldsymbol{p}\} = \frac{1}{2m}(\boldsymbol{p}_1^2 + \boldsymbol{p}_2^2 + \ldots + \boldsymbol{p}_N^2) . \tag{2.6.28}$$

For simplicity, we ignore the effects of quantum statistics here, namely Fermi or Bose statistics. The perturbation $\mathcal{H}_1$ has matrix elements only between the states connected by binary collisions. Since the potential $v(r_i - r_j)$ is dependent only on the relative position of particles, the momentum is conserved, i.e.,

$$(\boldsymbol{p}_i, \boldsymbol{p}_j) \rightarrow (\boldsymbol{p}_i + \hbar\boldsymbol{k}, \boldsymbol{p}_j - \hbar\boldsymbol{k}) , \tag{2.6.29}$$

in a collision exchanging a momentum $\hbar\boldsymbol{k}$. The corresponding matrix element of is

$$\langle \boldsymbol{p}_i, \boldsymbol{p}_j | v | \boldsymbol{p}_i + \hbar\boldsymbol{k}, \boldsymbol{p}_j - \hbar\boldsymbol{k} \rangle = \frac{1}{\Omega} \int \exp(\mathrm{i}\boldsymbol{k}\cdot\boldsymbol{r})\, v(\boldsymbol{r})\, d\boldsymbol{r} = \frac{1}{\Omega} v_{\boldsymbol{k}} , \tag{2.6.30}$$

where Ω is the volume of the box containing the particles. The energy change is given by

$$\begin{aligned} \Delta E(\boldsymbol{p}_i, \boldsymbol{p}_j, \boldsymbol{k}) &= \frac{1}{2m}[(\boldsymbol{p}_i + \hbar\boldsymbol{k})^2 + (\boldsymbol{p}_j - \hbar\boldsymbol{k})^2 - \boldsymbol{p}_i^2 - \boldsymbol{p}_j^2] \\ &= \frac{1}{m}[\hbar^2 \boldsymbol{k}^2 + \hbar\boldsymbol{k}\cdot(\boldsymbol{p}_i - \boldsymbol{p}_j)] . \end{aligned} \tag{2.6.31}$$

Defining

$$w(\boldsymbol{p}_i, \boldsymbol{p}_j | \boldsymbol{p}_i + \Delta\boldsymbol{p}, \boldsymbol{p}_j - \Delta\boldsymbol{p}) = \frac{2}{\hbar\Delta E} \sin\frac{\Delta E}{\hbar} t \cdot |v_k|^2 \frac{1}{\Omega^2} \tag{2.6.32}$$

$$\langle \boldsymbol{p}_1, \ldots, \boldsymbol{p}_N | \varrho(t) | \boldsymbol{p}_1, \ldots, \boldsymbol{p}_N \rangle = f(\boldsymbol{p}_1, \ldots, \boldsymbol{p}_N, t) \tag{2.6.33}$$

we rewrite (2.6.27) as

$$\begin{aligned} \frac{\partial}{\partial t} f(\boldsymbol{p}_1, \ldots, \boldsymbol{p}_N, t) = -\sum_{\langle ij \rangle} \sum_{\boldsymbol{k}} & w(\boldsymbol{p}_i, \boldsymbol{p}_j | \boldsymbol{p}_i + \Delta\boldsymbol{p}, \boldsymbol{p}_j - \Delta\boldsymbol{p}) \\ & \times [f(\boldsymbol{p}_1, \ldots, \boldsymbol{p}_N, t) - f(\boldsymbol{p}_1, \ldots, \boldsymbol{p}_i + \Delta\boldsymbol{p}, \ldots, \boldsymbol{p}_j - \Delta\boldsymbol{p}, \ldots, t)] , \end{aligned} \tag{2.6.34}$$

where

$$\Delta\boldsymbol{p} = \hbar\boldsymbol{k}$$

following (2.6.29). Equation (2.6.34) shows that the probability distribution function $f(\boldsymbol{p}_1, \boldsymbol{p}_2, \ldots, \boldsymbol{p}_N, t)$ evolves in time as a Markovian process induced by binary collisions. The latter are described by the transition rate

$w(\boldsymbol{p}_i, \boldsymbol{p}_j \,|\, \boldsymbol{p}_i + \Delta\boldsymbol{p}, \boldsymbol{p}_j - \Delta\boldsymbol{p})$ for the collision (2.6.29) or for its inverse. It is a function of time t in the form of (2.6.32) but becomes

$$w(\boldsymbol{p}_i, \boldsymbol{p}_j \,|\, \boldsymbol{p}_i + \Delta\boldsymbol{p}, \boldsymbol{p}_j - \Delta\boldsymbol{p}) = \frac{2\pi}{\hbar}\,\delta(\Delta E)\,|v_k|^2\,\frac{1}{\Omega^2} \tag{2.6.35}$$

through the approximation

$$\frac{2}{\Delta E}\sin\frac{\Delta E}{\hbar}t \rightarrow 2\pi\,\delta(\Delta E)\,. \tag{2.6.36}$$

In the form of (2.6.35), w is independent of time and defines the transition rate for the collision which conserves the energy, namely $\Delta E = 0$. In this sense, (2.6.34) is the *master equation* (or the Pauli equation [2.16] for a particle system with weak interactions).

If the particle system is macroscopic, the volume Ω and the particle number N are both very large. The sum over $\boldsymbol{k}$ in (2.6.34) is replaced by an integral

$$\sum_{\boldsymbol{k}} \rightarrow \frac{\Omega}{8\pi^3}\int d\boldsymbol{k}\,. \tag{2.6.37}$$

In such an integration, the energy change ΔE can take values from $-\infty$ to ∞. Let ε be a representative value of ΔE. For a nondegenerate particle system ε may be considered to be the average kinetic energy for a particle. According to (2.6.34), the rate of change of f with respect to a single variable p_j is of the order of

$$\frac{N}{\Omega}\frac{|v_k|^2}{\hbar\Delta E} \approx \frac{Na^3}{\Omega}\frac{v^2}{\hbar\varepsilon} \equiv \frac{1}{\tau_r}\,, \tag{2.6.38}$$

which represents the relaxation time of f. Here, a is the range of interaction $v(r)$ and v is the magnitude of the interaction. Therefore, the time t considered in (2.6.34) is of the order of τ_r

$$t = O(\tau_r)\,,$$

which is inversely proportional to the number density of particles N/Ω and to the square of the perturbation. On the other hand, the characteristic time rate for $\mathscr{H}_1(s)$ in (2.6.25) is of the order of

$$\frac{\hbar}{\varepsilon} = \tau_c\,. \tag{2.6.39}$$

If the perturbation is sufficiently weak, the condition

$$\tau_r \gg \tau_c \tag{2.6.40}$$

is fulfilled, which justifies the assumption (2.6.26). The approximation (2.6.35) is also justified since $t \approx \tau_r \gg \tau_c$ in the integration over k, corresponding to the well-known formula

$$\lim_{\xi \to \infty} \int \frac{\sin \xi x}{x} \phi(x)\, dx = \pi \phi(0)\,. \tag{2.6.41}$$

This situation arises more or less generally, not only for the above example of weakly interacting particles, but also for a macroscopic system with a great number of particles or large degrees of freedom, provided that the motion of the system can be considered in a perturbation scheme consisting of an unperturbed system and weak interactions. The time τ_c in (2.6.39) is the rate of change of the perturbation caused by the unperturbed motion of the system and characterizes the rate of microscopic motion of the system. This is determined by the structure of $\mathscr{H}_0$ and $\mathscr{H}_1$ and does not depend on the strength λ of the perturbation. On the other hand, τ_r measures the characteristic time rate of $\varrho_d(t)$ and is proportional to λ^{-2}. Therefore the condition (2.6.40) is fulfilled in the limit $\lambda \to 0$. The energy levels are almost continuously distributed in such a system so that the sum $\sum_l$ in (2.6.27) becomes an integral in which the replacement (2.6.35) is permissible. Thus the transitions occur only between states which conserve energy.

The eigenstates of the unperturbed Hamiltonian may be designated by the energy E and a set of other appropriate quantum numbers α. The density of state D is defined by the number of the states $D(E,\alpha)\, dE\, d\alpha$ in the interval $dE\, d\alpha$ and the distribution function f is defined by

$$\langle E\alpha | \varrho_d(t) | E\alpha \rangle = f(E,\alpha,t)\,.$$

Then (2.6.27) becomes

$$\frac{\partial}{\partial t} f(E,\alpha,t) = -\int w(E,\alpha \,|\, E,\alpha')\, D(E,\alpha')\, d\alpha' [f(E,\alpha,t) - f(E,a',t)]\,, \tag{2.6.42}$$

where

$$w(E,\alpha \,|\, E,\alpha') = \frac{2\pi}{\hbar} |\langle E,\alpha | \mathscr{H}_1 | E,\alpha' \rangle|^2 \tag{2.6.43}$$

is the transition rate between two states α and α' with equal energy. Equation (2.6.42) represents a uniform Markovian process. In other words, a non-Markovian process described by (2.6.21) recovers the Markovian property in the limit of weak perturbation and is described by a master equation. It is important to note that the conditions necessary to lead to this are essentially the same as those discussed in Sect. 2.1. If these conditions are not fulfilled, a master equation cannot be rigorous.

Equation (2.6.42) describes the long-time behavior of the diagonal part of the density matrix in the weak interaction limit, namely $t \to \infty$, $\lambda \to 0$, keeping $\lambda^2 t$ finite as $\lambda^2 t \sim \lambda^2 \tau_r \sim 1/\tau_c$ through (2.6.38, 39). This problem was treated in detail by *Van Hove* [2.17]. The limiting procedure is often called the *Van Hove limit* and, in fact, corresponds to the narrowing condition $\alpha \ll 1$ in (2.1.21) for a randomly modulated oscillator. If τ_c and γ^t in (2.1.21) are finite, this corresponds to the limit $\Delta \to 0$ and $t \to \infty$, keeping $\Delta^2 t \sim \Delta^2/\gamma \sim 1/\tau_c$ finite.

We treated (2.6.1) as a quantum-mechanical problem. Essentially the same treatment can be given to a classical problem considering this as a Liouville equation. Projection of the phase space distribution function onto a distribution in the momentum space corresponds to (2.6.33).

2.7 Brownian Motion of a Quantal System

The Kramers equation (2.2.34) describes Brownian motion of a classical system in contact with a heat bath at a constant temperature T. It is often necessary to extend this to a quantal system [2.18]. For example, Brownian motion of a quantal spin is one aspect of the problem of spin resonance and relaxation, and Brownian motion of a quantal oscillator is a fundamental problem in laser physics. Here we derive an equation corresponding to the Kramers equation by the method discussed in Sect. 2.5.

We consider a quantal system A in contact with a heat bath B and denote the density matrix of the whole system A + B by ϱ. Its temporal evolution is determined by the quantal Liouville equation (2.6.1)

$$\frac{\partial \varrho}{\partial t} = \frac{1}{\mathrm{i}\hbar}[\mathscr{H}_A + \mathscr{H}_B + \mathscr{H}_I, \varrho] \equiv \mathrm{i}\mathscr{L}\varrho\,, \qquad (2.7.1)$$

where $\mathscr{H}_A$ and $\mathscr{H}_B$ are the Hamiltonians of the systems A and B, respectively, and $\mathscr{H}_I$ is the interaction between the two. Denote the corresponding Liouville operators $\mathscr{L}_A$, $\mathscr{L}_B$, and $\mathscr{L}_I$ and define

$$\mathscr{L} = \mathscr{L}_A + \mathscr{L}_B + \mathscr{L}_I \equiv \mathscr{L}_0 + \mathscr{L}_I\,, \quad \mathscr{L}_0 = \mathscr{L}_A + \mathscr{L}_B\,. \qquad (2.7.2)$$

The density matrix for B at thermal equilibrium at temperature T is

$$\varrho_B = \exp(-\beta\mathscr{H}_B)/\mathrm{Tr}_B\{\exp(-\beta\mathscr{H}_B)\} \quad (\beta = 1/kT)\,, \qquad (2.7.3)$$

where Tr_B means the trace operation taking the diagonal sum of the following operator in the Hilbert space of B. Contraction of the density matrix ϱ of the whole system gives

$$\sigma(t) = \mathrm{Tr}_B\{\varrho(t)\}\,, \qquad (2.7.4)$$

which expresses the statistical state of system A.

Our problem is then to derive the equation of motion for $\sigma(t)$. We introduce the projection operator $\mathscr{P}$ by

$$\mathscr{P}f = \varrho_B \, \mathrm{Tr}_B\{f\} = \varrho_B \, g , \tag{2.7.5}$$

where f is an arbitrary operator in the Hilbert space of A + B. In particular, for the density matrix ϱ

$$\mathscr{P}\varrho = \varrho_B \, \sigma(t) . \tag{2.7.6}$$

Via (2.7.3) the operator $\mathscr{P}$ is in fact a projection satisfying the condition

$$\mathscr{P}^2 = \mathscr{P} .$$

The equation of motion for $\sigma(t)$ can be derived from (2.7.1) by using the formalism developed in Sect. 2.5. To rewrite (2.5.17) in a useful form, we note the following. For an arbitrary operator f

$$\mathrm{Tr}_B\{\mathrm{i}\mathscr{L}_B f\} = \frac{1}{\mathrm{i}\hbar}\mathrm{Tr}_B[\mathscr{H}_B f] = 0 .$$

Since

$$\mathrm{i}\mathscr{L}_B \varrho_B = \frac{1}{\mathrm{i}\hbar}[\mathscr{H}_B, \varrho_B] = 0$$

then also

$$\mathscr{P}\mathrm{i}\mathscr{L}_B = \mathrm{i}\mathscr{L}_B \mathscr{P} = 0$$

and therefore

$$\mathrm{i}\mathscr{L}_0\mathscr{P} = \mathscr{P}\mathrm{i}\mathscr{L}_0 = \mathscr{P}\mathrm{i}\mathscr{L}_0\mathscr{P} = \mathscr{P}\mathrm{i}\mathscr{L}_A\mathscr{P} . \tag{2.7.7}$$

Since $\mathscr{P}\mathscr{P}' = \mathscr{P}'\mathscr{P} = 0$, the above equation gives

$$\mathscr{P}'\mathrm{i}\mathscr{L}_0\mathscr{P} = \mathscr{P}\mathrm{i}\mathscr{L}_0\mathscr{P}' = 0 , \tag{2.7.8}$$

$$\mathscr{P}'\mathrm{i}\mathscr{L}\mathscr{P} = \mathscr{P}'\mathrm{i}\mathscr{L}_1\mathscr{P}, \quad \mathscr{P}\mathrm{i}\mathscr{L}\mathscr{P}' = \mathscr{P}\mathrm{i}\mathscr{L}_1\mathscr{P}' . \tag{2.7.9}$$

Now we define

$$\bar{\mathscr{H}}_1 = \mathrm{Tr}_B\{\varrho_B \mathscr{H}_1\} \tag{2.7.10}$$

which is the interaction averaged over the equilibrium distribution of the heat bath. Then

$$\begin{aligned}\mathscr{P}\mathrm{i}\mathscr{L}_1\mathscr{P}f &= \varrho_B \, \mathrm{Tr}_B[\mathscr{H}_1, \varrho_B \, \mathrm{Tr}_B\{f\}]/\mathrm{i}\hbar \\ &= \varrho_B[\mathrm{Tr}_B\{\varrho_B\mathscr{H}_1\}, \mathrm{Tr}_B\{f\}]/\mathrm{i}\hbar \\ &= [\bar{\mathscr{H}}_1, \mathscr{P}f]/\mathrm{i}\hbar = \mathrm{i}\bar{\mathscr{L}}_1\mathscr{P}f,\end{aligned} \tag{2.7.11}$$

where we used $\mathrm{Tr_B}\,[\mathscr{H}_1, \varrho_B] = 0$, and

$$\mathscr{P}'\,\mathrm{i}\,\mathscr{L}\mathscr{P} = \mathrm{i}(\mathscr{L}_1 - \bar{\mathscr{L}}_1)\,\mathscr{P}, \quad \mathscr{P}\,\mathrm{i}\,\mathscr{L}\mathscr{P}' = \mathscr{P}\,(\mathrm{i}\,\mathscr{L}_1 - \bar{\mathscr{L}}_1) \tag{2.7.12}$$

via (2.7.9). Corresponding to (2.5.17), thus

$$\frac{\partial}{\partial t}\sigma(t) = \mathrm{i}\,(\mathscr{L}_A + \bar{\mathscr{L}}_1)\,\sigma + \mathrm{Tr_B}\left\{\mathrm{i}\,(\mathscr{L}_1 - \bar{\mathscr{L}}_1)\int_{-\infty}^{t} \exp\,[(t-\tau)\,\mathscr{P}'\,\mathrm{i}\,\mathscr{L}\mathscr{P}']\,\mathrm{i}\,(\mathscr{L}_1 - \bar{\mathscr{L}}_1)\,\varrho_B\,\sigma(\tau)\,d\tau\right\} \tag{2.7.13}$$

where the contribution from the initial distribution is omitted. If initially it is of the form $\varrho = \varrho_B\,\sigma(0)$, this is obviously zero. If it is not, the contribution remains but decreases with time. In this sense the initial time was chosen to be in the infinite past in (2.7.13). If the average interaction is included in $\mathscr{H}_A$, $\bar{\mathscr{L}}_1$ can be omitted here.

Equation (2.7.13) is rigorous but not very useful in this form. The evolution operator $\exp\,[(t-\tau)\,\mathscr{P}'\,\mathrm{i}\,\mathscr{L}\mathscr{P}']$ in the integral is complex. If the interaction between A and B is weak enough, the operator $\mathrm{i}\,\mathscr{L}$ may be replaced by $\mathrm{i}\,\mathscr{L}_0$, meaning that the interaction is considered in second-order perturbation. Equation (2.7.8) gives

$$\mathscr{P}'\,\mathrm{i}\,\mathscr{L}_0\,\mathscr{P}' = \mathscr{P}'\,\mathrm{i}\,\mathscr{L}_0\,,$$

$$\therefore \exp\,[(t-\tau)\,\mathscr{P}'\,\mathrm{i}\,\mathscr{L}_0\,\mathscr{P}'] = \mathscr{P}'\exp\,[(t-\tau)\,\mathrm{i}\,\mathscr{L}_0]\,.$$

By including $\bar{\mathscr{H}}_1$ in $\mathscr{H}_A$ as noted above, then $\bar{\mathscr{L}}_1 = 0$ and $\mathscr{P}\,\mathrm{i}\,\mathscr{L}_1\,\mathscr{P} = 0$. Thus

$$\mathscr{P}\,\mathrm{i}\,\mathscr{L}_1\,\mathscr{P}' = \mathscr{P}\,\mathrm{i}\,\mathscr{L}_1\,.$$

Therefore (2.7.13) is reduced to

$$\frac{\partial}{\partial t}\sigma(t) = \mathrm{i}\,\mathscr{L}_A\,\sigma + \mathrm{Tr_B}\left\{\mathrm{i}\,\mathscr{L}_1\int_{-\infty}^{t}\exp\,[(t-\tau)\,\mathrm{i}\,\mathscr{L}_0]\,\mathrm{i}\,\mathscr{L}_1\,\varrho_B\,\sigma(\tau)\,d\tau\right\} \tag{2.7.14}$$

where the first term on the right-hand side represents the proper motion of A with its unperturbed Hamiltonian and the second term the thermal motion induced by the interaction with the heat bath.

For a Liouville operator $\mathrm{i}\,\mathscr{L}$ defined by

$$\mathrm{i}\,\mathscr{L}f = \frac{1}{\mathrm{i}\hbar}[\mathscr{H}, f]$$

we generally have

$$\exp(\mathrm{i}\,\mathscr{L}\,t)\,f = \exp(-\,\mathrm{i}\,\mathscr{H}\,t/\hbar)\,f\exp(\mathrm{i}\,\mathscr{H}\,t/\hbar)\,. \tag{2.7.15}$$

If $\sigma(t)$ in (2.7.14) is

$$\sigma(t) = \exp(\mathrm{i}\,\mathcal{L}_{\mathrm{A}}\,t)\,\hat{\sigma}(t) \equiv \exp(-\mathrm{i}\,\mathcal{H}_{\mathrm{A}}\,t/\hbar)\,\hat{\sigma}(t)\exp(\mathrm{i}\,\mathcal{H}_{\mathrm{A}}\,t/\hbar)\,, \tag{2.7.16}$$

the operator $\hat{\sigma}(t)$ represents the statistical motion of A observed in a reference system moving with the natural motion of A.

In many cases the interaction between A and B is expressed as

$$\mathcal{H}_{\mathrm{I}} = \sum_j X_j\,Y_j\,, \tag{2.7.17}$$

whre $X_j\,(j=1,2,\ldots)$ are dynamical quantities belonging to system A and $Y_j\,(j=1,2,\ldots)$ are those belonging to B. Their proper motion in the respective Hamiltonian is represented by the Heisenberg operators

$$X_j(t) = \exp(\mathrm{i}\,\mathcal{H}_{\mathrm{A}}\,t/\hbar)\,X_j\exp(-\mathrm{i}\,\mathcal{H}_{\mathrm{A}}\,t/\hbar) \equiv \exp(-\mathrm{i}\,\mathcal{L}_{\mathrm{A}}\,t)\,X_j\,, \tag{2.7.18}$$

$$Y_j(t) = \exp(\mathrm{i}\,\mathcal{H}_{\mathrm{B}}\,t/\hbar)\,Y_j\exp(-\mathrm{i}\,\mathcal{H}_{\mathrm{B}}\,t/\hbar) \equiv \exp(-\mathrm{i}\,\mathcal{L}_{\mathrm{B}}\,t)\,Y_j\,. \tag{2.7.19}$$

Using the above formulas we rewrite (2.7.14) as

$$\frac{\partial}{\partial t}\hat{\sigma}(t) = -\frac{1}{\hbar^2}\int_{-\infty}^{t} d\tau\,\mathrm{Tr}_{\mathrm{B}}\Big[\sum_j X_j(t)\,Y_j(0),\Big[\sum_l X_l(\tau)\,Y_l(-t+\tau),\,\varrho_{\mathrm{B}}\,\hat{\sigma}(\tau)\Big]\Big]\,. \tag{2.7.20}$$

Now we introduce correlation functions

$$\mathrm{Tr}_{\mathrm{B}}\,\varrho_{\mathrm{B}}\,Y_j(t)\,Y_l(0) = \Phi_{jl}(t) \tag{2.7.21}$$

for the dynamical quantities Y_j and Y_l evolving in time by the motion of B.

Disentangling the double commutator in (2.7.20) gives

$$\begin{aligned}\frac{\partial}{\partial t}\hat{\sigma}(t) = -\frac{1}{\hbar^2}\int_{-\infty}^{t} d\tau \sum_j\sum_l\,&[X_j(t)\,X_l(\tau)\,\hat{\sigma}(\tau)\,\Phi_{jl}(t-\tau)\\ &-X_j(t)\,\hat{\sigma}(\tau)\,X_l(\tau)\,\Phi_{lj}(-t+\tau)\\ &+\hat{\sigma}(\tau)\,X_l(\tau)\,X_j(t)\,\Phi_{lj}(-t+\tau)\\ &-X_l(\tau)\,\hat{\sigma}(\tau)\,X_j(t)\,\Phi_{jl}(t-\tau)]\,.\end{aligned} \tag{2.7.22}$$

This means that the forces acting on A fluctuate in time through the thermal motion of the heat bath B, causing the stochastic evolution of A. If the duration of the correlations $\Phi_{jl}(t)$ is characterized by τ_{c}, then the characteristic time of the temporal change of A, that is the relaxation time of A, is of the order of

$$\tau_{\mathrm{r}}^{-1} = O(\mathcal{H}_{\mathrm{I}}^2/\hbar^2)\,\tau_{\mathrm{c}} \tag{2.7.23}$$

as is seen from (2.7.22). If $\tau_{\mathrm{r}} \gg \tau_{\mathrm{c}}$, the change of $\hat{\sigma}(t)$ is slow so that $\hat{\sigma}(\tau)$ in the integral (2.7.22) may be replaced by $\hat{\sigma}(t)$. Going back to $\sigma(t)$ via the transformation (2.7.16), (2.7.22) becomes

$$\frac{\partial}{\partial t}\sigma(t) = \frac{1}{\mathrm{i}\hbar}[\mathscr{H}_{\mathrm{A}}, \sigma] + \Gamma\sigma(t), \tag{2.7.24}$$

where

$$\Gamma\sigma(t) = -\frac{1}{\hbar^2}\int_0^{\infty} dt' \int_{-\infty}^{\infty} d\omega\, \mathrm{e}^{\mathrm{i}\omega t'} \sum_j \sum_l \Phi_{jl}[\omega]\,[X_j(0)\, X_l(-t')\, \sigma(t)$$
$$- \mathrm{e}^{\beta\hbar\omega} X_j(0)\, \sigma(t)\, X_l(-t') + \mathrm{e}^{\beta\hbar\omega} \sigma(t)\, X_l(-t')\, X_j(0)$$
$$- X_l(-t')\, \sigma(t)\, X_j(0)]\,, \tag{2.7.25}$$

represents the Brownian motion of A induced by fluctuating forces exerted on A by the bath B. Here

$$\Phi_{jl}[\omega] = \frac{1}{2\pi}\int_{-\infty}^{\infty} dt\, \mathrm{e}^{-\mathrm{i}\omega t}\, \Phi_{jl}(t) \tag{2.7.26}$$

are the spectra of such fluctuating forces.

We have used

$$\frac{1}{2\pi}\int_{-\infty}^{\infty} dt\, \mathrm{e}^{-\mathrm{i}\omega t}\, \Phi_{lj}(-t) = \frac{\mathrm{e}^{\beta\hbar\omega}}{2\pi}\int_{-\infty}^{\infty} dt\, \mathrm{e}^{-\mathrm{i}\omega t}\, \Phi_{jl}(t) \tag{2.7.27}$$

in deriving (2.7.25) from (2.7.22). This is clearly evident by writing down explicit expressions of (2.7.21) in the representation diagonalizing $\mathscr{H}_{\mathrm{B}}$. If, further, we write the matrices of operators in (2.7.25) in the representation diagonalizing $\mathscr{H}_{\mathrm{A}}$ and carry the time integral using the formula

$$\int_0^{\infty} dt'\, \mathrm{e}^{\mathrm{i}(\omega-\nu)t'} = \pi\,\delta(\omega-\nu) + \frac{\mathrm{i}}{\omega-\nu}\,,$$

then

$$\sigma^0 = C\exp(-\beta\,\mathscr{H}_{\mathrm{A}}) \tag{2.7.28}$$

satisfies

$$\Gamma\sigma^0 = 0\,.$$

This means that (2.7.24) has the canonical density matrix (2.7.28) as the equilibrium solution, as is of course to be expected. Equation (2.7.24) is widely used as a fundamental equation for treating relaxation and fluctuation of quantal systems. If we concentrate our attention only on the diagonal part of the density matrix σ, which is the probabilities $\sigma(\alpha, t)$ of finding the system in its eigenstates, (2.7.24) is nothing but the usual Markovian equation in which the transition rates are calculated by the second-order perturbation calculation, or the so-called golden rule. Conventional equations of transports are of this type.

Equation (2.7.24) including off-diagonal elements is more general than such a conventional equation and is applied to spin or oscillator systems. For example, if A is an oscillator, then [2.19]

$$\frac{\partial}{\partial t}\sigma = \frac{1}{\mathrm{i}\hbar}[\omega_0 b^\dagger b, \sigma] + \nu\{[b, \sigma b^\dagger] + [b\sigma, b^\dagger\}$$
$$+ \nu'\{[b^\dagger, \sigma b] + [b^\dagger \sigma, b]\}, \qquad (2.7.29)$$

where $b^\dagger$ and b are the creation and destruction operators, respectively. To derive (2.7.29), assume (2.7.17) to be

$$\mathscr{H}' = b^\dagger Y + b Y^\dagger$$

and assume the spectra of the random forces $Y^\dagger$ and Y to be in suitable forms and apply (2.7.25). Details are left to the reader as an exercise.

The formalism explained here has a great variety of applications, but unfortunately the framework of this book does not allow these to be discussed.

We do, however, comment here on one point which is often misunderstood. Suppose that two systems A′ and A″ are in contact with a heat bath B and the Hamiltonian of the whole system is

$$\mathscr{H} = \mathscr{H}_{\mathrm{A}'} + \mathscr{H}_{\mathrm{A}''} + \mathscr{H}_{\mathrm{A}'\mathrm{A}''} + \mathscr{H}_{\mathrm{B}} + \mathscr{H}_{\mathrm{BA}'} + \mathscr{H}_{\mathrm{BA}''}.$$

One may be tempted to write

$$\frac{\partial \sigma}{\partial t} = \frac{1}{\mathrm{i}\hbar}(\mathscr{H}_{\mathrm{A}'} + \mathscr{H}_{\mathrm{A}''} + \mathscr{H}_{\mathrm{A}'\mathrm{A}''}, \sigma) + \Gamma_{\mathrm{A}'}\sigma + \Gamma_{\mathrm{A}''}\sigma$$

for the composite system A′ + A″ in which the stochastic operators $\Gamma_{\mathrm{A}'}$ and $\Gamma_{\mathrm{A}''}$ are calculated by treating the interactions $\mathscr{H}_{\mathrm{BA}'}$ and $\mathscr{H}_{\mathrm{BA}''}$ separately. In the classical limit this causes no trouble, but it is not correct, in general, for quantal systems and may lead to errors. The stochastic operators $\Gamma_{\mathrm{A}'}$ and $\Gamma_{\mathrm{A}''}$ drive the systems A′ and A″ separately to their equilibrium. If the interaction $\mathscr{H}_{\mathrm{A}'\mathrm{A}''}$ is not weak, however, the equilibrium of the composite system is not guaranteed by this treatment. In the Kramers equation for a classical system, the heat bath drives the kinetic energy of the system to thermal equilibrium, which in turn drives the potential energy to final equilibrium. This simple situation is not necessarily true for quantal systems.

2.8 Boltzmann Equation

In a dilute gas, molecules are almost free in motion. When two molecules by chance approach each other within the range of interaction, they collide and exchange momenta and energies. The probability that more than three molecules colliding together is proportional to the third power of the density, so that only binary collisions have to be considered in dilute gases. Such molecular collisions enable the gas to attain thermal equilibrium and govern the properties of the gas, such as hydrodynamic flow, heat conduction and diffusion.

A gas in thermal equilibrium is ideal and follows a very simple law. An equilibrium state is defined by only two variables, pressure and temperature. On the other hand, if the gas is not in equilibrium, there is no simple rule for the number of variables specifying such a state. However, for a dilute gas, the information contained in the one-particle distribution function may be supposed to be sufficient. Here the distribution function $f(\boldsymbol{r}, \boldsymbol{v}, t)$ gives the particle density and the velocity distribution of molecules as a function of space and time. It is, however, not very clear to what extent this information covers nonequilibrium states. It may be necessary to examine experimentally whether or not a real physical process belongs to the category which is well described in terms of the distribution function f. Even for a simple object like a dilute gas, the problem is very difficult since it essentially involves contraction of information from the microscopic level to the level of the one-particle distribution function.

For simplicity, we disregard the internal structure of molecules and treat them as mass points. We also disregard quantum-mechanical complications, and use a classical mechanics framework. Suppose that N particles in total are confined in a box with a volume Ω. We ask for the number of particles to be found in a volume element $d\boldsymbol{r}$ in the neighborhood of $\boldsymbol{r}$ with their momenta between $\boldsymbol{p}$ and $\boldsymbol{p} + d\boldsymbol{p}$. This number is denoted by

$$f(\boldsymbol{r}, \boldsymbol{p}, t)\, d\boldsymbol{r}\, d\boldsymbol{p}\,, \tag{2.8.1}$$

defining the distribution function, which is normalized as

$$\int f(\boldsymbol{r}, \boldsymbol{p}, t)\, d\boldsymbol{p} = n(\boldsymbol{r})\,, \quad \iint f(\boldsymbol{r}, \boldsymbol{p}, t)\, d\boldsymbol{r}\, d\boldsymbol{p} = N\,, \tag{2.8.2}$$

where $n(\boldsymbol{r})$ is the averaged number density at the position $\boldsymbol{r}$. In thermal equilibrium it holds that

$$f = f^0 = n(2\pi m k T)^{-3/2} \exp\left[-\frac{1}{kT}\left(\frac{\boldsymbol{p}^2}{2m} + V\right)\right]. \tag{2.8.3}$$

The distribution function $f(\boldsymbol{r}, \boldsymbol{p}, t)$ is analogous to that in the Kramers equation (2.2.34) and follows a similar equation

$$\frac{\partial}{\partial t} f(\boldsymbol{r}, \boldsymbol{p}, t) + \left(\frac{\boldsymbol{p}}{m} \cdot \frac{\partial}{\partial \boldsymbol{r}} - \frac{\partial V}{\partial \boldsymbol{r}} \cdot \frac{\partial}{\partial \boldsymbol{p}}\right) f = \Gamma[f] , \tag{2.8.4}$$

where the second term on the left-hand side is the drift term. If there is no collision, particles would move in phase space with

$$\frac{d\boldsymbol{r}}{dt} = \frac{\boldsymbol{p}}{m}, \quad \frac{d\boldsymbol{p}}{dt} = -\frac{\partial V}{\partial \boldsymbol{r}} .$$

The substantial change of f for an observer moving in phase space with this natural motion is

$$\begin{aligned} \frac{D}{Dt} f &= \lim \frac{1}{\Delta t} [f(\boldsymbol{r}, \boldsymbol{p}, t) - f(\boldsymbol{r} - \dot{\boldsymbol{r}} \Delta t, \boldsymbol{p} - \dot{\boldsymbol{p}} \Delta t, t - \Delta t)] \\ &= \frac{\partial f}{\partial t} + \left(\dot{\boldsymbol{r}} \cdot \frac{\partial}{\partial \boldsymbol{r}} + \dot{\boldsymbol{p}} \cdot \frac{\partial}{\partial \boldsymbol{p}}\right) f \end{aligned}$$

which must be balanced by the change due to collisions.

Suppose two particles with momenta $\boldsymbol{p}$ and $\boldsymbol{p}_1$ collide and change to $\boldsymbol{p}'$ and $\boldsymbol{p}_1'$. Both momentum and energy must be conserved so that

$$\boldsymbol{p} + \boldsymbol{p}_1 = \boldsymbol{p}' + \boldsymbol{p}_1' , \tag{2.8.5}$$

$$\varepsilon(\boldsymbol{p}) + \varepsilon(\boldsymbol{p}_1) = \varepsilon(\boldsymbol{p}') + \varepsilon(\boldsymbol{p}_1') . \tag{2.8.6}$$

In a nonrelativistic regime, the kinetic energy is usually assumed to be

$$\varepsilon(\boldsymbol{p}) = \frac{\boldsymbol{p}^2}{2m} ,$$

but the functional form may be different from this in general cases. The relative motion of colliding particles is shown in Fig. 2.3, where the vector $\boldsymbol{b}$ is the perpendicular from the target particle to the relative velocity before the collision. The magnitude of $\boldsymbol{b}$ is the impact parameter b. For a given set of $\boldsymbol{p}$ and $\boldsymbol{p}_1$, the conservation laws (2.8.5, 6) impose four conditions of

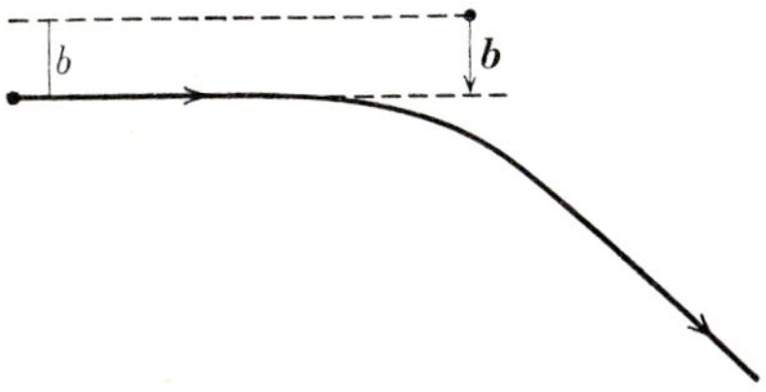

Fig. 2.3. The relative path of colliding particles. The lower dashed line is the asymptotic path of the incident particle relative to the scatterer. The distance b between it and a parallel line through the scatterer is the impact parameter

constraint, leaving two degrees of freedom for the momenta $\boldsymbol{p}'$ and $\boldsymbol{p}_1'$. These correspond to the impact parameter b and the angle of rotation around the vector $\boldsymbol{p}-\boldsymbol{p}_1$. These two parameters take various values in a probabilistic way so that the transition probability can be defined as a collision from $\boldsymbol{p}$ and $\boldsymbol{p}_1$ to $\boldsymbol{p}'$ and $\boldsymbol{p}_1'$. The relative probability is denoted by $(\boldsymbol{p},\boldsymbol{p}_1|\sigma|\boldsymbol{p}',\boldsymbol{p}_1')$ and is normalized as

$$\int (\boldsymbol{p},\boldsymbol{p}_1|\sigma|\boldsymbol{p}',\boldsymbol{p}_1')\, d\boldsymbol{p}'\, d\boldsymbol{p}_1' = \sigma(\boldsymbol{p},\boldsymbol{p}_1) \tag{2.8.7}$$

which is the total cross section for a collision between particles with momenta $\boldsymbol{p}$ and $\boldsymbol{p}_1$. The constraints (2.8.5, 6) can be included in the functional form of this transition probability as delta functions:

$$(\boldsymbol{p},\boldsymbol{p}_1|\sigma|\boldsymbol{p}',\boldsymbol{p}_1') \propto \delta(\boldsymbol{p}+\boldsymbol{p}_1-\boldsymbol{p}'-\boldsymbol{p}_1')\,\delta[\varepsilon(\boldsymbol{p})+\varepsilon(\boldsymbol{p}_1)-\varepsilon(\boldsymbol{p}')-\varepsilon(\boldsymbol{p}_1')]\,. \tag{2.8.8}$$

If a collision $(\boldsymbol{p},\boldsymbol{p}_1)\rightarrow(\boldsymbol{p}',\boldsymbol{p}_1')$ occurs in the volume element $d\boldsymbol{r}$ at position $\boldsymbol{r}$, the distribution function $f(\boldsymbol{r},\boldsymbol{p},t)$ suffers a loss. On the other hand, a collision from $(\boldsymbol{p}',\boldsymbol{p}_1')$ to $(\boldsymbol{p},\boldsymbol{p}_1)$ is a gain for f. Assuming by intuition that collisions are statistically independent, the collision term in (2.8.4) is

$$\begin{aligned}\Gamma[f] = &-\iiint d\boldsymbol{p}_1\, d\boldsymbol{p}'\, d\boldsymbol{p}_1' f(\boldsymbol{r},\boldsymbol{p},t) f(\boldsymbol{r},\boldsymbol{p}_1,t)(\boldsymbol{p},\boldsymbol{p}_1|\sigma|\boldsymbol{p}',\boldsymbol{p}_1')\\ &+\iiint d\boldsymbol{p}_1\, d\boldsymbol{p}'\, d\boldsymbol{p}_1' f(\boldsymbol{r},\boldsymbol{p}',t) f(\boldsymbol{r},\boldsymbol{p}_1',t)(\boldsymbol{p}',\boldsymbol{p}_1'|\sigma|\boldsymbol{p},\boldsymbol{p}_1)\,.\end{aligned} \tag{2.8.9}$$

When this is inserted, (2.8.4) becomes a nonlinear evolution equation for determining the distribution function f.

Equation (2.8.9) represents the *Stosszahl Ansatz* (assumption for the collision frequency) proposed by *Boltzmann* more than a hundred years ago. This was the basis of his monumental work on the kinetic theory of gases [2.20].

It is very difficult to prove the Stosszahl Ansatz from the more basic level of microscopic dynamics. Equation (2.6.34) represents a master equation for the probability distribution of the whole set of momentum variables of N particles. If this function is integrated over the variables, leaving us with functions of only one or two particles, a series of distribution functions arises such as

$$\int\ldots\int f(\boldsymbol{p}_1,\boldsymbol{p}_2,\ldots,\boldsymbol{p}_N,t)\, d\boldsymbol{p}_2\ldots d\boldsymbol{p}_N = f_1(\boldsymbol{p}_1,t)\,, \tag{2.8.10}$$

$$\int\ldots\int f(\boldsymbol{p}_1,\boldsymbol{p}_2,\ldots,\boldsymbol{p}_N,t)\, d\boldsymbol{p}_3\ldots d\boldsymbol{p}_N = f_2(\boldsymbol{p}_1,\boldsymbol{p}_2,t)\,. \tag{2.8.11}$$

Integrating both sides of (2.6.34) in the same way and changing the numbering of particles yields

$$\begin{aligned}\frac{\partial}{\partial t} f_1(\boldsymbol{p},t) = &-N\int d\boldsymbol{p}_1 \sum_{\boldsymbol{k}} w(\boldsymbol{p},\boldsymbol{p}_1|\boldsymbol{p}+\hbar\boldsymbol{k},\boldsymbol{p}_1-\hbar\boldsymbol{k})\\ &\times[f_2(\boldsymbol{p},\boldsymbol{p}_1,t)-f_2(\boldsymbol{p}+\hbar\boldsymbol{k},\boldsymbol{p}_1-\hbar\boldsymbol{k},t)]\,,\end{aligned} \tag{2.8.12}$$

where $f_2(\boldsymbol{p}, \boldsymbol{p}_1, t)$ is the two-particle distribution function. If statistical independence is assumed for the probability distribution of two particles, we may set

$$f_2(\boldsymbol{p}, \boldsymbol{p}_1, t) = f_1(\boldsymbol{p}, t) f_1(\boldsymbol{p}_1, t). \tag{2.8.13}$$

This makes the right-hand side of (2.8.12) essentially the same as the collision term in (2.8.9). The difference is that here the function f_1 is normalized to one and does not depend on the position. As is clear from the definition of w, (2.6.35, 37), the right-hand side of (2.8.12) is proportional to the number density of particles and gives the time rate of the change of the momentum distribution f_1 by binary collisions occurring everywhere in the container. Thus, (2.8.12) with the assumption of (2.8.13) is nothing but the Boltzmann equation given by (2.8.4, 9). We emphasize again that the master equation (2.6.34) is more general than the Boltzmann equation. It has not been made very clear what kind of conditions are needed for the solutions of these equations to coincide.

The master equation is already probabilistic. Derivation of the Boltzmann equation from the more basic dynamical equation, namely, from the Liouville equation, is very fundamental in statistical physics. Consider an ensemble of N-particle systems and define the phase space distribution function by

$$f_N(\boldsymbol{r}_1, \boldsymbol{p}_1; \boldsymbol{r}_2, \boldsymbol{p}_2; \ldots; \boldsymbol{r}_N, \boldsymbol{p}_N; t),$$

which is assumed to be symmetric in permutations of the particles. When this is integrated over the canonical variables, omitting only those for one particle, the one-particle distribution function (2.8.1) arises. If it is integrated, omitting variables of two or more particles, the hierarchy of reduced distribution functions for two or more particles is given. The original Liouville equation is concerned with N particles. When it is integrated over canonical variables, omitting one, two or more particles, the equation gives a hierarchy of equations for the reduced distribution functions. However, the temporal evolution of a one-particle distribution function contains a two-particle distribution function, and generally the n-particle distribution function contains a $(n+1)$-particle distribution function. In this sense, the chain of equations is not closed. Assuming that the higher-order distribution functions are functionals of the one-particle distribution function, the Boltzmann equation can be derived with a certain process of coarse graining. This was done by *Bogolyubov* [2.21] and has been examined in detail by other investigators. We shall not, however, go into this problem any further, Sect. 4.6 and [2.22]. If it were possible to start from the limit of dilute gases and to carry out a sort of cluster expansion as familiar from the theory of equilibrium of nonideal gases, the Boltzmann equation would be obtained in the limit of dilute gases and the corrections could be calculated at higher densities. It has turned out, however, that no simple expansion is possible in powers of density. Nonequilibrium statistical mechanics of dense gases still remains an outstanding problem.

The dynamical law is symmetric with respect to time reversal, so that for a collision process

$$(\boldsymbol{p}, \boldsymbol{p}_1) \rightarrow (\boldsymbol{p}', \boldsymbol{p}_1') \tag{2.8.14}$$

its reverse collision $(-\boldsymbol{p}', -\boldsymbol{p}_1') \rightarrow (-\boldsymbol{p}, -\boldsymbol{p}_1)$ exists. If, further, the law of force between particles is symmetric with respect to spatial inversion then the restitution collision

$$(\boldsymbol{p}', \boldsymbol{p}_1') \rightarrow (\boldsymbol{p}, \boldsymbol{p}_1) \tag{2.8.15}$$

exists and has the same probability as shown in Fig. 2.4. If this is the case, it has symmetry

$$\begin{aligned}(\boldsymbol{p}, \boldsymbol{p}_1 | \sigma | \boldsymbol{p}', \boldsymbol{p}_1') &= (\boldsymbol{p}', \boldsymbol{p}_1' | \sigma | \boldsymbol{p}, \boldsymbol{p}_1) \\ &\equiv \sigma(\boldsymbol{p}, \boldsymbol{p}_1 | \boldsymbol{p}', \boldsymbol{p}_1') .\end{aligned} \tag{2.8.16}$$

This symmetry seems to exist in (2.6.32) without the condition of spatial inversion, because the collision is treated in the second-order Born approximation. In general, the symmetry does not necessarily hold. However, let us assume this for simplicity. Then the two terms on the right-hand side of (2.8.9) can be combined into one. The Boltzmann equation now reads

$$\begin{aligned}&\frac{\partial}{\partial t} f(\boldsymbol{r}, \boldsymbol{p}, t) + \frac{\boldsymbol{p}}{m} \cdot \frac{\partial}{\partial \boldsymbol{r}} f - \frac{\partial V}{\partial \boldsymbol{r}} \cdot \frac{\partial}{\partial \boldsymbol{p}} f = \Gamma[f] \\ &\quad \equiv \iiint d\boldsymbol{p}_1 \, d\boldsymbol{p}' \, d\boldsymbol{p}_1' \, \sigma(\boldsymbol{p}, \boldsymbol{p}_1 | \boldsymbol{p}', \boldsymbol{p}_1')(f f_1 - f' f_1')\end{aligned} \tag{2.8.17}$$

with the abbreviation

$$f_1 = f(\boldsymbol{r}, \boldsymbol{p}_1, t), \quad f' = f(\boldsymbol{r}, \boldsymbol{p}', t), \quad f_1' = f(\boldsymbol{r}, \boldsymbol{p}_1', t) . \tag{2.8.18}$$

Equation (2.8.4, 9) or (2.8.17) is Markovian in a sense because the evolution of the distribution function f is determined by the same function at that time point. However, the evolution equation is not linear – an essential difference from the Markovian property in the ordinary sense. The

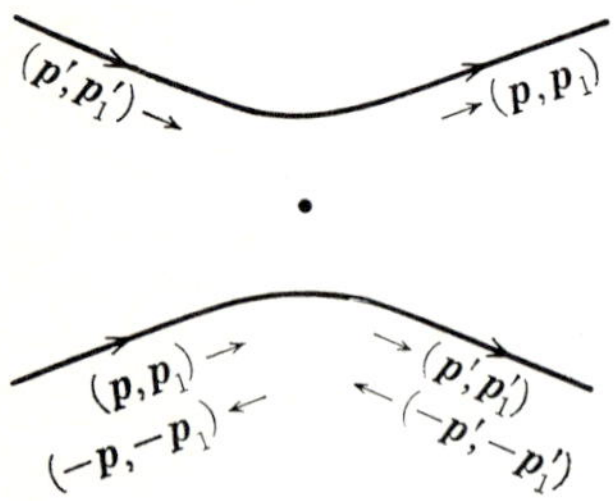

Fig. 2.4. A collision and its restitution

nonlinearity causes mathematical difficulties which have not yet been completely solved.

The H theorem explained in Sect. 2.3 can be proved for the Boltzmann equation. In fact, Boltzmann showed this to prove the approach to equilibrium. Let us define the entropy density by

$$s(\boldsymbol{r}, t) = -k \int f(\boldsymbol{r}, \boldsymbol{p}, t) \ln f(\boldsymbol{r}, \boldsymbol{p}, t)\, d\boldsymbol{p}. \tag{2.8.19}$$

Multiplying (2.8.17) by $(1 + \ln f)$ and integrating it with respect to $\boldsymbol{p}$ gives

$$\int d\boldsymbol{p} \left(\frac{\partial}{\partial t} + \frac{\boldsymbol{p}}{m} \cdot \frac{\partial}{\partial \boldsymbol{r}} - \frac{\partial V}{\partial \boldsymbol{r}} \cdot \frac{\partial}{\partial \boldsymbol{p}} \right) f \ln f = \int d\boldsymbol{p} (1 + \ln f)\, \Gamma[f]. \tag{2.8.20}$$

Since we may assume that f vanishes for $|p| \to \infty$, then

$$\int d\boldsymbol{p} \frac{\partial}{\partial \boldsymbol{p}} f \ln f = 0.$$

Note also that

$$\int \Gamma[f]\, d\boldsymbol{p} = 0$$

as seen from (2.8.17), and define the density of entropy flow by

$$\boldsymbol{j}_s(\boldsymbol{r}, t) = -k \int d\boldsymbol{p} \frac{\boldsymbol{p}}{m} f \ln f. \tag{2.8.21}$$

Then (2.8.20) is written as

$$\frac{\partial}{\partial t} s(\boldsymbol{r}, t) + \operatorname{div} \boldsymbol{j}_s(\boldsymbol{r}, t) = \left(\frac{ds}{dt} \right)_{\text{irr}}, \tag{2.8.22}$$

where the right-hand side is

$$\left(\frac{ds}{dt} \right)_{\text{irr}} = -k \int \ln f\, \Gamma[f]\, d\boldsymbol{p}.$$

This integration is made over $\boldsymbol{p}$, $\boldsymbol{p}_1$, $\boldsymbol{p}'$, and $\boldsymbol{p}_1'$ so that f in $\ln f$ may be changed to f_1. The integral changes its sign if primed and unprimed functions are interchanged. Therefore

$$\left(\frac{ds}{dt} \right)_{\text{irr}} = \frac{k}{4} \iiiint d\boldsymbol{p}\, d\boldsymbol{p}_1\, d\boldsymbol{p}'\, d\boldsymbol{p}_1'\, \sigma(\boldsymbol{p}, \boldsymbol{p}_1 | \boldsymbol{p}', \boldsymbol{p}_1') (f f_1 - f' f_1') \ln \frac{f f_1}{f' f_1'}.$$

Here, σ is positive because it is the cross section of collisions, and the rest is also positive because of the inequality

$$(x - y) \ln(x/y) > 0$$

for $x \neq y$. Thus it is proved that

$$\left(\frac{ds}{dt}\right)_{\text{irr}} \geqq 0. \tag{2.8.23}$$

This is the irreversible production of entropy, which becomes zero when and only when

$$ff_1 = f'f_1' \tag{2.8.24}$$

holds. Integrated over the whole space, the total entropy is

$$S(t) = \int s(\boldsymbol{r}, t)\, d\boldsymbol{r} = -k \iint d\boldsymbol{r}\, d\boldsymbol{p}\, f \ln f, \tag{2.8.25}$$

for which (2.8.22) gives

$$\frac{dS}{dt} \geqq 0. \tag{2.8.26}$$

Boltzmann's H function is related to S by

$$S = -kH. \tag{2.8.27}$$

Through collisions, H decreases in time and S increases. In the state where H or S no longer changes, equilibrium is attained.

It is not difficult to derive the Maxwellian distribution (2.8.3) from the condition (2.8.24), written as

$$\ln f(\boldsymbol{p}) + \ln f(\boldsymbol{p}_1) = \ln f(\boldsymbol{p}') + \ln f(\boldsymbol{p}_1').$$

Generally a function $\varphi(\boldsymbol{p})$ is called an invariant of the collision if it satisfies

$$\varphi(\boldsymbol{p}) + \varphi(\boldsymbol{p}_1) = \varphi(\boldsymbol{p}') + \varphi(\boldsymbol{p}_1') \tag{2.8.28}$$

for a collision process $(\boldsymbol{p}, \boldsymbol{p}_1) \to (\boldsymbol{p}', \boldsymbol{p}_1')$. Obviously

$$\varphi = 1, p_x, p_y, p_z, \boldsymbol{p}^2$$

are such invariants. There is no other invariant because the degrees of freedom for a collision would be limited to less than two if one existed. Since the function $\ln f$ satisfies the same functional equation as (2.8.28), it must have the form

$$\ln f = a + \boldsymbol{b} \cdot \boldsymbol{p} + c\boldsymbol{p}^2.$$

If the gas is flowing, then $\boldsymbol{b} \neq 0$. If the gas is confined in a container, $\boldsymbol{b} = 0$ and the distribution becomes Maxwellian. Thus, the Boltzmann equation has equilibrium distribution as its solution and any solution presumably approaches it in the course of time.

If the state is not very far from equilibrium, we may set

$$f = f_0(1+g) \quad (|g| \ll 1)$$

and the Boltzmann equation may be linearized in the deviation g as

$$\left(\frac{\partial}{\partial t} + \frac{\boldsymbol{p}}{m} \cdot \frac{\partial}{\partial \boldsymbol{r}} - \frac{\partial V}{\partial \boldsymbol{r}} \cdot \frac{\partial}{\partial \boldsymbol{p}}\right) g = -\mathscr{D} g .$$

Derivation of the collision operator $\mathscr{D}$ from (2.8.9 or 17) is left to the reader as an exercise.

The basic idea of the Boltzmann equation is not limited to dilute gases. In metals and semiconductors, conduction electrons are scattered by crystalline defects or phonons or they collide with each other. The Boltzmann equation is applied to these cases with the scattering probabilities calculated by quantum mechanics. At low temperatures, degenerate Fermi particles can be treated by Landau's Fermi-liquid theory [2.23]. This is a bold extension of the Boltzmann equation which is made possible by Pauli's exclusion principle that greatly suppresses collisions between particles.

Fundamental problems regarding the Boltzmann equation remain even today. Nevertheless, it has played an extremely important role in the history of statistical physics.

2.9 Generalized Langevin Equation and the Damping Theory

In Sect. 1.6 we generalized the Langevin equation from a phenomenological point of view. Now we discuss a microscopic derivation of such a generalized Langevin equation, due to *Mori* [2.24], whereby the damping formalism is applied to an equation of motion of dynamical variables, transforming it to a form which corresponds to the Langevin equation. The formalism is the same in classical and quantum mechanics, but the classical case elucidates the essential points better. For this purpose we need to discuss some preliminaries regarding equations of motion of dynamical variables.

We denote a point in phase space by (p, q), which is an abbreviation of the whole set of canonical variables. In the following a sum over all the degrees of freedom is simply written as $\sum$. By the motion of the system, a phase point moves from (p, q) to (p_t, q_t) in a time interval $(0, t)$. The evolution of a dynamical quantity a in the same time interval is denoted by

$$a_t \equiv a(p_t, q_t) = a(p, q; t) . \tag{2.9.1}$$

Its value a_t at time t is determined by p_t and q_t, which are functions of the initial phase (p,q) and t. The evolution is expressed by the equation of motion

$$\begin{aligned}\frac{d}{dt}a_t &= \sum\left(\frac{\partial a_t}{\partial p_t}\dot{p}_t + \frac{\partial a_t}{\partial q_t}\dot{q}_t\right)\\ &= \sum\left(-\frac{\partial \mathscr{H}_t}{\partial q_t}\frac{\partial a_t}{\partial p_t} + \frac{\partial \mathscr{H}_t}{\partial p_t}\frac{\partial a_t}{\partial q_t}\right),\end{aligned} \tag{2.9.2}$$

where $\mathscr{H}_t = \mathscr{H}(p_t, q_t)$ is the Hamiltonian of the system. Alternatively the motion is expressed by

$$\begin{aligned}\frac{\partial}{\partial t}a(p,q;t) &= \sum\left(-\frac{\partial \mathscr{H}}{\partial q}\frac{\partial}{\partial p} + \frac{\partial \mathscr{H}}{\partial p}\frac{\partial}{\partial q}\right)a(p,q;t)\\ &\equiv -\mathrm{i}\mathscr{L}a,\end{aligned} \tag{2.9.3}$$

which defines a_t as a function of (p,q) and t. Here $\mathrm{i}\mathscr{L}$ is the Liouville operator. It should be noted that (2.9.3) has a negative sign in contrast to the Liouville equation describing the evolution of the phase space distribution function. The situation is the same as the relation between the equation of motion of a density matrix and the Heisenberg equation of motion for a dynamical variable,

$$\mathrm{i}\hbar\frac{dA_t}{dt} = A_t\mathscr{H} - \mathscr{H}A_t\,. \tag{2.9.4}$$

We now show that (2.9.3) in fact gives the motion of a. A formal solution of this equation is written as

$$a(p,q;t) = \exp(-\mathrm{i}\mathscr{L}t)\,a(p,q)\,, \tag{2.9.5}$$

where $a(p,q)$ is the initial value of a. Note that the commutator of the operator $\mathrm{i}\mathscr{L}$ and an arbitrary function $G(p,q)$ is simply a function of (p,q) as may be inferred from

$$\begin{aligned}[\mathrm{i}\mathscr{L},G] &= \sum\left(-\frac{\partial \mathscr{H}}{\partial q}\left[\frac{\partial}{\partial p},G\right] + \frac{\partial \mathscr{H}}{\partial p}\left[\frac{\partial}{\partial q},G\right]\right)\\ &= \sum\left(-\frac{\partial \mathscr{H}}{\partial q}\frac{\partial G}{\partial p} + \frac{\partial \mathscr{H}}{\partial p}\frac{\partial G}{\partial q}\right) = \mathrm{i}\mathscr{L}G\,.\end{aligned}$$

Repeated use of this expression gives

$$[\mathrm{i}\mathscr{L},[\mathrm{i}\mathscr{L},\dots[\mathrm{i}\mathscr{L},G]\dots] = (\mathrm{i}\mathscr{L})^n G\,.$$

Furthermore, note a general formula for two operators

$$e^X Y e^{-X} = X + [X, Y] + \frac{1}{2}[X, [X, Y]] + \ldots + \frac{1}{n!}[X, [X, \ldots [X, Y] \ldots] + \ldots .$$

Therefore

$$\exp(-i\mathscr{L}t)\, a(p,q) \exp(i\mathscr{L}t) = \exp(-i\mathscr{L}t)\, a(p,q) . \tag{2.9.6}$$

For example, if a is $p^n q^m$, then

$$\begin{aligned}&\exp(-i\mathscr{L}t)\, p^n q^m \exp(i\mathscr{L}t)\\ &\quad = \{\exp(-i\mathscr{L}t)\, p \exp(i\mathscr{L}t)\}^n \{\exp(-i\mathscr{L}t)\, q \exp(i\mathscr{L}t)\}^m\end{aligned}$$

[note that $\exp(-i\mathscr{L}t)$ and $\exp(i\mathscr{L}t)$ are inverse]. If we set

$$\begin{aligned} p_t &= \exp(-i\mathscr{L}t)\, p \exp(i\mathscr{L}t) = \exp(-i\mathscr{L}t)\, p \\ q_t &= \exp(-i\mathscr{L}t)\, q \exp(i\mathscr{L}t) = \exp(-i\mathscr{L}t)\, q , \end{aligned} \tag{2.9.7}$$

(2.9.6) shows that

$$a(p,q;t) = \exp(-i\mathscr{L}t)\, a(p,q) \exp(i\mathscr{L}t) = a(p_t, q_t)$$

holds for an arbitrary function $a(p,q)$. From (2.9.7) it follows that

$$\frac{dp_t}{dt} = \exp(-i\mathscr{L}t)(-i\mathscr{L}p) = \exp(-i\mathscr{L}t)\left(-\frac{\partial \mathscr{H}}{\partial q}\right) = -\frac{\partial}{\partial q_t}\mathscr{H}(p_t, q_t)$$

(with a similar equation for dq_t/dt), so that the functions p_t and q_t are the solution of Hamilton's equation of motion for the initial phase (p,q). Therefore, (2.9.5) represents the motion of a dynamical quantity a and (2.9.3) is its evolution equation. This equation has the same form as (2.5.1), so that the damping theory can be applied by introducing a properly defined projection operator.

Consider a dynamical system in equilibrium and study the fluctuation of a dynamical quantity (the velocity of a Brownian particle, for example) of the system. For an arbitrary dynamical quantity $g(p,q)$, its projection is defined by

$$\mathscr{P}g = \frac{a\langle a, g\rangle}{\langle a, a\rangle} , \tag{2.9.8}$$

where the inner product is defined by

$$\langle a, g\rangle = C \iint dp\, dq \exp[-\mathscr{H}(p,q)/kT]\, a(p,q)\, g(p,q) . \tag{2.9.9}$$

Here C is the normalization for the canonical distribution function and the integration is made over the entire phase space. This projection obviously satisfies the condition $\mathscr{P}^2=\mathscr{P}$. For $g=a$

$$a_0=a\,,\qquad \mathscr{P}a=a\,,\qquad \mathscr{P}'a=0\qquad (\mathscr{P}'=1-\mathscr{P})\,. \tag{2.9.10}$$

For two phase functions $g\,(p,q)$ and $h\,(p,q)$

$$\langle g_t,h_t\rangle=\langle g_0,h_0\rangle\equiv\langle g,h\rangle \tag{2.9.11}$$

holds and hence

$$\frac{d}{dt}\langle g_t,h_t\rangle=\langle \dot{g}_t,h_t\rangle+\langle g_t,\dot{h}_t\rangle=0\,,$$
$$\langle \dot{g},h\rangle=-\langle g,\dot{h}\rangle\,, \tag{2.9.12}$$

$$\langle \dot{g},g\rangle=\langle g,\dot{g}\rangle=0\,. \tag{2.9.13}$$

Equation (2.9.11) is found to be true by

$$\begin{aligned}\langle g_t,h_t\rangle&=C\textstyle\iint dp\,dq\,\exp[-\mathscr{H}(p,q)/kT]\,g\,(p_t,q_t)\,h\,(p_t,q_t)\\&=C\textstyle\iint dp_t\,dq_t\,\exp[-\mathscr{H}(p_t,q_t)/kT]\,g\,(p_t,q_t)\,h\,(p_t,q_t)\\&=\langle g,h\rangle\,,\end{aligned}$$

where the invariance $\mathscr{H}(p_t,q_t)=\mathscr{H}(p,q)$ and $dp\,dq=dp_t\,dq_t$ is used. The second line is an integral over the whole phase space, so that the integration variables p_t and q_t can be replaced by the initial phase p and q. Differentiating the above equality by t and setting $t=0$ yields (2.9.12).

Now we apply (2.5.17), inserting a for f and $-\mathrm{i}\mathscr{L}$ for Γ. Since

$$-\mathrm{i}\mathscr{L}a=\dot{a}(0)\equiv\dot{a}\,, \tag{2.9.14}$$

the first term on the right-hand side vanishes because of $(a,\dot{a})=0$, and the third term via (2.9.10). The second term is found to be

$$\begin{aligned}&a\left\langle a,-\mathrm{i}\mathscr{L}\int_0^t d\tau\,\exp[-\mathrm{i}\,(t-\tau)\mathscr{P}'\mathscr{L}]\mathscr{P}'(-\mathrm{i}\mathscr{L})\,a\right\rangle\frac{\langle a,a_\tau\rangle}{(a,a)^2}\\&=-\left\langle \dot{a},\int_0^t d\tau\,\exp[-\mathrm{i}\,(t-\tau)\mathscr{P}'\mathscr{L}]\mathscr{P}'\dot{a}\right\rangle\frac{\mathscr{P}a_\tau}{\langle a,a\rangle}\,,\end{aligned} \tag{2.9.15}$$

where (2.9.12) was used for

$$\langle \mathrm{i}\mathscr{L}g,h\rangle=-\langle g,\mathrm{i}\mathscr{L}h\rangle\,.$$

We can further write $\mathscr{P}'\dot{a}=\dot{a}$ because $\mathscr{P}\dot{a}=0$. Equation (2.5.17) is thus reduced to

$$\frac{d}{dt}\mathscr{P}a_t=-\int_0^t \gamma(t-\tau)\,\mathscr{P}a_\tau\,d\tau \tag{2.9.16}$$

or

$$\frac{d}{dt}\langle a,a_t\rangle=-\int_0^t \gamma(t-\tau)\,\langle a,a_\tau\rangle\,d\tau\,, \tag{2.9.17}$$

where $\gamma(t)$ is defined by

$$\gamma(t)=\frac{\langle\dot{a},\exp(-\mathrm{i}t\mathscr{P}'\mathscr{L})\,\dot{a}\rangle}{\langle a,a\rangle} \tag{2.9.18}$$

as seen from (2.9.15).

Equation (2.5.16) gives

$$\mathscr{P}'a_t=\int_0^t d\tau\,\exp[-\mathrm{i}(t-\tau)\mathscr{P}'\mathscr{L}]\mathscr{P}'(-\mathrm{i}\mathscr{L}a)\,\frac{\langle a,a_\tau\rangle}{\langle a,a\rangle}\,. \tag{2.9.19}$$

Equation (2.9.17) now suggests that the equation of motion for a_t is transformed into

$$\frac{d}{dt}a_t=-\int_0^t \gamma(t-\tau)\,a_\tau\,d\tau+\bar{R}_t\,, \tag{2.9.20}$$

where $\bar{R}_t$ corresponds to the random force in the Langevin equation of the type (1.6.20). In fact, $\bar{R}_t$ is defined by

$$\bar{R}_t=\exp(-\mathrm{i}t\mathscr{P}'\mathscr{L})\,\dot{a}\,,\qquad \bar{R}_0=\dot{a} \tag{2.9.21}$$

which makes (2.9.20) equivalent to (2.9.16, 19).

This is seen as follows. The random force $\bar{R}_t$ defined by (2.9.21) satisfies the condition

$$\mathscr{P}\bar{R}_t=0\,,\quad\text{or}\quad\langle\bar{R}_t,a\rangle=0\,. \tag{2.9.22}$$

Projection of (2.9.20) by $\mathscr{P}$ gives (2.9.16). To see that (2.9.19) is really contained in (2.9.21), it is convenient to Laplace transform (2.9.20) into

$$a[\omega]=\frac{1}{\mathrm{i}\omega+\gamma[\omega]}\,(a+\bar{R}[\omega])\,, \tag{2.9.23}$$

where $\gamma[\omega]$ is that quantity which was defined by (1.6.5) and $a[\omega]$ and $\bar{R}[\omega]$ are similar Laplace transforms of a_t and $\bar{R}_t$. Similarly, (2.9.17) is transformed into

$$\langle a, a[\omega]\rangle = \frac{1}{\mathrm{i}\omega + \gamma[\omega]}\langle a, a\rangle . \tag{2.9.24}$$

We thus find from (2.9.23) that

$$\mathscr{P}' a[\omega] = \frac{1}{\mathrm{i}\omega + \gamma[\omega]}\bar{R}[\omega] = \frac{\langle a, a[\omega]\rangle}{\langle a, a\rangle}\bar{R}[\omega] . \tag{2.9.25}$$

This is equivalent to (2.9.19).

One may wonder if the transformation (2.9.20) of the equation of motion is uniquely determined. It is indeed unique as long as the Langevin equation is assumed in the form of (2.9.20) with the condition (2.9.22) imposed. This leads to (2.9.21) for $\bar{R}_t$ as shown in the following.

First, from (2.9.20, 22)

$$\frac{d}{dt}\langle a, a_t\rangle = -\int_0^t \gamma(t-\tau)\langle a, a_\tau\rangle\, d\tau \tag{2.9.26}$$

and hence (2.9.24). Next we derive

$$\langle \bar{R}_0, \bar{R}[\omega]\rangle = \gamma[\omega]\langle a, a\rangle \tag{2.9.27}$$

using $\bar{R}_0 = \dot{a}$ and

$$\bar{R}_t = \dot{a}_t + \int_0^t \gamma(t-\tau)\, a_\tau\, d\tau .$$

By repeated partial integrations as done in (1.6.24, 25) using (2.9.12). Therefore

$$\gamma(t) = \langle \bar{R}_0, \bar{R}_t\rangle / \langle a, a\rangle . \tag{2.9.28}$$

Now, differentiating (2.9.20) gives

$$\frac{d}{dt}\bar{R}_t = -\mathrm{i}\,\mathscr{L}\bar{R}_t + \gamma(t)\, a .$$

Inserting (2.9.28) and using $\bar{R}_0 = \dot{a} = -\mathrm{i}\mathscr{L}a$ and again (2.9.12), we find that the above equation is nothing but

$$\frac{d}{dt}\bar{R}_t = -\mathrm{i}\mathscr{P}'\mathscr{L}\bar{R}_t \tag{2.9.29}$$

which gives (2.9.21).

Using (2.9.28), (2.9.24) can be written explicitly as

$$\int_0^\infty \langle a_0, a_t \rangle \, \mathrm{e}^{-\mathrm{i}\omega t} \, dt = \frac{\langle a, a \rangle}{\mathrm{i}\omega + \int_0^\infty \langle \bar{R}_0, \bar{R}_t \rangle \, \mathrm{e}^{-\mathrm{i}\omega t} \, dt / \langle a, a \rangle}, \tag{2.9.30}$$

This expresses the correlation function of a dynamical quantity a at two time points 0 and t in terms of that of the random force $\bar{R}_t$. Calculation of correlation functions using this type of equation is often called Mori's method. Equation (2.9.24) corresponds to (1.6.17), and (2.9.27) to (1.6.18). The former corresponds to the first kind of fluctuation-dissipation theorem and the latter to the second kind. It should be kept in mind that the above is a purely formal transformation of the equation of motion and provides us only with a formalism convenient to fit to, say, the Langevin equation. This does not say anything about how a response to an external perturbation, for example, the mobility of a Brownian particle under the influence of an external force, is related to the correlation function of the velocity fluctuation. Such a relation was incorporated in (1.6.1) which contains an external force K, but this was only an assumption. The relationship between response and fluctuation is proved only by the linear response theory discussed in Chap. 4.

The random force $\bar{R}_t$ is the remainder of the true force, which is equal to $\dot{a}_t$ subtracted by the systematic part; namely

$$\dot{a}_t \equiv F_t = -\int_0^t \gamma(t-\tau) \, a_\tau \, dt + \bar{R}_t \, . \tag{2.9.31}$$

The time evolution of $\bar{R}_t$ is not natural as shown by (2.9.21 or 29). This makes it difficult to see intuitively its physical meaning. On the other hand, the time evolution of the true force is governed by the natural dynamics. A simple relation exists between the correlation functions of F_t and $\bar{R}_t$ [2.25],

$$\frac{1}{\gamma_t[\omega]} - \frac{1}{\gamma[\omega]} = \frac{1}{\mathrm{i}\omega}, \tag{2.9.32}$$

which can be written as

$$\gamma_t[\omega] = \frac{\mathrm{i}\omega\gamma[\omega]}{\mathrm{i}\omega + \gamma[\omega]} \tag{2.9.33}$$

or

$$\gamma[\omega] = \frac{\mathrm{i}\omega\gamma_t[\omega]}{\mathrm{i}\omega - \gamma_t[\omega]}, \tag{2.9.34}$$

where

$$\gamma_t[\omega] = \int_0^\infty e^{-i\omega t} \langle F_0, F_t \rangle \, dt / \langle a, a \rangle \tag{2.9.35}$$

is the Laplace transform of the true force correlation. This can be easily derived by partially integrating the Laplace transform of the correlation function $\langle a_0, a_t \rangle$ twice with respect to the factor $\exp(-i\omega t)$, yielding

$$\langle a, a[\omega] \rangle = \frac{\langle a, a \rangle}{i\omega} - \frac{\langle F_0, F[\omega] \rangle}{(i\omega)^2} , \tag{2.9.36}$$

which, with (2.9.24), leads to the above equations.

Equation (2.9.33) shows that

$$\int_0^\infty \langle F_0, F_t \rangle \, dt = 0 \quad \text{or} \quad \gamma_t[0] = 0 . \tag{2.9.37}$$

This is an important theorem which holds generally for any $F_t = \dot{a}_t$ as long as a_t is a bounded random variable, as seen by

$$\begin{aligned} \int_0^\infty \langle \dot{a}_0, \dot{a}_t \rangle \, dt &= \lim_{T \to \infty} \frac{1}{2T} \int_0^T dt_1 \int_0^T dt_2 \langle \dot{a}_{t_1}, \dot{a}_{t_2} \rangle \\ &= \lim_{T \to \infty} \langle a_T - a_0, a_T - a_0 \rangle / 2T = 0 . \end{aligned}$$

Thus the time integral of the true force correlation ought to vanish if integrated to infinity. On the other hand, the corresponding integral

$$\frac{1}{\langle a, a \rangle} \int_0^\infty \langle \bar{R}, \bar{R}_t \rangle \, dt = \gamma[0] \tag{2.9.38}$$

of the random force $\bar{R}_t$ (or R_t) is finite. Figure 2.5 illustrates this characteristic difference of the functions $\gamma(t)$ and $\gamma_t(t)$. If the random force is white noise

$$\langle \bar{R}_0, \bar{R}_t \rangle / \langle a, a \rangle = \gamma \, \delta_+(t) \quad \text{with} \quad \gamma = \text{const} , \tag{2.9.39}$$

where $\delta_+(t)$ is a delta function normalized to

$$\int_0^\infty \delta_+(t) \, dt = 1 ,$$

then

$$\gamma_t[\omega] = \frac{i\omega\gamma}{i\omega + \gamma} . \tag{2.9.40}$$

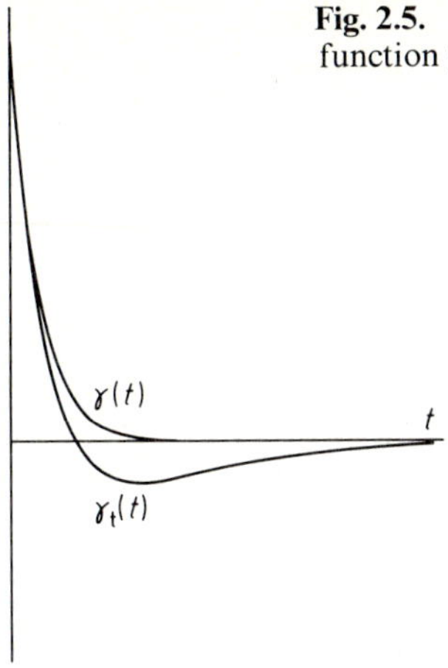

Fig. 2.5. Characteristic behavior of the random-force correlation function $\gamma(t)$ and the true force correlation function $\gamma_t(t)$

This means that the true force correlation has the form

$$\frac{\langle F_0, F_t \rangle}{\langle a, a \rangle} = \gamma[\delta_+(t) - \gamma\, \mathrm{e}^{-\gamma t}], \tag{2.9.41}$$

which consists of a delta-type peak at $t=0$ and a negative long tail compensating the positive peak when integrated.

A delta function is only an idealization, which can be a reasonable approximation when the time scale is well separated for the variables a_t and $\dot{a}_t$. If the characteristic times are $\tau_r = 1/\gamma$ and τ_c, respectively, a nearly ideal case arises when the condition $\tau_r \gg \tau_c$ is fulfilled. Then it is possible to approximate (2.9.38) by

$$\gamma \cong \frac{1}{\langle a, a \rangle} \int_0^\tau \langle F_0, F_t \rangle \, dt \tag{2.9.42}$$

by choosing τ such that

$$\tau_c \ll \tau \ll \tau_r .$$

Equation (2.9.42) was first obtained by *Kirkwood* [2.26] as an expression of the friction coefficient of a Brownian particle. It is an approximation and becomes exact only in the ideal limit of $\tau_c/\tau_r \to 0$.

A simple example is the resistivity of conduction electrons independently scattered by short-ranged scatterers. Even if collisions are assumed to occur completely randomly, the force acting on an electron in successive collisions must have long-lived correlation over the time between collisions. The projection $\mathscr{P}'$ in (2.9.29) removes this correlation and extracts only a single collision process. If the scattering interaction is long ranged, the duration of collision is no longer short in comparison with the mean free time and so the projection $\mathscr{P}'$ is not simple. It becomes simple if the random force is regarded as a white noise so that the process a_t is Markovian in a good approximation. Then the projection $\mathscr{P}'$ is simply the extraction of a single scattering process. Equation (2.9.38) then becomes a resistivity formula in terms of the correlation function of force in a single scattering process in the Born approximation. Such an equation is often used in calculations of resistivity in liquid and amorphous conductors. The approximate nature of the formula must be kept in mind. The exact expression is given by (2.9.34) [2.27].

So far we have considered a single quantity $a_t(p,q)$ in a classical system. We now conclude this chapter by showing that the above formalism is easily generalizable to the case of many variables in a classical or quantal system. To make the expressions concise, we introduce vector notations such as

$$A=\begin{pmatrix} a_1 \\ \vdots \\ a_n \end{pmatrix}, \qquad A^+=(a_1,\ldots,a_n)$$

for a set of n real quantities a_j $(j=1,2,\ldots,n)$. Dynamical evolution is governed by the equation of motion

$$\dot{A}_t=-\mathrm{i}\,\mathscr{L}A_t \tag{2.9.43}$$

which is the same as that in (2.9.3) in a classical case, and is the Heisenberg equation of motion (2.9.4) in a quantal case, where $\mathrm{i}\mathscr{L}$ is a commutator operation with the Hamiltonian of the system. We shall write A for A_0 for brevity. The projection operator $\mathscr{P}$ is defined by

$$\mathscr{P}X=\langle X,A^+\rangle\langle A,A^+\rangle^{-1}A \tag{2.9.44}$$

for a vector quantity X. Note that $\langle X,A^+\rangle$ and $\langle A,A^+\rangle$ are matrices whose elements are inner products of the form $(f,g\rangle$ of two dynamical quantities. In a classical case these quantities are defined by (2.9.9). In a quantal case, the most appropriate definition of $\langle f,g\rangle$ is the *canonical correlation*

$$\langle f;g\rangle=\frac{1}{\beta}\int_0^\beta d\lambda\,\mathrm{Tr}\{\mathrm{e}^{-\beta H}\,\mathrm{e}^{\lambda H}\,f\,\mathrm{e}^{-\lambda H}\,g\}/\mathrm{Tr}\{\mathrm{e}^{-\beta H}\} \tag{2.9.45}$$

to be introduced by (4.1.12). This quantal definition has all the properties required for the formal parallelism between classical and quantal formulations.

Equation (2.9.20) is generalized to

$$\dot{A}_t=\mathrm{i}\Omega A_t-\int_0^t \Gamma(t-s)\,A_s\,ds+\bar{R}_t \tag{2.9.46}$$

or equivalently to

$$\dot{A}_t^+ = -A_t^+ \mathrm{i}\Omega^+ - \int_0^t A_s^+ \Gamma(t-s)\, ds + \bar{R}_t^+ , \tag{2.9.47}$$

where

$$\begin{aligned} \mathrm{i}\Omega &= \langle \dot{A}, A^+ \rangle \langle A, A^+ \rangle^{-1} \\ -\mathrm{i}\Omega^+ &= \langle A, A^+ \rangle^{-1} \langle A, \dot{A}^+ \rangle = -\langle A, A^+ \rangle^{-1} \mathrm{i}\Omega \langle A, A^+ \rangle \\ \bar{R}_t &= \exp(-\mathrm{i}t\mathscr{P}'\mathscr{L})\, \bar{R}_0 , \qquad \bar{R}_0 = \dot{A} - \mathrm{i}\Omega A \end{aligned} \tag{2.9.48}$$

with

$$\mathscr{P}' = 1 - \mathscr{P},$$

and

$$\Gamma(t) = \langle \bar{R}_t, \bar{R}_0^+ \rangle \langle A, A^+ \rangle^{-1} . \tag{2.9.49}$$

Corresponding to (2.9.31) we introduce

$$F_t = \dot{A}_t - \mathrm{i}\Omega A_t \tag{2.9.50}$$

and call it the *true force* to distinguish it from the random force $\bar{R}_t$ appearing in (2.9.46). These two are related by

$$F_t = \bar{R}_t - \int_0^t \Gamma(t-s) A_s\, ds . \tag{2.9.51}$$

The above equations are proved by a little generalization of the previous equations for a single variable. It is illuminating, however, to prove them by a different method. Here we regard (2.9.48) as the definition of $\bar{R}_t$ and derive (2.9.46) from this. First note that

$$\langle \bar{R}_t, A^+ \rangle = 0 . \tag{2.9.52}$$

Then, using the operator identity

$$\mathrm{e}^{t(x+y)} = \mathrm{e}^{tx} + \int_0^t \mathrm{e}^{sx}\, y\, \mathrm{e}^{(t-s)(x+y)}\, ds ,$$

we expand (2.9.48) as

$$\bar{R}_t = \left(\exp(-\mathrm{i}t\mathscr{L}) + \int_0^t ds \exp(-\mathrm{i}s\mathscr{L}) \mathscr{P} \mathrm{i}\mathscr{L} \exp[-\mathrm{i}(t-s)\mathscr{P}'\mathscr{L}] \right) \bar{R}_0 . \tag{2.9.53}$$

From (2.9.43) we have the general relation

$$\exp(-\mathrm{i}t\mathscr{L})\, X_0 = X_t$$

for arbitrary X. So the first term on the right-hand side of (2.9.53) is

$$\exp(-\mathrm{i}t\mathscr{L})\,\bar{R}_0 = F_t\,,$$

since $\bar{R}_0 = F_0$, from (2.9.51). The second term becomes

$$\int_0^t ds\,\langle \mathrm{i}\mathscr{L}\bar{R}_{t-s}, A^+\rangle \langle A, A^+\rangle^{-1} A_s\,,$$

using (2.9.44). This is equal to

$$\int_0^t ds\,\langle \bar{R}_{t-s}, \dot{A}^+\rangle \langle A, A^+\rangle^{-1} A_s\,.$$

Using (2.9.48, 52), we thus find that (2.9.53) gives

$$\bar{R}_t = F_t + \int_0^t ds\,\langle \bar{R}_{t-s}, \bar{R}^+\rangle \langle A, A^+\rangle^{-1} A_s\,,$$

which is (2.9.51) with $\Gamma(t)$ defined by (2.9.49). Using (2.9.52),

$$\frac{d}{dt}\langle A_t, A^+\rangle = \mathrm{i}\Omega\langle A_t, A^+\rangle - \int_0^t ds\,\Gamma(t-s)\langle A_s, A^+\rangle$$

corresponding to (2.9.26).

The relationship (2.9.32) between the random and true forces is generalized to

$$\Gamma[\omega]^{-1} = \Gamma_t[\omega]^{-1} - \frac{1}{\mathrm{i}(\omega-\Omega)}\,, \qquad (2.9.54)$$

where $\Gamma[\omega]$ is the Laplace transform of $\Gamma(t)$ and correspondingly

$$\Gamma_t[\omega] = \int_0^\infty \mathrm{e}^{-\mathrm{i}\omega t}\langle F_t, F_0^+\rangle\,dt\,. \qquad (2.5.55)$$

Equation (2.9.54) is proved as follows. From (2.9.50)

$$A_s = \mathrm{e}^{\mathrm{i}\Omega s} A_0 + \int_0^s \mathrm{e}^{\mathrm{i}\Omega(s-s')} F_{s'}\,ds'$$

and from (2.9.51)

$$\langle \bar{R}_t, \bar{R}_0^+\rangle = \langle F_t, F_0^+\rangle + \int_0^t ds\,\Gamma(t-s)\langle A_s, F_0^+\rangle\,.$$

Inserting the first equation into the second and Laplace-transforming the expression yields an equation equivalent to (2.5.54).

3. Relaxation and Resonance Absorption

In Chaps. 1 and 2 we have treated temporal developments of physical phenomena as stochastic processes, and examined in detail how our understanding of them depends on the degree of crudeness of our description, or in other words, on alterations in the spatial and temporal precision of discernment and also on projections of processes. We have introduced various concepts and physical quantities as tools for our description of phenomena. In Chaps. 4 and 5 we shall sketch general statistical-mechanical formulations for these physical quantities and the method of their evaluation when atomic structures of substances are assumed.

This chapter is intended to link Chaps. 1, 2 with Chaps. 4, 5. As a preliminary to Chaps. 4, 5, we shall systematically introduce physical quantities such as those often used in phenomenological theories, and determine the mutual relationship among those quantities. Statistical-mechanical quantities like the correlation functions are discussed in Chap. 4. Because a phenomenological theory is advanced to picture phenomenological theory is advanced to picture phenomena which are observed in substances, the concepts introduced inevitably reflect the characteristics of the means of observation. Thermodynamic concepts are apt when the phenomena are almost static. On the other hand, if the phenomena are dynamic with higher frequencies, we must resort to other concepts unfamiliar in thermodynamics, such as those used in the theory of electric circuits. We should note that in discussing high-frequency phenomena we have to alter the precision of our description, shifting it to a more microscopic level. The simple microscopic theories providing a basis for the phenomenological theories are referred to in this chapter only in so far as they help our understanding of the relation to Chaps. 1 and 2.

3.1 Linear Irreversible Processes

This section and the following develop phenomenological theory in a most general form. The examples given in Sect. 3.4 illustrate the present treatment.

3.1.1 Mechanical and Thermal Forces vs Displacements and Currents

In dielectrics and paramagnets, an external electric or magnetic field of force induces electric polarization or magnetic moment, respectively. In conductors, the electric current is induced when their shapes or the boundary conditions are compatible with the flow of electricity, or when the electric or magnetic field varies in time. If the external force is weak enough, the electric polarization or current or the magnetization may be considered proportional to the force. When the temperature gradient or the concentration gradient, which is not a mechanical (or electromagnetic) force, exists and is weak enough, a heat or diffusion current is induced, respectively. According to irreversible thermodynamics [3.1], the negative gradient of the logarithm of the absolute temperature or the negative gradient of the chemical potential is the thermal internal force corresponding to nonuniformity of temperature or of concentration, respectively. Generally, there is no clear-cut distinction between mechanical and thermal forces. For an electric field, the externally applied field differs, in general, from the internal field within the system. The internal field is composed of the applied field and a reaction field produced by other parts of the system [3.2]. The latter is affected by the thermal motion within the substance.

Electric polarization or magnetization can appear in thermal equilibrium under an electric or magnetic static field, respectively. It is connected with irreversible processes under a time-dependent field. Other examples mentioned above are instances of irreversible processes.

The electric polarization appears as the result of changes in the spatial distribution of electric charges or in the orientational distribution of permanent electric dipoles, and the magnetization is produced by the change in the orientational distribution of magnetic dipoles and so on. Let us call these and similar changes *generalized displacements,* after the example of the displacement of electric charges and the rotation of magnetic dipoles. We shall call the electric, heat, and other currents *generalized currents.* The external or internal forces inducing the generalized displacements or currents are called *generalized forces.* Hereafter, we shall omit the adjective generalized for brevity.

3.1.2 Linear Relations

In the phenomenological theory of irreversible processes we have to assume some relations connecting the displacements or the currents with the forces, in addition to the equations of the conversation of energy and matter. The latter equations represent the universal laws common to all substances, while the former relations are connected with the special properties of each substance and can differ depending on the substance or the phenomenon under consideration. Therefore they are sometimes called the *phenomenological relations.* If the forces are weak enough, the displacements or the currents may be expanded in powers of the forces. Whenever we can neglect powers higher than the second in the phenomenological relations for an irreversible process, that process is called a *linear irreversible process*.

Phenomenological relations linear in the forces are called *linear relations*. Ohm's law of electric conduction, Fourier's law of thermal conduction, and Fick's law of diffusion are well-known examples of current. Hooke's law of elasticity and the magnetization equation for the paramagnet are instances of displacements. Let us denote the displacements or the currents at a time t by $B_\mu(t)$, and their corresponding forces by $X_\mu(t)$. Then the linear relations for an isotropic system have the form

$$B_\mu(t) - B_\mu^{\text{eq}} = L_{\mu\mu} X_\mu(t) .$$

The subscript μ distinguishes the kinds and the components of vectors or tensors. For a nonuniform system, the spatial coordinates should be included in the subscript μ (Sect. 3.5). Here B_μ^{eq} denotes the equilibrium value when there are no external forces. For the current $B_\mu^{\text{eq}} = 0$.

In the phenomenological theory, we endeavor to generalize the relations as far as possible. By doing so we can extend the range of their applicability and increase the usefulness of the theory.

First, let us consider the easy generalization to the case in which various irreversible processes coexist. In general, for the linear phenomenon we may use the superposition principle. Summing up the effects of various kinds of forces gives linear relations of the form

$$B_\mu(t) - B_\mu^{\text{eq}} = \sum_\nu L_{\mu\nu} X_\nu(t) \tag{3.1.1}$$

with the so-called *kinetic coefficients* $L_{\mu\nu}$. The appearance of the coefficients $L_{\mu\nu}$, in which μ is related to a kind of irreversible process different from that concerning ν, shows the interference effect among irreversible processes, and the symmetry of the coefficients represents the Onsager relation of reciprocity [3.1, 3].

Phenomenological relations of the form (3.1.1) can be applied only to slowly varying linear irreversible processes. The fact that the quantities appear on both sides of the equation at the same instant means that the displacements or the currents respond instantaneously to the forces. When the time variation of the forces exceeds a certain limit, the response of the displacements or of the currents generally lags the forces. Only when we approximate the description of the phenomena to such a crude degree that we may neglect the delay may we use (3.1.1). If events for which the delay cannot be neglected follow each other within the precision used, we must superpose all aftereffects due to forces prior to time t. That the forces in the past enter but those of the future do not is due to *causality*, according to which the present state results from the past cause. By this correction, (3.1.1) is generalized, in the terminology of Sect. 2.5, into the *non-*

Markovian form

$$B_\mu(t) - B_\mu^{\rm eq} = \sum_\nu \chi_{\mu\nu}^\infty X_\nu(t) + \int_{-\infty}^{t} dt' \sum_\nu \Phi_{\mu\nu}(t-t')\, X_\nu(t') . \tag{3.1.2}$$

The first term on the right-hand side denotes a contribution whose delay can be neglected, and the second term one whose delay cannot be neglected. We have assumed that the coefficients $\Phi_{\mu\nu}$ depend on the time difference $t-t'$, because the properties of the system, except the time dependence of the forces, are considered to be independent of the choice of the origin of the time axis.

For a macroscopic system with a huge number of degrees of freedom, the effect of past values of the forces usually vanishes with time; phenomena such as plastic deformation are not linear irreversible processes and are not taken into account in this chapter. This vanishing effect is represented by the condition

$$\lim_{t\to+\infty} \Phi_{\mu\nu}(t-t') = 0 . \tag{3.1.3}$$

Convergence to zero means that the function $\Phi_{\mu\nu}(s)$ is almost zero if s exceeds a certain time τ characteristic of the system.

Let us return to the case when the phenomena can be described with such crude precision that the time interval τ may be neglected. In other words, the temporal variation of the forces is supposed to be so slow that the forces may be regarded as constant during a time interval of the order of τ. Changing the integration variable from t' to $s=t-t'$ in the second term of (3.1.2) yields

$$B_\mu(t) - B_\mu^{\rm eq} = \sum_\nu \chi_{\mu\nu}^\infty X_\nu(t) + \int_0^\infty ds \sum_\nu \Phi_{\mu\nu}(s)\, X_\nu(t-s) \tag{3.1.4}$$

in which the main contribution to the integral comes from the range of s less than τ. According to the assumption, $X_\mu(t-s)$ may be replaced by $X_\mu(t)$, and (3.1.4) reduces to (3.1.1), in which the kinetic coefficients are found to be given by

$$L_{\mu\nu} = \chi_{\mu\nu}^\infty + \int_0^\infty ds\, \Phi_{\mu\nu}(s) . \tag{3.1.5}$$

The coarse-graining procedure is necessary to obtain the Markovian law (3.1.1).

3.1.3 Response to a Pulsed Force

To understand the physical meaning of the functions $\Phi_{\mu\nu}$ that appeared in place of the kinetic coefficients when the linear relations were generalized so as to be applicable to rapidly varying phenomena, let us assume a pulsed force of kind $\varkappa$, or in the $\varkappa$ direction, acting at a time t_1. If we regard the action as instantaneous in our time scale, we may represent the time dependence of the force by the Dirac δ function:

$$X_\nu(t) = \delta_{\nu\varkappa}\,\delta(t-t_1)\,. \tag{3.1.6}$$

We have chosen the unit pulse to obtain unity for the total impulse $\int_{-\infty}^{\infty} X_\varkappa(t)\,dt$. Substituting (3.1.6) into (3.1.2), and introducing the Heaviside unit step function

$$\theta(x) = \int_{-\infty}^{x} dx'\,\delta(x') = \begin{cases} 1 & (x>0) \\ 0 & (x<0) \end{cases} \tag{3.1.7}$$

yields

$$B_\mu(t) - B_\mu^{\rm eq} = \chi_{\mu\varkappa}^{\infty}\,\delta(t-t_1) + \Phi_{\mu\varkappa}(t-t_1)\,\theta(t-t_1)\,. \tag{3.1.8}$$

Before time t_1 there was no force, so that this system was in equilibrium: $B_\mu(t) = B_\mu^{\rm eq}\,(t<t')$. The first term in (3.1.8) represents the instantaneous response to the pulsed force, while the second is the effect of the pulsed force remaining after time t_1:

$$B_\mu(t) - B_\mu^{\rm eq} = \Phi_{\mu\varkappa}(t-t_1) \quad (t>t_1)\,.$$

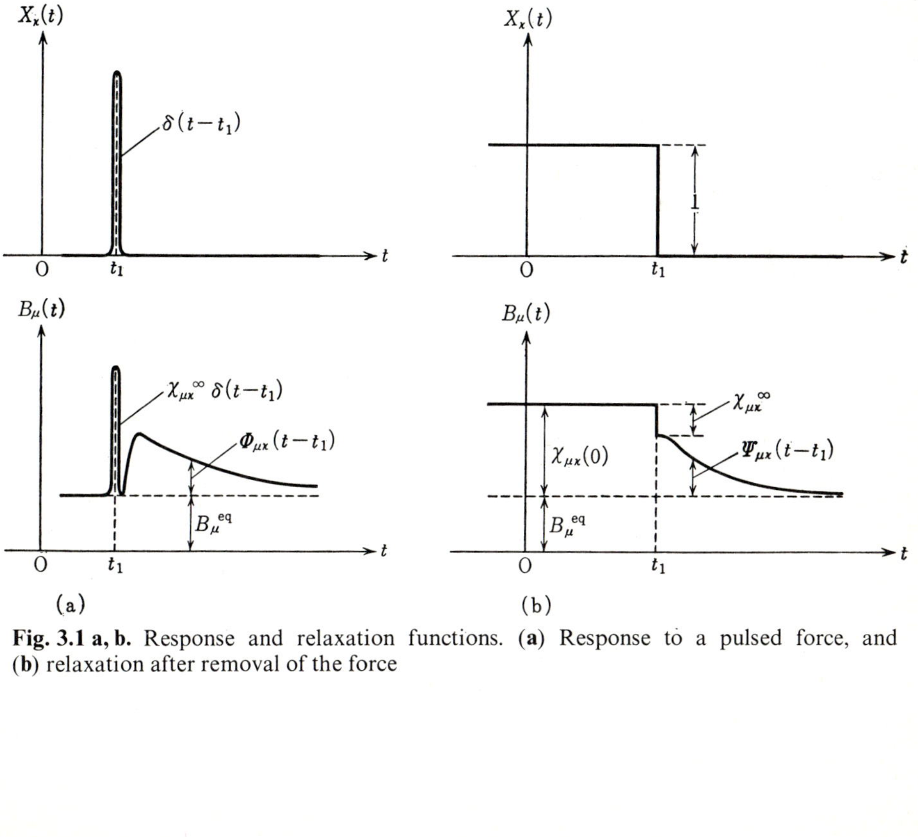

Fig. 3.1 a, b. Response and relaxation functions. **(a)** Response to a pulsed force, and **(b)** relaxation after removal of the force

By virtue of (3.1.3) our system will return to thermal equilibrium in a sufficiently long time after t_1 (Fig. 3.1a). Thus we call the function Φ the *aftereffect function* or the *response function.*

3.1.4 Relaxation Phenomena

Let us consider the case of a sudden removal of force kept constant until time t_1. Our system is in equilibrium under the action of the force till t_1, but it proceeds to a new equilibrium state without the action of the force. A phenomenon observed in such a case is called the *relaxation phenomenon.* Assume that the force has its component only in the $\varkappa$ direction with a strength of unity

$$X_\nu(t) = \delta_{\nu\varkappa}\,\theta(t_1 - t)\,, \tag{3.1.9}$$

where θ denotes the unit step function (3.1.7):

$$X_\nu(t) = \delta_{\nu\varkappa} \quad \text{for} \quad t < t_1 \quad \text{and} \quad X_\nu(t) = 0 \quad \text{for} \quad t > t_1\,.$$

Substituting this expression into (3.1.4) gives

$$B_\mu(t) = \begin{cases} B_\mu^{\mathrm{eq}} + \chi_{\mu\varkappa}^{\infty} + \Psi_{\mu\varkappa}(0) & \text{for} \quad t < t_1\,, \\ B_\mu^{\mathrm{eq}} + \Psi_{\mu\varkappa}(t - t_1) & \text{for} \quad t > t_1 \end{cases} \tag{3.1.10}$$

where the function

$$\Psi_{\mu\nu}(t) = \int_t^\infty ds\, \Phi_{\mu\nu}(s) \tag{3.1.11}$$

is introduced. According to (3.1.10), B_μ has the equilibrium value $B_\mu^{\mathrm{eq}} + \chi_{\mu\varkappa}^{\infty} + \Psi_{\mu\varkappa}(0)$ under the action of the force in the $\varkappa$ direction before time t_1. After time t_1 it changes its value toward the equilibrium value B_μ^{eq} in a way determined by the function $\Psi_{\mu\varkappa}(t - t_1)$ (Fig. 3.1b). In this sense the function Ψ defined by (3.1.11) is called the *relaxation function.* By the definition (3.1.11)

$$\Phi_{\mu\nu}(t) = -\frac{d\Psi_{\mu\nu}(t)}{dt}\,. \tag{3.1.12}$$

Equations (3.1.11, 12) show that if either the response Φ or the relaxation function Ψ is given, the other can be derived: the property of our system can be described by either of these functions. When the convergence in (3.1.3) is sufficiently rapid, and the integral in (3.1.11) converges, then

$$\lim_{t\to+\infty} \Psi_{\mu\nu}(t) = 0\,, \tag{3.1.13}$$

which guarantees the convergence of $B_\mu(t)$ toward B_μ^{eq}. If this relation holds, Ψ is zero for a sufficiently large t, so that by virtue of (3.1.12) Φ converges to zero. For a system moving toward equilibrium the effect of a force should vanish after sufficient elapsed time when the force is removed.

To see the relaxation when the inital condition is specified, the force X must be eliminated. Let us consider a more general case

$$X_\nu(t) = X_\nu\,\theta(t_1 - t) \tag{3.1.14}$$

in place of (3.1.9). Then

$$B_\mu(t) = \begin{cases} B_\mu^{\mathrm{eq}} + \sum_\nu [\chi_{\mu\nu}^\infty + \Psi_{\mu\nu}(0)]\,X_\nu & \text{for} \quad t < t_1\,, \\ B_\mu + \sum_\nu \Psi_{\mu\nu}(t - t_1)\,X_\nu & \text{for} \quad t > t_1 \end{cases} \tag{3.1.15}$$

in place of (3.1.10). When the initial value is given at time $t = t_1 - 0$,

$$B_\mu(t_1 - 0) = B_\mu^{\mathrm{eq}} + \sum_\nu [\chi_{\mu\nu}^\infty + \Psi_{\mu\nu}(0)]\,X_\nu \tag{3.1.16}$$

from (3.1.15). Solving for X and substituting it into (3.1.15) gives

$$B_\mu(t) - B_\mu^{\mathrm{eq}} = \sum_\nu \Psi_{\mu\nu}^{\mathrm{m}}(t - t_1)\,[B_\nu(t_1 - 0) - B_\nu^{\mathrm{eq}}] \tag{3.1.17}$$

where the function

$$\Psi_{\mu\nu}^{\mathrm{m}}(t) = \sum_\varkappa [\chi_{\mu\varkappa}^\infty\,\theta(-t) + \Psi_{\mu\varkappa}(t)]\,\{[\chi^\infty + \Psi(0)]^{-1}\}_{\varkappa\nu} \tag{3.1.18}$$

describes the relaxation phenomenon. In a matrix form

$$\begin{aligned} \Psi^{\mathrm{m}}(t) &= 1 \quad \text{for} \quad t = 0-\,, \\ \Psi^{\mathrm{m}}(t) &= \Psi(t)\,[\chi^\infty + \Psi(0)]^{-1} \quad \text{for} \quad t > 0\,. \end{aligned} \tag{3.1.19}$$

Since the knowledge of $\chi^\infty + \Psi(0)$ is lost by virtue of the normalization $\Psi^{\mathrm{m}}(0) = 1$, measuring the relaxation (3.1.19) from an initial value does not give knowledge equivalent to that contained in the response or the relaxation function except when it is combined with the static response measurement (3.1.16).

3.2 Complex Admittance

For linear phenomena, as already mentioned, the superposition principle is valid: we need not examine the response of our system to a force having a general time dependence since it suffices to examine the response to a harmonically oscillating force. The response to the harmonic force is described by a new function different from either the response or the relaxation function.

3.2.1 Harmonic (Fourier) Analysis

The force and the displacement or the current are expressed as the superposition of harmonic oscillations[1]:

$$X_\nu(t) = \int_{-\infty}^{\infty} \frac{d\omega}{2\pi} \, e^{-i\omega t} X_{\nu,\omega} \,,$$
$$B_\mu(t) - B_\mu^{eq} = \int_{-\infty}^{\infty} \frac{d\omega}{2\pi} \, e^{-i\omega t} B_{\mu,\omega} \,, \tag{3.2.1}$$

$$X_{\nu,\omega} = \int_{-\infty}^{\infty} dt \, e^{i\omega t} X_\nu(t) \,,$$
$$B_{\mu,\omega} = \int_{-\infty}^{\infty} dt \, e^{i\omega t} \{B_\mu(t) - B_\mu^{eq}\} \,. \tag{3.2.2}$$

Substituting the linear relation (3.1.4) into the right-hand side of the second equation of (3.2.2) and the first of (3.2.1) into the right-hand side of (3.1.4) gives the Fourier transform of the linear relation.

$$B_{\mu,\omega} = \sum_\nu \chi_{\mu\nu}(\omega) \, X_{\nu,\omega} \tag{3.2.3}$$

where the formula for the Dirac δ-function was used

$$\delta(x) = \int_{-\infty}^{\infty} \frac{d\omega}{2\pi} \, e^{i\omega x} \tag{3.2.4}$$

and the complex function was introduced

$$\chi_{\mu\nu}(\omega) = \chi_{\mu\nu}^{\infty} + \int_0^{\infty} dt \, e^{i\omega t} \, \Phi_{\mu\nu}(t) \,. \tag{3.2.5}$$

Linear relations of the form (3.2.3) are well known in the theory of electric circuits. Accordingly, we shall call the coefficient χ the *complex admittance.*

The relationship of the complex admittance to the response function is given by definition (3.2.5), while that to the relaxation function is derived by substituting (3.1.12) into (3.2.5), integrating by parts, and using (3.1.13):

$$\chi_{\mu\nu}(\omega) = \chi_{\mu\nu}^{\infty} + \Psi_{\mu\nu}(0) + i\omega \int_0^{\infty} dt \, e^{i\omega t} \, \Psi_{\mu\nu}(t) \,. \tag{3.2.6}$$

---

1 Note the change of sign in the exponential function in the integrals. Unlike $\exp(i\omega t)$ used in Chaps. 1 and 2, we use $\exp(-i\omega t)$ hereafter. This convention is common in literature dealing with many-particle theories. The standard form of a plane wave with frequency ω and wave number $\boldsymbol{k}$ is chosen as $\exp[i(\boldsymbol{k}\boldsymbol{r} - \omega t)]$, which fits the conventional choice of the sign of i in the Schrödinger equation

In particular, setting $\omega \to 0$ gives

$$\chi_{\mu\nu}(0) = \chi_{\mu\nu}^{\infty} + \Psi_{\mu\nu}(0) \tag{3.2.7}$$

provided that the integral in the third term of (3.2.6) converges. This static admittance is real. Equation (3.2.5) shows that the Fourier-Laplace transform of the response function is equal to $\chi(\omega) - \chi^{\infty}$, while (3.2.6) indicates that the Fourier-Laplace transform of the relaxation function gives

$$\frac{\chi_{\mu\nu}(\omega) - \chi_{\mu\nu}(0)}{\mathrm{i}\omega} = \int_0^{\infty} dt\, \mathrm{e}^{\mathrm{i}\omega t}\, \Psi_{\mu\nu}(t)\,. \tag{3.2.8}$$

For the normalized relaxation function (3.1.18)

$$\frac{[\chi(\omega)\,\chi^{-1}(0) - 1]_{\mu\nu}}{\mathrm{i}\omega} = \int_0^{\infty} dt\, \mathrm{e}^{\mathrm{i}\omega t}\, \Psi_{\mu\nu}^{\mathrm{m}}(t) \tag{3.2.9}$$

as is seen from (3.1.19) and (3.2.8).

Thus it is clear that describing a linear irreversible process in terms of the response or the relaxation function is equivalent to that in terms of the complex admittance. Since the admittance is complex, it is decomposed into a real and an imaginary part. Section 3.6 shows that only one of these parts contains independent information while the other does not give us any more knowledge. Experimentally, preparing a pulsed force is not easy: it is easier to measure either the relaxation or the response to harmonic forces, i.e., the complex admittance. In particular, of the two real functions composing the complex admittance, it is convenient to study the one relating to energy dissipation.

3.2.2 Energy Dissipation

When the force $X(t)$ is periodic in time, we can express in terms of the complex admittance the rate at which the work done by the external force on the system is changed by the system into thermal energy and is dissipated. Assume a periodic force of the form

$$X_{\nu}(t) = x_{\nu} \cos(\omega t + \delta_{\nu})\,. \tag{3.2.10}$$

The phase δ_{ν} enables circular polarization (3.4.7), Sect. 3.4, to be included. The corresponding response is given by

$$B_{\mu}(t) - B_{\mu}^{\mathrm{eq}} = \sum_{\nu} [\chi_{\mu\nu}'(\omega)\, x_{\nu} \cos(\omega t + \delta_{\nu}) + \chi_{\mu\nu}''(\omega)\, x_{\nu} \sin(\omega t + \delta_{\nu})] \tag{3.2.11}$$

according to (3.1.4). On the right-hand side the first term is the in-phase component, while the second denotes the quadrature (out-of-phase) component. The coefficients of both terms are

$$\begin{aligned} \chi'_{\mu\nu}(\omega) &= \chi^{\infty}_{\mu\nu} + \int_0^{\infty} dt\, \Phi_{\mu\nu}(t) \cos\omega t\,, \\ \chi''_{\mu\nu}(\omega) &= \int_0^{\infty} dt\, \Phi_{\mu\nu}(t) \sin\omega t\,. \end{aligned} \tag{3.2.12}$$

These are both real functions and, on comparison with (3.2.5), are found to be the real and imaginary parts of the complex admittance

$$\chi_{\mu\nu}(\omega) = \chi'_{\mu\nu}(\omega) + \mathrm{i}\chi''_{\mu\nu}(\omega)\,. \tag{3.2.13}$$

If A_ν denotes the system's displacement corresponding to the external force X_ν, the work done by the force on the system is

$$d'W = \sum_{\nu} X_\nu\, dA_\nu\,. \tag{3.2.14}$$

The energy conservation law then is

$$dU = d'Q + d'W\,, \tag{3.2.15}$$

where dU is the change of the internal energy of the system and $d'Q$ is the heat transferred to the system during the process of change, whereby we assume that no other sort of energy transfer takes place. When integrated over a cycle, the heat evolved from the system into the environment per cycle amounts to

$$-\oint d'Q = \oint d'W = \sum_{\nu} \oint X_\nu(t) \frac{dA_\nu(t)}{dt}\, dt\,. \tag{3.2.16}$$

If the system is considered to follow a succession of thermodynamic states during the periodic process, the entropy change dS in the change $\{dA_\nu\}$ is conditioned by the inequality

$$dS - \frac{d'Q}{T} > 0\,, \tag{3.2.17}$$

where $-d'Q/T$ is the entropy change of the reservoir, T being the temperature of the reservoir. Integrated over a cycle, this yields

$$-\oint d'Q > 0\,, \tag{3.2.18}$$

since $\oint dS = 0$. Thus the work done on the system must be dissipated as heat into the environment. The heat dissipated per unit of time

$$L = -\oint \frac{d'Q}{2\pi/\omega} = \overline{\sum_{\mu} X_\mu(t) \frac{dA_\mu(t)}{dt}} > 0 \tag{3.2.19}$$

is called the *power loss*. Here the horizontal bar means taking the time average per period. This dissipation may be considered as irreversible generation of entropy [3.1]. Then the rate of entropy production is

$$\left(\frac{dS}{dt}\right)_{\rm irr} = \frac{L}{T}.$$

It should be noted, however, that the above argument cannot be generally applied to the irreversible processes with which we shall be concerned, because these processes can be so fast that the states of the system are no longer thermodynamic. The positivity of heat dissipation is still expected to hold generally. As Chap. 4 shows the fluctuation-dissipation theorem as proved by the linear response theory shows that the heat dissipation (3.2.19) is expressed in terms of the power spectrum of a certain correlation function of a relevant physical quantity with the desired positivity.

Now let us examine how the dissipation is related to the general admittance. First consider the case in which B_μ is the displacement A_μ itself. Then the time derivative appearing on the right-hand side of (3.2.19) becomes

$$\frac{dA_\mu(t)}{dt} = \sum_\nu [-\omega\chi'_{\mu\nu}(\omega)\, x_\nu \sin(\omega t + \delta_\nu) + \omega\chi''_{\mu\nu}(\omega)\, x_\nu \cos(\omega t + \delta_\nu)],$$

which, if inserted together with (3.2.10) into (3.2.19), gives in turn

$$L = \sum_{\mu,\nu} \omega [\chi'_{\mu\nu}(\omega) \sin(\delta_\mu - \delta_\nu) + \chi''_{\mu\nu}(\omega) \cos(\delta_\mu - \delta_\nu)] \frac{x_\mu}{\sqrt{2}} \frac{x_\nu}{\sqrt{2}} > 0. \quad (3.2.20)$$

Since we may interchange the subscripts μ and ν in the sum of the above expression, then only the coefficients

$$\begin{aligned} \chi'^{a}_{\mu\nu}(\omega) &= \tfrac{1}{2}[\chi'_{\mu\nu}(\omega) - \chi'_{\nu\mu}(\omega)], \\ \chi''^{s}_{\mu\nu}(\omega) &= \tfrac{1}{2}[\chi''_{\mu\nu}(\omega) + \chi''_{\nu\mu}(\omega)] \end{aligned} \quad (3.2.21)$$

relate to the energy dissipation.

Next let us proceed to the case in which B_μ is the current J_μ corresponding to the force X_μ. For a uniform system, the current is the time derivative itself of the corresponding displacement:

$$J_\mu(t) = \frac{dA_\mu(t)}{dt}. \quad (3.2.22)$$

For a nonuniform system, the current is connected by an equation, say, of continuity with the time derivative of the displacement; discussed in Sect. 3.5. Noticing that $B_\mu^{\rm eq} = 0$, we can rewrite (3.2.11) as

$$J_\mu(t) = \sum_\nu [\chi'_{\mu\nu}(\omega)\, x_\nu \cos(\omega t + \delta_\nu) + \chi''_{\mu\nu}(\omega)\, x_\nu \sin(\omega t + \delta_\nu)], \quad (3.2.23)$$

so that the power loss is given by

$$L = \sum_{\mu,\nu} [\chi'_{\mu\nu}(\omega) \cos(\delta_\mu - \delta_\nu) - \chi''_{\mu\nu}(\omega) \sin(\delta_\mu - \delta_\nu)] \frac{x_\mu}{\sqrt{2}} \frac{x_\nu}{\sqrt{2}} > 0 . \quad (3.2.24)$$

The quantities relating to the energy dissipation are found to be

$$\begin{aligned} \chi''^{a}_{\mu\nu}(\omega) &= \tfrac{1}{2}[\chi''_{\mu\nu}(\omega) - \chi''_{\nu\mu}(\omega)] , \\ \chi'^{s}_{\mu\nu}(\omega) &= \tfrac{1}{2}[\chi'_{\mu\nu}(\omega) + \chi'_{\nu\mu}(\omega)] . \end{aligned} \quad (3.2.25)$$

3.3 Debye Relaxation

The preceding two sections describe the phenomenological theory of linear irreversible processes in a general form common to most processes. Before proceeding further, it is convenient for the descriptions in Sects. 3.6, 7 to give concrete examples here.

3.3.1 Dielectric Relaxation

Consider a plane parallel condenser filled with a uniform dielectric. For simplicity, assume an isotropic dielectric. The electric displacement is defined as

$$\boldsymbol{D}(t) = \boldsymbol{E}(t) + 4\pi \boldsymbol{P}(t) . \quad (3.3.1)$$

Regarding the electric field $\boldsymbol{E}$ as the force X in Sect. 3.1 and the electric displacement as the displacement B, we write the linear relation (3.1.2). The quantity corresponding to B^{eq} vanishes because we are not concerned with ferroelectrics. The part of the electric polarization $\boldsymbol{P}$ which follows the electric field $\boldsymbol{E}$ instantaneously within our time precision here combines with the first term of (3.3.1): $\varepsilon_\infty \boldsymbol{E}(t)$. Since the dielectric is assumed to be isotropic, the response function becomes scalar. Thus we may write (3.1.2) in the form

$$\boldsymbol{D}(t) = \varepsilon_\infty \boldsymbol{E}(t) + \int_{-\infty}^{t} dt' \varphi(t-t') \boldsymbol{E}(t') \quad (3.3.2)$$

or, by making the Fourier transformation (3.2.2),

$$\boldsymbol{D}_\omega = \varepsilon(\omega) \boldsymbol{E}_\omega . \quad (3.3.3)$$

In the static case

$$\boldsymbol{D} = \varepsilon_s \boldsymbol{E} , \quad (3.3.4)$$

the proportionality coefficient ε_s is called the *static dielectric constant.* In (3.3.3) the quantity

$$\varepsilon(\omega) = \varepsilon'(\omega) + \mathrm{i}\,\varepsilon''(\omega) = \varepsilon_\infty + \int_0^\infty dt\, \mathrm{e}^{\mathrm{i}\omega t}\, \varphi(t) \tag{3.3.5}$$

corresponding to the complex admittance (3.2.5) is called the *complex dielectric function.* The static dielectric constant is given by

$$\varepsilon_s = \varepsilon(0) = \varepsilon_\infty + \int_0^\infty dt\, \varphi(t)\,. \tag{3.3.6}$$

According to thermodynamics, the work done by a slight change in the polarization is given as $\boldsymbol{E} \cdot d\boldsymbol{P}$, whence the electric polarization and, in turn, the electric displacement is the displacement corresponding to the force exerted by $\boldsymbol{E}$. The power loss due to the oscillating electric field relates to the imaginary part of the complex dielectric function by means of (3.2.20), and the inequality

$$\omega\varepsilon''(\omega) > 0 \tag{3.3.7}$$

must hold. This inequality indicates that the sign of $\varepsilon''(\omega)$ should be the same as that of ω and thus $\varepsilon''(\omega)$ must be an odd function of ω.

3.3.2 Response Functions with Exponential Damping

In phenomenological theory either the response function, the relaxation function or the complex admittance must be given in order to specify dynamical properties of the system. In nature there are many examples in which the relaxation function and, in turn, the response function decay exponentially in time by virtue of (3.1.12). Let us call this type of relaxation phenomena *Debye-type relaxation phenomena* [3.2, 4].

Assuming Debye-type dielectric relaxation in our example, we are led to use the response function of the form

$$\varphi(t) = \frac{\varepsilon_s - \varepsilon_\infty}{\tau}\, \mathrm{e}^{-t/\tau} \quad (t > 0)\,, \tag{3.3.8}$$

the proportionality constant being determined by (3.3.6). According to (3.1.11) the relaxation function is

$$\psi(t) = (\varepsilon_s - \varepsilon_\infty)\, \mathrm{e}^{-t/\tau} \quad (t > 0)\,. \tag{3.3.9}$$

The time necessary for the relaxation is of the order of τ, so that τ is called the *relaxation time.* The complex dielectric function becomes

$$\varepsilon(\omega) = \varepsilon_\infty + \frac{\varepsilon_s - \varepsilon_\infty}{1 - \mathrm{i}\omega\tau} \tag{3.3.10}$$

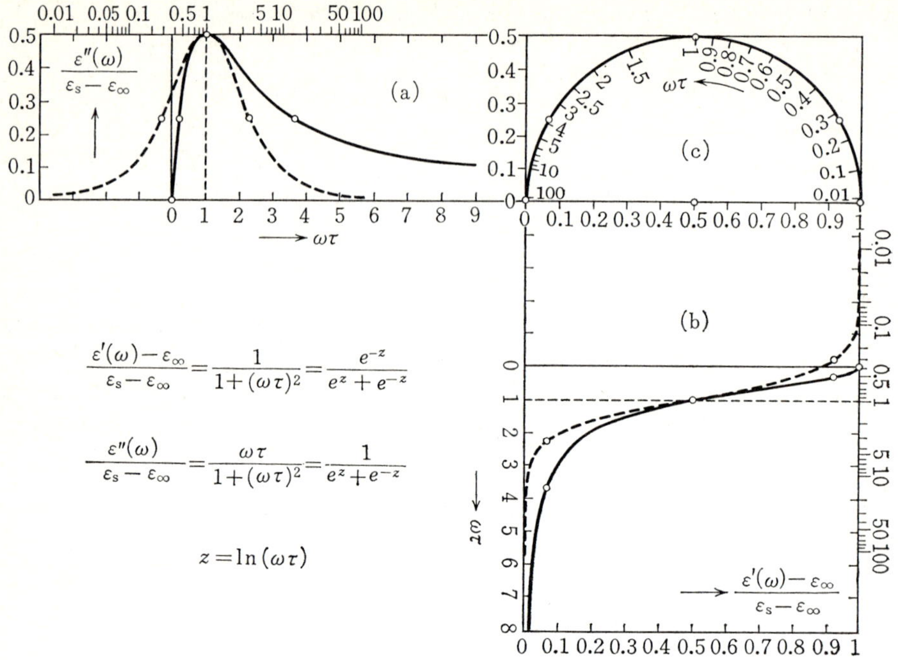

Fig. 3.2 a–c. Complex Debye type, dielectric function. (**a**) the imaginary part, (**b**) the real part, and (**c**) the Cole-Cole diagram. The broken lines illustrate the real and imaginary parts plotted against the logarithm of $\omega\tau$, the scale given on top of (***a***) and to the right of (***b***)

by virtue of (3.3.5), and its real and imaginary parts are

$$\varepsilon'(\omega) - \varepsilon_\infty = (\varepsilon_s - \varepsilon_\infty)\,\frac{1}{1+(\omega\tau)^2}\,,$$
$$\varepsilon''(\omega) = (\varepsilon_s - \varepsilon_\infty)\,\frac{\omega\tau}{1+(\omega\tau)^2}\,, \qquad (3.3.11)$$

respectively (Fig. 3.2a, b). Note that, as is obvious from these equations, the real part changes from ε_s to ε_∞ as ω increases from 0 to $+\infty$.

A diagram drawn with the real part ε' as the abscissa and the imaginary part ε'' as the ordinate and with the angular frequency ω as the parameter is called a Cole-Cole diagram (Fig. 3.2c). As is clear from (3.3.11), characteristic of Debye relaxation is that its Cole-Cole diagram is a semicircle centered at $(\varepsilon_s + \varepsilon_\infty)/2$ on the ε' axis and with a radius of $(\varepsilon_s - \varepsilon_\infty)/2$. We can read the relaxation time τ from the Cole-Cole diagram, because $\varepsilon''(\omega)$ becomes maximum at $\omega = 1/\tau$.

We may draw the graph with the set of coordinates $(\varepsilon', \omega\varepsilon'')$ or $(\varepsilon', \varepsilon''/\omega)$ in place of $(\varepsilon', \varepsilon'')$, which gives a straight line, as is obvious from

(3.3.11), and determine τ or $1/\tau$, respectively, by measuring the slope. To find out from experiments whether the relaxation is Debye or not, it is suitable to check the validity of

$$\varepsilon''(\omega) = \sqrt{[\varepsilon_s - \varepsilon'(\omega)][\varepsilon'(\omega) - \varepsilon_\infty]} \, .$$

Since $\varepsilon_s - \varepsilon_\infty \ll 1$ for ordinary dielectrics, we need very accurate measurements.

3.3.3 Solution of Polar Molecules

Let us consider a uniform dilute solution composed of polar molecule solute and nonpolar molecule solvent as an example showing Debye relaxation phenomena. A polar molecule is a molecule with a permanent electric dipole. The electric polarization of this solution may approximately be supposed to consist of three parts:

$$\boldsymbol{P}(t) = \boldsymbol{P}_d(t) + \boldsymbol{P}_a(t) + \boldsymbol{P}_e(t) \, . \tag{3.3.12}$$

The second term on the right-hand side appears because of the change in interionic distances of each molecule induced by the electric field. It lags behind the field only when the frequency of the field becomes higher than that of the infrared region. The third term due to the change in the electronic distribution in each molecule shows a time lag at frequencies higher than that of the optical region. If we consider a field of frequency sufficiently lower than these frequencies, i.e., if we develop our phenomenological theory within such a crude time precision that is applicable to sufficiently slowly varying electric fields, we may include these two terms in $\varepsilon_\infty \boldsymbol{E}(t)$ of (3.3.2).

The first term on the right-hand side of (3.3.12) is the part produced by the change in the orientation of permanent dipoles due to the rotational motion of molecules, and it ought to give the integral term of (3.3.2). Let us examine the orientational distribution of dipoles, supposing that the solution contains n polar molecules per unit volume and that the direction of the electric field does not change. Let us denote by $f(\theta, \phi)\, d\Omega$ the number of polar molecules having their permanent dipoles oriented within the infinitesimal solid angle $d\Omega = \sin\theta \, d\theta \, d\phi$ in the direction (θ, ϕ), the polar axis being taken in the fixed direction of the electric field. In thermal equilibrium, the orientational distribution is given by the Boltzmann distribution function

$$f_{eq}(\theta, \varphi) = n \exp\left(\frac{\mu E \cos\theta}{kT}\right) \Big/ \int \exp\left(\frac{\mu E \cos\theta}{kT}\right) d\Omega \, , \tag{3.3.13}$$

where μ denotes the magnitude of the electric dipole. Since $|\mu E| \ll kT$ holds for ordinary temperatures T and electric fields E, we may use the linear approximation with respect to the electric field

$$f_{\rm eq}(\theta, \varphi) = \frac{n}{4\pi}\left(1 + \frac{\mu E}{kT}\cos\theta\right). \tag{3.3.14}$$

The first component $\boldsymbol{P}_{\rm d}$ of the electric polarization, which is evaluated as

$$\boldsymbol{P}_{\rm d} = \int \mu\cos\theta\,\frac{\boldsymbol{E}}{E} f(\theta, \varphi)\, d\Omega\,, \tag{3.3.15}$$

becomes

$$\boldsymbol{P}_{\rm d} = \frac{n\mu^2}{3kT}\boldsymbol{E} \tag{3.3.16}$$

according to (3.3.14). Then the static dielectric function is given by the *Langevin-Debye equation*

$$\varepsilon_{\rm s} - \varepsilon_\infty = 4\pi\frac{n\mu^2}{3kT}. \tag{3.3.17}$$

Here we ignored the local field correction [3.2], which would replace the left-hand side of (3.3.17) by $3(\varepsilon_{\rm s} - \varepsilon_\infty)/(\varepsilon_{\rm s} + 2\varepsilon_\infty)$. This equation guarantees the inequality $\varepsilon_{\rm s} - \varepsilon_\infty > 0$, so that ε'' of (3.3.11) certainly satisfies the requirement (3.3.7) of the second law of thermodynamics.

The simplest assumption leading to the response function (3.3.8) is that the polarization $\boldsymbol{P}_{\rm d}(t)$ satisfies the equation of motion

$$\frac{d\boldsymbol{P}_{\rm d}(t)}{dt} = -\frac{1}{\tau}\left[\boldsymbol{P}_{\rm d}(t) - \frac{n\mu^2}{3kT}\boldsymbol{E}(t)\right], \tag{3.3.18}$$

according to which the polarization approaches with relaxation time τ the equilibrium value (3.3.16) corresponding to the instantaneous value of the electric field $\boldsymbol{E}(t)$. In fact, the solution of (3.3.18) is

$$\boldsymbol{P}_{\rm d}(t) = \int_{t_0}^{t} dt'\,\frac{\exp[-(t-t')/\tau]}{\tau}\,\frac{n\mu^2}{3kT}\boldsymbol{E}(t') + \exp[-(t-t_0)/\tau]\,\boldsymbol{P}_{\rm d}(t_0) \tag{3.3.19}$$

and, moving the initial time t_0 to $-\infty$, it leads to the linear relation (3.3.2) including the Debye-type response function (3.3.8).

To give a molecular kinetic theoretical basis to the equation of motion (3.3.18), we must build up a kinetic equation controlling the temporal variation of the orientational distribution function $f(\theta, \phi, t)$. Considering only the orientation of an electric dipole of a polar molecule, we may assume that the molecule undergoes rotational Brownian motion. According to the theory of Brownian motion [3.4], it is proved that the orientational distribution function satisfies an equation similar to the Fokker-Planck equation described in Sect. 2.4

$$\frac{\partial f}{\partial t} = D_{\mathrm{r}} \left\{ \frac{1}{\sin\theta} \frac{\partial}{\partial\theta} \left[\sin\theta \left(\frac{\partial f}{\partial\theta} + \frac{\mu E(t) \sin\theta}{kT} f \right) \right] + \frac{1}{\sin^2\theta} \frac{\partial^2 f}{\partial\varphi^2} \right\} . \tag{3.3.20}$$

It is obvious that the Boltzmann distribution (3.3.13) is the stationary solution of this equation for a static electric field. To obtain the general solution of (3.3.20), the orientational distribution function must be expanded in terms of spherical harmonics. It is necessary and sufficient, however, to obtain only an approximate solution of the form

$$f(\theta, \varphi, t) = \frac{n}{4\pi} \left(1 + \frac{\mu F(t)}{kT} \cos\theta \right) \tag{3.3.21}$$

because the electric polarization (3.3.15) is determined only by a term corresponding to a first-order Legendre polynomial, $\cos\theta$. The expansion coefficient is taken in a form similar to (3.3.14). As the equilibrium solution (3.3.14) gave (3.3.16), the solution (3.3.21) gives

$$\boldsymbol{P}_{\mathrm{d}}(t) = \frac{n\mu^2}{3kT} F(t) \frac{\boldsymbol{E}(t)}{E(t)} \tag{3.3.22}$$

if it is substituted into (3.3.15). By using (3.3.21), (3.3.20) is rewritten into an equation for the unknown coefficient $F(t)$ to the first order of $E(t)$

$$\frac{dF(t)}{dt} = -2D_{\mathrm{r}} [F(t) - E(t)] . \tag{3.3.23}$$

By virtue of (3.3.22), this equation is nothing but the equation of motion for $\boldsymbol{P}_{\mathrm{d}}$ (3.3.18), provided that the relaxation time τ is determined from the rotational diffusion coefficient D_{r} by

$$\tau = \frac{1}{2D_{\mathrm{r}}} . \tag{3.3.24}$$

Clearly τ is the relaxation time of $F(t)$ and thus that of the orientational distribution approaching the Boltzmann distribution.

As seen from (3.3.22), the solution of the form (3.3.21) varies in time synchronized with the quantity governed by the macroscopic equation (3.3.18). This type of solution is often assumed to derive macroscopic equations.

3.4 Resonance Absorption

For the case of Debye-type relaxation described above, the relaxation function is monotonically decaying (3.3.18, 20), because only such mechanisms act within the system, so making the system approach a thermal equilibrium state corresponding to an instantaneous value of the external force. In nature there are many cases where other mechanisms are built in the system, making it oscillate harmonically with certain characteristic frequencies. In such cases the relaxation function is of the damped oscillation type, and phenomena relating to *resonance absorption of external energy* are observed. These phenomena can provide knowledge of the characteristic oscillation, exemplified below.

3.4.1 Van Vleck-Weisskopf-Fröhlich Type Resonance Absorption

Let us call the resonance absorption described by an exponential type of relaxation function, such as (3.3.9), multiplied by the cosine function with a characteristic frequency ω_0

$$\psi(t) = \Delta\varepsilon\, \mathrm{e}^{-t/\tau} \cos \omega_0 t \quad (t > 0) \tag{3.4.1}$$

resonance absorption of the Van Vleck-Weisskopf-Fröhlich type [3.2, 5]. The reason why we write $\Delta\varepsilon$ in place of $\varepsilon_s - \varepsilon_\infty$ is to indicate that the real part of the complex dielectric function $\varepsilon'(\omega)$ decreases by $\Delta\varepsilon$, as that of the Debye type does by $\varepsilon_s - \varepsilon_\infty$, when the angular frequency increases from zero to infinity.

The response function corresponding to the relaxation function (3.4.1) is

$$\varphi(t) = \frac{\Delta\varepsilon}{\tau}\, \mathrm{e}^{-t/\tau} [\cos \omega_0 t + \omega_0 \tau \sin \omega_0 t] \quad (t > 0) \tag{3.4.2}$$

from (3.1.12), and the complex dielectric function is

$$\varepsilon(\omega) = \varepsilon_\infty + \frac{\Delta\varepsilon}{2} \left(\frac{1 - \mathrm{i}\omega_0 \tau}{1 - \mathrm{i}(\omega + \omega_0)\tau} + \frac{1 + \mathrm{i}\omega_0 \tau}{1 - \mathrm{i}(\omega - \omega_0)\tau} \right) \tag{3.4.3}$$

by virtue of (3.3.5). Here ε_∞ denotes the value of $\varepsilon'(\omega)$ at ω sufficiently far from the resonance point. Therefore, the value of $\varepsilon'(\omega)$ on the low-frequency side becomes $\varepsilon_\infty + \Delta\varepsilon$. Separating the function (3.4.3) into its real and imaginary parts gives (Fig. 3.3a, b)

$$\begin{aligned} \varepsilon'(\omega) - \varepsilon_\infty &= \frac{\Delta\varepsilon}{2} \left(\frac{1 + \omega_0(\omega + \omega_0)\tau^2}{1 + (\omega + \omega_0)^2\tau^2} + \frac{1 - \omega_0(\omega - \omega_0)\tau^2}{1 + (\omega - \omega_0)^2\tau^2} \right) \\ &= \Delta\varepsilon \frac{\omega_0^2}{\omega_0^2 - \omega^2} \left(1 - \frac{(1/2)(\omega/\omega_0) + (1/2)(\omega/\omega_0)^2}{\tau^2[(\omega - \omega_0)^2 + (1/\tau)^2]} \right. \\ &\quad \left. + \frac{(1/2)(\omega/\omega_0) - (1/2)(\omega/\omega_0)^2}{\tau^2[(\omega + \omega_0)^2 + (1/\tau)^2]} \right), \\ \varepsilon''(\omega) &= \frac{\Delta\varepsilon}{2} \left(\frac{\omega\tau}{1 + (\omega + \omega_0)^2\tau^2} + \frac{\omega\tau}{1 + (\omega - \omega_0)^2\tau^2} \right). \end{aligned} \tag{3.4.4}$$

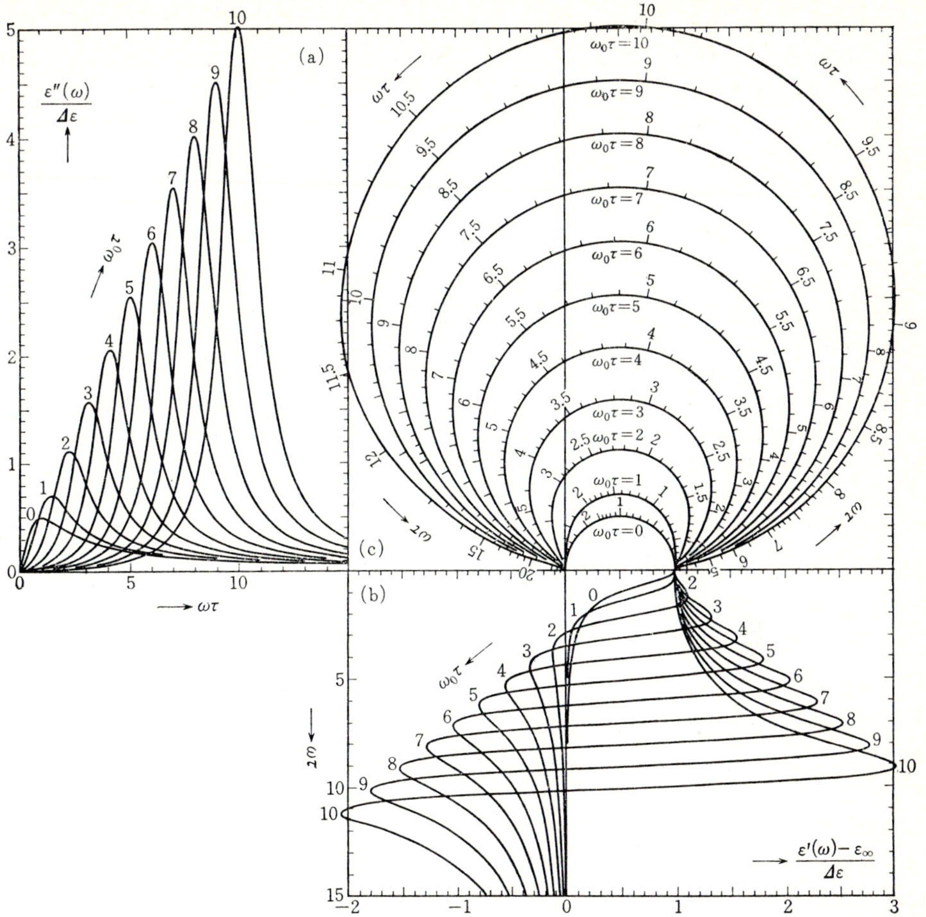

Fig. 3.3 a – c. Complex Van Vleck-Weisskopf-Fröhlich type, dielectric function. (**a**) the imaginary part, (**b**) the real part, and (**c**) the Cole-Cole diagram. When $\omega_0 \tau = 0$, this dielectric function coincides with that of the Debye type (Fig. 3.2)

The imaginary part $\varepsilon''(\omega)$ relating to the energy absorption has its maximum value $\sqrt{1+(\omega_0 \tau)^2}\,(\Delta\varepsilon/2)$ at the frequency $\omega = \sqrt{1+(\omega_0 \tau)^2}/\tau$ which is larger than ω_0 and $1/\tau$. The half-width measured in the logarithmic scale of the angular frequency is given by

$$\Delta(\ln \omega) = 2\cosh^{-1}\left(\frac{1+\sqrt{1+(1+\omega_0^2\tau^2)\,\omega_0^2\tau^2}}{1+\omega_0^2\tau^2}\right).$$

The Cole-Cole diagram differs from the semicircle (Fig. 3.3c). If $\omega_0 \tau$ is sufficiently large, the maximum occurs nearly at $\omega = \omega_0$, and the half-width decreases as $\Delta\omega = (\omega_0 \tau^2)^{-1}$, so the peak becomes very sharp. Of course, the Debye type arises again for $\omega_0 \to 0$ as expected from (3.4.1).

The dielectric function as given above was introduced to explain the ionic polarization $\boldsymbol{P}_{\mathrm{a}}(t)$ in (3.3.12) for a certain class of molecules with their resonance points in the microwave region. We shall not enter into this problem now, but show an example taken from the nuclear-magnetic-resonance area.

3.4.2 Nuclear Magnetic Resonance

Let us consider the problem of nuclear magnetic resonance in a substance composed of molecules which carry nuclear magnetic moments. For simplicity, assume that there is only one kind of nuclei, but generalizations to more complex systems are not difficult. The treatment also applies to similar sorts of physical phenomena, for example, paramagnets carried by electronic spins, and optical excitation of resonant atoms or molecules, with suitable modifications.

The magnetization vector of the whole system is denoted by $\boldsymbol{M}(t)$ which is exposed to a uniform magnetic field $\boldsymbol{H}(t)$. The equation of motion of $\boldsymbol{M}(t)$ is assumed phenomenologically to be

$$\frac{d\boldsymbol{M}(t)}{dt} = \gamma \boldsymbol{M}(t) \times \boldsymbol{H}(t) - \tau^{-1}[\boldsymbol{M}(t) - \chi \boldsymbol{H}(t)], \tag{3.4.5}$$

where γ is the gyromagnetic ratio and the second term on the right-hand side represents relaxation as explained below. This equation corresponds to (3.3.18) for the electric polarization of a dielectric. We consider the situation in which $\boldsymbol{H}$ is composed of a static field $\boldsymbol{H}_0$ in the z direction and an oscillating field $\boldsymbol{H}_1(t)$ perpendicular to $\boldsymbol{H}_0$:

$$\boldsymbol{H}(t) = \boldsymbol{H}_0 + \boldsymbol{H}_1(t). \tag{3.4.6}$$

In usual nuclear resonance experiments, $\boldsymbol{H}_0$ is of the order of 10^3 or 10^4 Gauss and the frequency of $\boldsymbol{H}_1(t)$ is in the radio wave region. On the right-hand side of (3.4.5) the last term indicates relaxation toward the equilibrium magnetization $\chi \boldsymbol{H}(t)$ corresponding to the field $\boldsymbol{H}(t)$, where χ is the isothermal susceptibility. This assumption should hold as long as $\boldsymbol{H}_1(t)$ is weak and changes slowly. In general cases where $\boldsymbol{H}_1(t)$ is not weak and changes in time with higher frequencies, there is, however, no guarantee for this [3.6]. Commonly, therefore, the end relaxation is assumed to be $\chi \boldsymbol{H}_0$ rather $\chi \boldsymbol{H}(t)$. In this form, the equation is called the *Bloch equation* [3.7].

We treat here the generalized Bloch equation of the form (3.4.5) which makes it possible to go over smoothly to the limits of $\boldsymbol{H}_0 \to 0$ and $\omega \to 0$. The relaxation is anisotropic because of the presence of $\boldsymbol{H}_0$. Accordingly, τ is a tensor, the zz component being the longitudinal relaxation time T_1 and the xx and yy components the transverse relaxation time T_2. The simple exponential relaxation assumed here is an idealization which may not always be realized. Transverse relaxation is caused by random local fields due to the interactions of magnetic spins. A simple model of this is the random frequency modulation, Sect. 2.1. In the limit of motional narrowing, the relaxation becomes simple exponential decay with a well-defined relaxation time T_2. Longitudinal relaxation is caused by quantum jumps of spins between the Zeeman levels. It becomes simple also if the static field $\boldsymbol{H}_0$ is sufficiently strong in comparison with the random local fields. Within these limitations, (3.4.5) may be reasonably assumed.

First let us consider the case of a circularly polarized magnetic field

$$\boldsymbol{H}_1(t) = (H_1 \cos \omega t, -H_1 \sin \omega t, 0). \tag{3.4.7}$$

Then (3.4.5) is written as

$$\begin{aligned}
\frac{dM_x}{dt} &= \gamma [M_y H_0 + M_z H_1 \sin \omega t] - \frac{M_x - \chi H_1 \cos \omega t}{T_2}, \\
\frac{dM_y}{dt} &= \gamma [-M_x H_0 + M_z H_1 \cos \omega t] - \frac{M_y + \chi H_1 \sin \omega t}{T_2}, \\
\frac{dM_z}{dt} &= \gamma [-M_x H_1 \sin \omega t - M_y H_1 \cos \omega t] - \frac{M_z - \chi H_0}{T_1}.
\end{aligned} \tag{3.4.8}$$

In the preceding section the transient phenomenon was removed by setting $t_0 \to -\infty$ in the solution (3.3.19). Here also we are not interested in transient phenomena and seek a solution that is stationary in the coordinate system rotating about the z axis together with $\boldsymbol{H}_1(t)$. After some lengthy calculations we get

$$\begin{aligned}
M_x &= \frac{1 + \omega_0(\omega_0 - \omega) T_2^2 + \omega_1^2 T_1 T_2}{1 + (\omega_0 - \omega)^2 T_2^2 + \omega_1^2 T_1 T_2} \chi H_1 \cos \omega t \\
&\quad + \frac{\omega T_2}{1 + (\omega_0 - \omega)^2 T_2^2 + \omega_1^2 T_1 T_2} \chi H_1 \sin \omega t, \\
M_y &= \frac{\omega T_2}{1 + (\omega_0 - \omega)^2 T_2^2 + \omega_1^2 T_1 T_2} \chi H_1 \cos \omega t \\
&\quad - \frac{1 + \omega_0(\omega_0 - \omega) T_2^2 + \omega_1^2 T_1 T_2}{1 + (\omega_0 - \omega)^2 T_2^2 + \omega_1^2 T_1 T_2} \chi H_1 \sin \omega t, \\
M_z &= \frac{1 + (\omega_0 - \omega)^2 T_2^2 + (\omega_0 - \omega)\,\omega_0^{-1} \omega_1^2 T_1 T_2}{1 + (\omega_0 - \omega)^2 T_2^2 + \omega_1^2 T_1 T_2} \chi H_0,
\end{aligned} \tag{3.4.9}$$

setting

$$\omega_0 = \gamma H_0, \quad \omega_1 = \gamma H_1.$$

If the transverse field $\boldsymbol{H}_1$ is sufficiently weak and we may assume $(\gamma H_1)^2 T_1 T_2 \ll 1$, we may neglect those terms that include the square of ω_1, so obtaining the linear response solution with respect to the oscillating field

$\boldsymbol{H}_1(t)$. In this solution, the longitudinal relaxation time T_1 drops out completely. To measure T_1, we either have to observe a nonlinear phenomenon, such as the saturation phenomenon, or examine a transient phenomenon, e.g., the paramagnetic relaxation.

In linear phenomena the superposition principle can be used. The solution for a linearly polarized field

$$\boldsymbol{H}_1(t) = (H_1 \cos \omega t, 0, 0) \tag{3.4.10}$$

can be derived by averaging the solution for the clockwise circular polarization (3.4.7) and that for the counterclockwise circular polarization, obtained by replacing ω by $-\omega$. The result has the form (3.2.11)

$$M_\mu - \delta_{\mu z} \chi H_0 = \chi'_{\mu x}(\omega) H_1 \cos \omega t + \chi''_{\mu x}(\omega) H_1 \sin \omega t\,, \tag{3.4.11}$$

where the corresponding complex magnetic susceptibility is

$$\begin{aligned} \chi_{xx}(\omega) &= \chi + \frac{\mathrm{i}}{2} \chi \omega T_2 \left(\frac{1}{1 - \mathrm{i}(\omega + \omega_0) T_2} + \frac{1}{1 - \mathrm{i}(\omega - \omega_0) T_2} \right) \\ &= \frac{\chi}{2} \left(\frac{1 - \mathrm{i}\omega_0 T_2}{1 - \mathrm{i}(\omega + \omega_0) T_2} + \frac{1 + \mathrm{i}\omega_0 T_2}{1 - \mathrm{i}(\omega - \omega_0) T_2} \right), \\ \chi_{yx}(\omega) &= \frac{\chi}{2} \left(\frac{-\omega T_2}{1 - \mathrm{i}(\omega + \omega_0) T_2} + \frac{\omega T_2}{1 - \mathrm{i}(\omega - \omega_0) T_2} \right), \\ \chi_{zx}(\omega) &= 0\,. \end{aligned} \tag{3.4.12}$$

The xx component of the complex magnetic susceptibility has the same form as the Van Vleck-Weisskopf-Fröhlich complex dielectric function (3.4.3). Here $\chi_{xx}(\omega)$ approaches χ as $\omega \to 0$ and zero at a sufficiently large $|\omega|$.

The preceding section gives a basis for the equation of motion of the electric polarization (3.3.18) with the kinetic equation (3.3.20), derived by considering the rotational Brownian motion of the permanent electric dipole of each molecule. In the present problem, the magnetic moments are carried by nuclear or electronic spins, which must be treated by quantum mechanics. In spite of this, an ensemble of spins can be represented by that of classical spins, each member of which is considered to be in rotational Brownian motion. The kinetic equation is shown [3.8] to be

$$\begin{aligned} \left(\frac{\partial}{\partial t} + \omega_0 \frac{\partial}{\partial \varphi} \right) f = &\left\{ \frac{1}{\sin\theta} \frac{\partial}{\partial \theta} \left[\sin\theta\, D_\perp \left(\frac{\partial}{\partial \theta} + \frac{M H_0 \sin\theta}{kT} \right) \right] \right. \\ &\left. + \frac{1}{\sin^2\theta} \frac{\partial}{\partial \varphi} \left[(D_\perp \cos^2\theta + D_\parallel \sin^2\theta) \frac{\partial}{\partial \varphi} \right] \right\} f\,. \end{aligned} \tag{3.4.13}$$

This is a generalization of (3.3.20) in that firstly the longitudinal and transverse diffusion constants are distinguished, and secondly the Larmor precession is taken into account. Note that the Larmor frequency ω_0 is not equal to γH_0 but is shifted from it as a result of the random field acting on a spin to cause its Brownian motion. The relation (3.3.24) is generalized to

$$\frac{1}{T_1}=\frac{D_\perp}{2}, \quad \frac{1}{T_2}=D_\perp + D_\parallel, \tag{3.4.14}$$

where T_1 and T_2 are the longitudinal and transverse relaxation times, respectively.

3.4.3 Failure at High Frequencies

To describe high-frequency phenomena we have to use a correspondingly high time precision. Then, as was mentioned previously in connection with the electric polarization (3.3.12), new mechanisms will successively be called into play. They transfer contributions to the first term on the right-hand side of the linear relation (3.1.2), which follow the force instantaneously, the second term, which shows a time lag behind the force. Accordingly, the functional form of the response function at small $|t|$ and hence the form of the complex admittance at large $|\omega|$ ought to be modified. However, if these new mechanisms are not yet know, we can only check the degree of the approximation by our assumed expression of the response function or of the complex admittance at high frequencies by using the sum rules given in Sect. 3.7, and improve the expression by using the sum rules. We discuss the details in Sect. 3.7, and here examine only the total energy loss, i.e. the sum of the power loss over all angular frequencies.

For dielectrics, the power loss was found to be proportional to $\omega\varepsilon''(\omega)$. Assuming that the Debye-type expression (3.3.11) is valid over all frequencies, the total energy loss would diverge:

$$\int_{-\infty}^{\infty} \omega\varepsilon''(\omega)\, d\omega = \infty . \tag{3.4.15}$$

The divergence arises because the assumed $\omega\varepsilon''(\omega)$ becomes constant at large $|\omega|$ and does not vanish sufficiently rapidly. The situation does not change for the Van Vleck-Weisskopf-Fröhlich type expression (3.4.4). Since the total energy loss for a finite system should essentially be finite even for an infinitely short unit pulse (3.1.6), the divergence means that the approximation is poor at high frequencies.

3.5 Wave Number-Dependent Complex Admittance

So far, we have assumed that the state of a system is uniform in space. This assumption of uniformity is not generally true, because an external action and the response produced within the system essentially propagate with finite velocity, forming a sort of wave. If the dimension of the system is far smaller than the wavelengths, the spatial change may be unimportant. In general, however, the action and response are nonuniform in space, even if the system is structurally homogeneous. In a nonuniform system the force and the displacement or the current become functions of the spatial coordinate $\boldsymbol{r}$ in addition to time t. This section discusses the generalization of the theory developed in Sects. 3.1, 2 for the nonuniform case.

3.5.1 Non-Markovian Nonlocal Linear Relations

As noted in Sect. 3.1, the subscript μ attached to the force and the displacement or the current should be understood to include the spatial coordinate in the case of a nonuniform system, or rather, the set $(\mu, \boldsymbol{r})$ should be used in place of the single μ. Then the linear relation (3.1.2) is written in the form

$$\begin{aligned} B_\mu(\boldsymbol{r}, t) - B_\mu^{\mathrm{eq}} &= \int d\boldsymbol{r}' \sum_\nu \chi_{\mu\nu}^\infty(\boldsymbol{r}-\boldsymbol{r}') X_\nu(\boldsymbol{r}', t) \\ &\quad + \int_{-\infty}^{t} dt' \int d\boldsymbol{r}' \sum_\nu \Phi_{\mu\nu}(\boldsymbol{r}-\boldsymbol{r}', t-t') X_\nu(\boldsymbol{r}', t') . \end{aligned} \tag{3.5.1}$$

Besides the non-Markovian property, i.e., the retarded effect of the force, we have taken into account the nonlocal property, i.e., the effect of forces at different points. Here, as for the case of time, we have assumed that the system is invariant under the spatial translation except for the coordinate dependence of the force so that the response and other functions depend only on the difference of the space coordinates $\boldsymbol{r}-\boldsymbol{r}'$.

In ordinary cases, the response function rapidly converges to zero as the distance $|\boldsymbol{r}-\boldsymbol{r}'|$ increases:

$$\lim_{|\boldsymbol{r}-\boldsymbol{r}'|\to\infty} \Phi_{\mu\nu}(\boldsymbol{r}-\boldsymbol{r}', t-t') = 0 . \tag{3.5.2}$$

Thus we may assume the existence of a distance l at which the function Φ may be regarded to have converged to zero already.

The nonuniformity arises when the difference of values of force at two points at a distance of the order l can no longer be neglected. This is exemplified by superfluidity or superconductivity, in which the coherence length of the quantum effect appears, and the anomalous skin effect of metals, in which the mean free paths of conduction electrons are comparable to or longer than the wavelengths of electromagnetic waves. The characteristic distance l becomes very long at temperatures near the critical point of a phase transition of the second kind. When we have to take nonlocality into account, it should be noted that the equation of state and related expressions ought ot have nonlocal forms, as exemplified in Sect. 4.5.

If the spatial precision of our phenomenological theory is sufficiently crude in comparison with the distance l, we may neglect the nonlocality in the response function and make the approximation

$$\Phi_{\mu\nu}(\boldsymbol{r}-\boldsymbol{r}', t-t') = \Phi_{\mu\nu}(t-t')\,\delta(\boldsymbol{r}-\boldsymbol{r}')\,. \tag{3.5.3}$$

Similarly, if the first term on the right-hand side of (3.5.1) χ^{∞} includes the factor $\delta(\boldsymbol{r}-\boldsymbol{r}')$, the linear relation (3.5.1) reduces to (3.1.2), but with space-dependent B and X given by

$$B_\mu(\boldsymbol{r}, t) - B_\mu^{\text{eq}} = \sum_\nu \chi_{\mu\nu}^{\infty} X_\nu(\boldsymbol{r}, t) + \int_{-\infty}^{t} dt' \sum_\nu \Phi_{\mu\nu}(t-t')\, X_\nu(\boldsymbol{r}, t')\,. \tag{3.5.4}$$

This is the linear relation commonly used for a nonuniform system, ignoring the nonlocal effects.

When the time precision is crude enough so that the non-Markovian property can be disregarded, (3.5.1) reduces further to the same form as (3.1.1)

$$B_\mu(\boldsymbol{r}, t) - B_\mu^{\text{eq}} = \sum_\nu L_{\mu\nu} X_\nu(\boldsymbol{r}, t)\,, \tag{3.5.5}$$

where the kinetic coefficients are given by

$$L_{\mu\nu} = \int d\boldsymbol{r}' \left[\chi_{\mu\nu}^{\infty}(\boldsymbol{r}') + \int_0^{\infty} ds\, \Phi_{\mu\nu}(\boldsymbol{r}', s) \right]. \tag{3.5.6}$$

The linear relation (3.5.5) is the relation corresponding to the Fourier law of heat conduction and the Fick law of diffusion, Sect. 3.1.

The relaxation function is given by the generalization of (3.1.11)

$$\Psi_{\mu\nu}(\boldsymbol{r}, t) = \int_t^{\infty} ds\, \Phi_{\mu\nu}(\boldsymbol{r}, s) \tag{3.5.7}$$

and satisfies

$$\Phi_{\mu\nu}(\boldsymbol{r}, t) = -\frac{\partial \Psi_{\mu\nu}(\boldsymbol{r}, t)}{\partial t}\,. \tag{3.5.8}$$

as the function (3.1.11) satisfies (3.1.12).

To obtain the complex admittance, we Fourier-transform with respect to the space coordinate, in addition to that with respect to time, i.e., we Fourier-analyze into the superposition of running plane waves:

$$\begin{aligned} X_\nu(\boldsymbol{r}, t) &= \int_{-\infty}^{\infty} \frac{d\omega}{2\pi} \int \frac{d\boldsymbol{k}}{(2\pi)^3} \exp[\mathrm{i}(\boldsymbol{k}\cdot\boldsymbol{r} - \omega t)]\, X_{\nu;\, \boldsymbol{k}, \infty}\,, \\ B_\mu(\boldsymbol{r}, t) - B_\mu^{\text{eq}} &= \int_{-\infty}^{\infty} \frac{d\omega}{2\pi} \int \frac{d\boldsymbol{k}}{(2\pi)^3} \exp[\mathrm{i}(\boldsymbol{k}\cdot\boldsymbol{r} - \omega t)]\, B_{\mu;\, \boldsymbol{k}, \omega}\,. \end{aligned} \tag{3.5.9}$$

Following this transformation, the linear relation (3.5.1) is rewritten as

$$B_{\mu;\,\boldsymbol{k},\omega} = \sum_{\nu} \chi_{\mu\nu}(\boldsymbol{k},\omega)\, x_{\nu;\,\boldsymbol{k},\omega} \tag{3.5.10}$$

in the same way as for (3.2.3), introducing the complex admittance depending on the wave vector $\boldsymbol{k}$, too,

$$\chi_{\mu\nu}(\boldsymbol{k},\omega) = \int d\boldsymbol{r} \exp(-\mathrm{i}\,\boldsymbol{k}\cdot\boldsymbol{r}) \left[\chi^{\infty}_{\mu\nu}(\boldsymbol{r}) + \int_0^{\infty} dt\, \mathrm{e}^{\mathrm{i}\omega t}\, \Phi_{\mu\nu}(\boldsymbol{r},t)\right]. \tag{3.5.11}$$

We also obtain

$$\frac{\chi_{\mu\nu}(\boldsymbol{k},0) - \chi_{\mu\nu}(\boldsymbol{k},\omega)}{\mathrm{i}\omega} = \int_0^{\infty} dt \int d\boldsymbol{r} \exp[-\mathrm{i}(\boldsymbol{k}\cdot\boldsymbol{r} - \omega t)]\, \Psi_{\mu\nu}(\boldsymbol{r},t) \tag{3.5.12}$$

by using the relaxation function (3.5.7).

3.5.2 Complex Admittance for the Diffusion Phenomenon

As an example let us consider the diffusion occurring in a two-component fluid. The heat conduction can be treated similarly. Let $c(\boldsymbol{r},t)$ denote the concentration of one of the components and assume that there is no chemical reaction between the components. Then the equation of continuity

$$\frac{\partial c(\boldsymbol{r},t)}{\partial t} = -\operatorname{div} \boldsymbol{j}(\boldsymbol{r},t) \tag{3.5.13}$$

holds as the law of conservation of mass. We shall assume the Fick law for the diffusion current $\boldsymbol{j}(\boldsymbol{r},t)$:

$$\boldsymbol{j}(\boldsymbol{r},t) = -D \operatorname{grad} c(\boldsymbol{r},t)\,. \tag{3.5.14}$$

A generalization to the non-Markovian law is discussed in Sect. 3.7.2. If there is no external force, the thermal equilibrium value of concentration c^{eq} is a constant and that of the diffusion current vanishes. Therefore, we may suppose without losing generality that c represents the shift from its equilibrium value. Further, D denotes the diffusion coefficient, for which we may assume a value at $c = c^{\mathrm{eq}}$ in the linear approximation.

Substituting (3.5.14) into (3.5.13), yields the diffusion equation

$$\frac{\partial c(\boldsymbol{r},t)}{\partial t} = D \frac{\partial^2 c(\boldsymbol{r},t)}{\partial \boldsymbol{r}^2}\,. \tag{3.5.15}$$

As an example of how to obtain the complex admittance from the relaxation function, let us solve this equation for an initial concentration distribution $c(\boldsymbol{r},t_0)$ at time t_0. To impose the initial condition at time t_0, we do not take the inverse of the transformation (3.5.9), but instead the

Fourier-Laplace transformation

$$c_{k,\omega} = \int_{t_0}^{\infty} dt \int d\boldsymbol{r} \exp[-\mathrm{i}(\boldsymbol{k} \cdot \boldsymbol{r} - \omega t)]\, c(\boldsymbol{r}, t)\,. \tag{3.5.16}$$

Since the plane wave $\exp(\mathrm{i}\boldsymbol{k} \cdot \boldsymbol{r})$ is the eigenfunction of the Laplacian operator, the diffusion equation (3.5.15) is changed into

$$(-\mathrm{i}\omega + Dk^2)\, c_{k,\omega} = \mathrm{e}^{\mathrm{i}\omega t_0} \int d\boldsymbol{r} \exp(-\mathrm{i}\boldsymbol{k} \cdot \boldsymbol{r})\, c(\boldsymbol{r}, t_0)\,. \tag{3.5.17}$$

On the other hand, if we use the normalized relaxation function Ψ^{m}, an equation corresponding to (3.1.17)

$$c(\boldsymbol{r}, t) = \int d\boldsymbol{r}'\, \Psi^{\mathrm{m}}(\boldsymbol{r} - \boldsymbol{r}', t - t_0)\, c(\boldsymbol{r}', t_0) \tag{3.5.18}$$

holds. Substituting this into (3.5.16), the expected form of the solution is obtained as

$$c_{k,\omega} = \left[\int_0^{\infty} ds \int d\boldsymbol{R} \exp[-\mathrm{i}(\boldsymbol{k} \cdot \boldsymbol{R} - \omega s)]\, \Psi^{\mathrm{m}}(\boldsymbol{R}, s)\right] \times \mathrm{e}^{\mathrm{i}\omega t_0} \int d\boldsymbol{r}' \exp(-\mathrm{i}\boldsymbol{k} \cdot \boldsymbol{r}')\, c(\boldsymbol{r}', t_0)\,.$$

Comparing this with (3.5.17), then

$$\frac{1}{-\mathrm{i}\omega + Dk^2} = \int_0^{\infty} ds \int d\boldsymbol{R} \exp[-\mathrm{i}(\boldsymbol{k} \cdot \boldsymbol{R} - \omega s)]\, \Psi^{\mathrm{m}}(\boldsymbol{R}, s)\,. \tag{3.5.19}$$

Since the right-hand side of the equation thus obtained must have the form of the right-hand side of (3.5.12) normalized in the same way as in (3.2.9), we arrive at the complex admittance

$$\chi(\boldsymbol{k}, \omega) = \frac{Dk^2}{-\mathrm{i}\omega + Dk^2}\, \chi(\boldsymbol{k}, 0)\,. \tag{3.5.20}$$

The concentration variation $c(\boldsymbol{r}, t)$ can be induced, in principle, by applying an external potential field $\phi^{\mathrm{e}}(\boldsymbol{r}, t)$ [3.9]. The response of the Fourier component $c_{k,\omega}$ to the Fourier component $\phi^{\mathrm{e}}_{k,\omega}$ of the potential is defined by the admittance $\chi(\boldsymbol{k}, \omega)$. The static admittance $\chi(\boldsymbol{k}, 0)$ is obtained by solving an equilibrium problem. In particular, its limiting value for $\boldsymbol{k} \to 0$ ought to be determined by thermodynamics, Sect. 4.5. The inverse transformation of (3.5.19) yields the fundamental solution of the diffusion equation

$$\Psi^{\mathrm{m}}(\boldsymbol{r}, t) = \frac{1}{(4\pi D t)^{3/2}} \exp\left(-\frac{r^2}{4Dt}\right) \quad (t > 0)\,. \tag{3.5.21}$$

3.6 Dispersion Relations

As stated in Sect. 3.2, the real and imaginary parts of the complex admittance are not independent of each other. We shall now prove this. Of the various proofs, we shall use one based on the theory of complex functions to lay the foundations for the next section. Another proof is given in Sect. 4.2. We treat only uniform systems.

3.6.1 Proof of the Dispersion Relations

If complex admittance exists for all values of the angular frequency ω, the integral on the right-hand side of (3.2.5) converges. According to the theory of Fourier-Laplace transformation, the function by replacing ω with a complex number z

$$\Xi(z) = \int_0^\infty dt\, e^{izt}\,\Phi(t) \tag{3.6.1}$$

is analytic in the upper half complex plane $\mathrm{Im}\,\{z\} > 0$ and behaves as

$$\Xi(z) = 0(1/z) \quad \text{for large } |z|, \tag{3.6.2}$$

if the function $\Phi(t)$ is differentiable for $t > 0$ and is continuous at $t = 0$. the complex admittance is its limiting value at the real axis from the upper side, and (3.2.5) should be understood more precisely as

$$\chi(\omega) - \chi^\infty = \lim_{\varepsilon \to +0} \Xi(\omega + i\varepsilon) = \lim_{\varepsilon \to +0} \int_0^\infty dt\, e^{i\omega t - \varepsilon t}\,\Phi(t)\,. \tag{3.6.3}$$

According to the Cauchy integral formula, the following equation arises for an arbitrary closed contour Γ placed in the complex region, within which the complex function $\Xi(z)$ is analytic:

$$\oint_\Gamma \frac{dz'}{2\pi i} \frac{\Xi(z')}{z' - z} = \begin{cases} \Xi(z) & (\text{if } z \text{ is enclosed with } \Gamma), \\ 0 & (\text{if } z \text{ is not enclosed with } \Gamma). \end{cases}$$

As the contour Γ, we choose the one that goes on the real axis from $-\infty$ to $+\infty$ and returns on the upper semicircle with an infinite radius from $+\infty$ to $-\infty$. Then the contribution from the integral of the function (3.6.1) on the semicircle vanishes by virtue of (3.6.2), so that

$$\int_{-\infty}^{\infty} \frac{d\omega}{2\pi i} \frac{\chi(\omega) - \chi^\infty}{\omega - z} = \begin{cases} \Xi(z) & \mathrm{Im}\,\{z\} > 0, \\ 0 & \mathrm{Im}\,\{z\} < 0. \end{cases}$$

If we choose z so that $\operatorname{Im}\{z\} > 0$, then $\operatorname{Im}\{z^*\} < 0$, and hence we may add or subtract a term of zero value as in

$$\Xi(z) = \int_{-\infty}^{\infty} \frac{d\omega}{2\pi \mathrm{i}} [\chi(\omega) - \chi^{\infty}] \left(\frac{1}{\omega - z} \pm \frac{1}{\omega - z^*} \right).$$

Thus

$$\begin{aligned} \Xi(z) &= \int_{-\infty}^{\infty} \frac{d\omega}{\pi \mathrm{i}} [\chi(\omega) - \chi^{\infty}] \operatorname{Re} \left\{ \frac{1}{\omega - z} \right\}, \\ & \qquad\qquad\qquad\qquad\qquad\qquad \operatorname{Im}\{z\} > 0 . \\ \Xi(z) &= \int_{-\infty}^{\infty} \frac{d\omega}{\pi} [\chi(\omega) - \chi^{\infty}] \operatorname{Im} \left\{ \frac{1}{\omega - z} \right\}, \end{aligned} \tag{3.6.4}$$

Adding the real part of the first equation to the imaginary part of the second equation multiplied by i gives

$$\Xi(z) = \int_{-\infty}^{\infty} \frac{d\omega}{\pi} \frac{\chi''(\omega)}{\omega - z}, \quad \operatorname{Im}\{z\} > 0, \tag{3.6.5}$$

and adding the real part of the second equation to the imaginary part of the first equation multiplied by i,

$$\Xi(z) = -\mathrm{i} \int_{-\infty}^{\infty} \frac{d\omega}{\pi} \frac{\chi'(\omega) - \chi^{\infty}}{\omega - z}, \quad \operatorname{Im}\{z\} > 0 . \tag{3.6.6}$$

Taking the limit of (3.6.3), and using

$$\lim_{\varepsilon \to +0} \frac{1}{x \pm \mathrm{i}\varepsilon} = \frac{\mathscr{P}}{x} \mp \mathrm{i}\pi\delta(x), \tag{3.6.7}$$

we arrive at the final equations

$$\begin{aligned} \chi'(\omega) - \chi^{\infty} &= \int_{-\infty}^{\infty} \frac{d\omega'}{\pi} \frac{\mathscr{P}}{\omega' - \omega} \chi''(\omega'), \\ \chi''(\omega) &= -\int_{-\infty}^{\infty} \frac{d\omega'}{\pi} \frac{\mathscr{P}}{\omega' - \omega} [\chi'(\omega') - \chi^{\infty}] \end{aligned} \tag{3.6.8}$$

or

$$\Xi(\omega) = \frac{1}{\pi \mathrm{i}} \int_{-\infty}^{\infty} d\omega' \frac{\mathscr{P}}{\omega' - \omega} \Xi(\omega') . \tag{3.6.9}$$

The notation $\mathscr{P}$ indicates taking the Cauchy principal value. The last result can also be obtained more directly by choosing the contour Γ comprising an infinite upper semicircle closed by a straight line along the real axis detouring the point ω by an infinitely small semicircle. Equations (3.6.8) show that the real and imaginary parts of the complex admittance are connected to each other by the Hilbert transformation, and are called the *Kramers-Kronig relations* or sometimes simply the *dispersion relations.*

3.6.2 Dispersion Relations and Causality

In the proof given above we have not directly used the representation (3.6.1) of the response function. The dispersion relations hold between the real and imaginary parts of any function analytic in the upper half-plane and which vanishes sufficiently rapidly as $|z| \to \infty$.

For complex admittance, the analyticity in the upper half-plane is guaranteed because the limits of the time integral in (3.6.1) are $(0, +\infty)$. This in turn originates in the limits of the time integral $(-\infty, t)$ in the linear relation (3.1.2) and hence in causality, according to which only the past values of force have an effect but not the future values. Therefore, the position of the minus sign in the dispersion relations (3.6.8) originates from causality.

According to the dispersion relations (3.6.8), the function which contains the same amount of knowledge as the imaginary part of the complex admittance $\chi''(\omega)$ is not the real part $\chi'(\omega)$, but the real part minus χ^∞. The quantity equivalent to the response or the relaxation function comprises these parts: the knowledge contained in χ^∞ should be supplemented separately. However, normally χ^∞ is introduced owing to the crudeness of the time precision in our observations and, as the next chapter shows, by its nature can not exist in the microscopic theory based on statistical mechanics. Exceptions whereby χ^∞ remains intrinsic are the cases where the forces are explicitly contained in the observed quantities. For example, the electric displacement $\boldsymbol{D}$ contains the electric field $\boldsymbol{E}$ as is seen in (3.3.1), so that the response of $\boldsymbol{D}$ to $\boldsymbol{E}$ naturally has a χ^∞ term. Another example is a diamagnetic current carried by electrons in a varying magnetic field, in which case the current operator explicitly depends on the vector potential deriving the magnetic field.

3.6.3 Analytical Continuation into the Complex Plane

Expression (3.6.5) gives us a method of defining a single function $\Xi(z)$ both in the upper and lower half-planes of z, namely

$$\Xi(z) = \int_{-\infty}^{\infty} \frac{d\omega}{\pi} \frac{\chi''(\omega)}{\omega - z}. \tag{3.6.10}$$

In the upper half-plane it is identical with the original function (3.6.1) or the function $\chi(\omega) - \chi^\infty$, with ω replaced by the complex variable z:

$$\Xi(z) = \chi(z) - \chi^\infty, \quad \mathrm{Im}\{z\} > 0.$$

In the lower half-plane, it is the analytical continuation of the original function by means of the integral expression (3.6.10) and is equal to the function $\chi^*(z)-\chi^\infty$, where $\chi^*(\omega)$ for real ω is the complex conjugate to $\chi(\omega)$. Therefore, when z approaches the real axis from the upper or the lower side,

$$\lim_{\varepsilon\to+0} \bar{\Xi}(\omega+\mathrm{i}\,\varepsilon)=\chi(\omega)-\chi^\infty \quad \text{or}$$

$$\lim_{\varepsilon\to+0} \bar{\Xi}(\omega-\mathrm{i}\,\varepsilon)=\chi^*(\omega)-\chi^\infty .$$

Accordingly, the function $\bar{\Xi}(z)$ has the discontinuity

$$\lim_{\varepsilon\to+0}\,[\bar{\Xi}(\omega+\mathrm{i}\,\varepsilon)-\bar{\Xi}(\omega-\mathrm{i}\,\varepsilon)]=2\,\mathrm{i}\,\chi''(\omega) \tag{3.6.11}$$

across the real axis. In the integral expression (3.6.10) $\chi''(\omega)$ may be called the spectral function of $\bar{\Xi}(z)$. Chapter 5, which treats the Green's functions, also deals with this subject.

As an example, let us consider the complex Debye dielectric function (3.3.10). The integral (3.6.10) is easily evaluated with the Cauchy theorem to give

$$\bar{\Xi}(z)=\int_{-\infty}^{\infty}\frac{d\omega}{\pi}\,\frac{\varepsilon''(\omega)}{\omega-z}=\frac{\varepsilon_\mathrm{s}-\varepsilon_\infty}{1\mp\mathrm{i}\,z\,\tau}\,,\quad \mathrm{Im}\,\{z\}\gtrless 0\,. \tag{3.6.12}$$

If $\mathrm{Im}\,\{z\}>0$, this coincides with $\varepsilon(\omega)-\varepsilon_\infty$ of (3.3.10) in which the complex number z replaces ω formally. This function $\varepsilon(z)-\varepsilon_\infty$ has a first-order pole at $z=-\mathrm{i}/\tau$, while $\bar{\Xi}(z)$ obtained above has no pole in the lower half-plane.

Similarly, the reader can easily check that for the Van Vleck-Weisskopf-Fröhlich type absorption (3.4.4)

$$\bar{\Xi}(z)=\frac{1\pm\mathrm{i}\,\omega_0\,\tau}{1\mp\mathrm{i}\,(z+\omega_0)\,\tau}+\frac{1\pm\mathrm{i}\,\omega_0\,\tau}{1\mp\mathrm{i}\,(z-\omega_0)\,\tau}\,,\quad \mathrm{Im}\,\{z\}\gtrless 0\,. \tag{3.6.13}$$

3.7 Sum Rules and Interpolation Formulas

An equation requiring that the real or imaginary part of the complex admittance $\chi'(\omega)$ or $\chi''(\omega)$ multiplied by certain powers of ω and integrated over all values of ω should be equal to a certain finite value, is generally called a *moment sum rule*.

3.7.1 Moment Sum Rules

As is obvious from (3.2.12), $\chi'(\omega)-\chi^{\infty}$ is an even function of ω, while $\chi''(\omega)$ is an odd function. Setting $\omega=0$ in the first equation of (3.6.8) gives

$$\int_{-\infty}^{\infty}\frac{d\omega}{\pi}\frac{\mathscr{P}}{\omega}\chi''(\omega)=\chi'(0)-\chi^{\infty}. \tag{3.7.1}$$

By this method, we obtain only the identity $0=0$ from the second equation of (3.6.8), but if we multiply both sides by ω and take the limit $\omega\to\infty$, then

$$\int_{-\infty}^{\infty}\frac{d\omega}{\pi}[\chi'(\omega)-\chi^{\infty}]=\lim_{\omega\to\infty}\omega\chi''(\omega)=\Phi(+0). \tag{3.7.2}$$

To get the last equality we used the Abel theorem. Namely, by partially integrating in (3.6.1), for $\mathrm{Im}\{z\}>0$

$$\Xi(z)=-\frac{\Phi(+0)}{\mathrm{i}z}-\frac{1}{\mathrm{i}z}\int_{0}^{\infty}dt\,\mathrm{e}^{\mathrm{i}zt}\,\dot{\Phi}(t), \tag{3.7.3}$$

where the dot $\cdot$ denotes the time derivative, and by taking the limit $|z|\to\infty$ after multiplying both sides by $\mathrm{i}z$ then

$$\lim_{|z|\to\infty}\mathrm{i}z\,\Xi(z)=-\Phi(+0), \tag{3.7.4}$$

because the integral in (3.7.3) vanishes in this limit as that in (3.6.2). Remembering (3.6.3) and the reality of $\Phi(+0)$, we have the required equation.

It is instructive to confirm by elementary calculations that the example of a dielectric (3.3.11, 4.4) and of the paramagnetic resonance (3.4.12) actually satisfy the sum rules (3.7.1, 2). However, more remarkably, the sum rules are valid in more general forms. As stated in Sect. 3.2, χ' and χ'' relate to dispersion and energy dissipation. Although the sum rules mentioned above are valid whether they relate to χ' or χ'', they have an especially important physical meaning when their left-hand sides represent energy dissipation. When χ'' represents, for instance, the dielectric loss, the magnetic resonance absorption, or the elastic loss, the integral on the left-hand side of (3.7.1), which relates to the frequency spectrum of the power loss, must be equal to the quantity on the right-hand side which is determined as a corresponding thermodynamic response such as the static dielectric constant, the static magnetic susceptibility, or the elastic constants.

When $\chi(\omega)$ is the complex electric conductivity $\sigma(\omega)$ in (3.7.2), the imaginary part σ'' coincides with the real part of the complex dielectric function $\varepsilon(\omega)=1+\mathrm{i}4\pi\sigma(\omega)/\omega$. In the limit of a very high frequency ω, forces acting upon an electron due to scatterers, for example, may be neglected and only the inertial term contributes, so that

$$\sigma(\omega) \approx \mathrm{i}\frac{ne^2}{m\omega} \quad (\omega \to \infty) \tag{3.7.5}$$

and hence (3.7.2) yields

$$\frac{2}{\pi}\int_0^\infty d\omega\, \sigma'(\omega) = \frac{ne^2}{m}\,. \tag{3.7.6}$$

This is a very general representation of the *oscillator strength sum rule* for a system of electrons.

The sum rule of the form (3.7.2) can be generalized further [3.10]. Rewriting (3.7.3) as

$$-\mathrm{i}z\,\Xi(z) - \Phi(+0) = \int_0^\infty dt\, \mathrm{e}^{\mathrm{i}zt}\dot{\Phi}(t)\,, \tag{3.7.7}$$

its right-hand side is the same as that of (3.6.1) except that $\dot{\Phi}$ is substituted in place of Φ. Therefore (3.7.1) may be applied, which is valid between the real and imaginary parts of the complex function $\Xi(\omega + \mathrm{i}0)$ defined by (3.6.1), for the real and imaginary parts of the complex function

$$[-\mathrm{i}z\,\Xi(z) - \Phi(+0)]_{z=\omega+\mathrm{i}0} = \omega\left\{\chi''(\omega) - \frac{\Phi(+0)}{\omega} - \mathrm{i}\,[\chi'(\omega) - \chi^\infty]\right\}$$

to obtain

$$\int_{-\infty}^{\infty} \frac{d\omega}{\pi}\,\omega\left[\chi''(\omega) - \frac{\Phi(+0)}{\omega}\right] = \dot{\Phi}(+0)\,. \tag{3.7.8}$$

By adopting the same method with regard to the equation obtained by repeating once more the partial integration

$$\Xi(z) + \frac{\Phi(+0)}{\mathrm{i}z} - \frac{\dot{\Phi}(+0)}{(\mathrm{i}z)^2} = \frac{1}{(\mathrm{i}z)^2}\int_0^\infty dt\, \mathrm{e}^{\mathrm{i}zt}\ddot{\Phi}(t)\,, \tag{3.7.9}$$

we can derive

$$\int_{-\infty}^{\infty} \frac{d\omega}{\pi}\,\omega^2\left[\chi'(\omega) - \chi^\infty + \frac{\dot{\Phi}(+0)}{\omega^2}\right] = -\,\ddot{\Phi}(+0)\,. \tag{3.7.10}$$

Since $\Phi(+0)$ is the limiting value of $\omega\,\chi''(\omega)$, as seen in (3.7.2), the quantity within the square brackets of the integrand in (3.7.8) is nothing but $\chi''(\omega)$ with its asymptotic form subtracted. The same can be said about (3.7.10). In general, we can repeat the same technique and find that the integral that appears on the left-hand side is the moment of ω^p $(p=1, 2, \ldots)$

evaluated by using the spectrum χ' or χ'' as the weight suitably modified by subtracting its asymptotic form. We are thus led to sum rules stating that such moments are equal to the values of some derivatives of $\Phi(t)$ at $t=+0$.

In Sect. 4.2 we shall extend $\Phi(t)$ to $t<0$. If the function $\Phi(t)$ thus obtained is an even function of t, all the odd derivatives of Φ vanish at $t=0$, and therefore in the even order moments the supplementary terms giving the asymptotic forms mentioned above also vanish, so that

$$\int_{-\infty}^{\infty} \frac{d\omega}{\pi} \omega^{2n} [\chi'(\omega) - \chi^{\infty}] = (-1)^n \Phi^{(2n)}(0). \tag{3.7.11}$$

If $\Phi(t)$ is odd in t,

$$\int_{-\infty}^{\infty} \frac{d\omega}{\pi} \omega^{2n+1} \chi''(\omega) = (-1)^{n+1} \Phi^{(2n+1)}(0). \tag{3.7.12}$$

As Sect. 4.2, $\Phi(t)$ is the correlation function of a suitable physical quantity, and the quantity on the right-hand side of (3.7.11 or 12) is derived from that function and consequently determined as a statistical-mechanics quantity in thermal equilibrium. This is the important meaning of the sum rules.

Equations (3.7.11, 12) can be derived by differentiating the equations obtained by taking the inverse Fourier transformation from the first and second equations of (3.2.12), provided that $\Phi(t)$ is *analytic* at $t=0$. As long as the response function $\Phi(t)$ is determined by the dynamical motion, strictly it ought to be analytic at $t=0$, but the approximate response function, e.g., that of the Debye type (3.3.8) extended into $t<0$ by assuming evenness, has a cusp at $t=0$ and is not analytic. Corresponding to such discontinuities of the approximate response functions at $t=0$, the integrals in the sum rules (3.7.11, 12) diverge. Sum rules generalized to the form of (3.7.8, 10) are, however, also valid for such approximate response functions and the corresponding admittances.

We can also use these moment sum rules to ascertain how poor a given approximate expression of the complex admittance is, substituting accurate values of the response function and its derivatives calculated at $t=0$ by the statistical mechanics of thermal equilibrium states; we examine up to which order of the sum rules the approximate complex admittance in question satisfies. We can use them not only to make the degree of poorness in approximation at high frequencies clear, but also to improve the degree of approximation. Namely, an approximate expression in the phenomenological theory is in many cases valid at low frequencies, so that by inserting in advance a certain number of parameters into the approximate expression and determining them so as to satisfy some moment sum rules, we get an interpolation formula which is consistent at high frequencies. An example is given below.

3.7.2 Non-Markovian Law of Diffusion

We have discussed the dielectric relaxation as a typical example of Debye relaxation in Sect. 3.3. In that case, the rotating electric dipoles were concerned, but the dipoles cannot follow high-frequency external fields because of molecular inertia. The behavior in a short-time interval is determined from the dynamical equation, so that normalized relaxation function ought to be expanded to

$$\psi^{\mathrm{m}}(t) = 1 - \alpha t^2 + \dots, \quad \therefore \lim_{t\to 0} \frac{d\psi^{\mathrm{m}}(t)}{dt} = 0 \tag{3.7.13}$$

and, in turn, the response function should satisfy

$$\lim_{t\to 0} \varphi(t) = 0, \quad \lim_{t\to 0} \frac{d\varphi(t)}{dt} = 2\alpha \text{ being finite} \tag{3.7.14}$$

according to (3.1.12). The approximate expression of the Debye type (3.3.8) does not satisfy these. The corresponding complex admittance (3.3.11) does not satisfy the sum rules (3.7.2, 8). In particular, the latter leads to the divergence of the total loss, as stated by (3.4.15).

For the example of diffusion described in Sect. 3.5.2, the complex admittance (3.5.20) is of the Debye type

$$\chi(\boldsymbol{k}, \omega) = \frac{\chi(\boldsymbol{k}, 0)}{1 - \mathrm{i}\omega\tau_{\mathrm{D}}}, \quad \tau_{\mathrm{D}} = \frac{1}{D\boldsymbol{k}^2}, \tag{3.7.15}$$

if we fix the wave vector $\boldsymbol{k}$, and it is expected to be a poor approximation at high frequencies. Indeed, equations of the same form as (3.5.13, 14) have been applied to heat conduction and spin diffusion phenomena, and there have been attempts to modify the linear law (3.5.14) by comparison with experiments. Since the idea is the same for either of these phenomena, we shall describe it in the form of a modification of the Fick law of diffusion. For thermal conduction and diffusion phenomena we can hardly expect entirely new mechanisms as in the case of dielectric relaxation (3.3.12), and therefore have to modify the linear relation anyway.

The Fick law (3.5.14) corresponds to the linear law (3.5.5), and should be transformed to a law represented in (3.5.1), in which the delay is taken into account, when the time presicision is increased to describe high-frequency phenomena. The law of the type (3.5.5) is valid when the change of state reaches local stationarity and the delay is supposed to appear because it takes a relaxation time τ_{F} to reach stationarity. For dielectric relaxation we have generalized the stationary law (3.3.16) into (3.3.18). In the same way, generalizing the Fick law, let us assume [3.11]

$$\frac{\partial \boldsymbol{j}(\boldsymbol{r}, t)}{\partial t} = -\frac{1}{\tau_{\mathrm{F}}}[\boldsymbol{j}(\boldsymbol{r}, t) + D \operatorname{grad} c(\boldsymbol{r}, t)]. \tag{3.7.16}$$

Although this line of thinking seems very natural because the stationary law appears explicitly in the generalized one, it gives a poor approximation as the frequency increases further, since in (3.7.16) the Debye-type response is assumed with respect to the diffusion current $\boldsymbol{j}$, as for the electric polarization $\boldsymbol{P}_{\mathrm{d}}$ in dielectric relaxation. However, this deterioration in approximation can be reduced appreciably by determining the parameter τ_{F} based on the sum rules.

The improvement of behavior of the concentration $c(\boldsymbol{r}, t)$ at high frequencies due to using (3.7.16) in place of the Fick law (3.5.14) is evident in the fact that the order of the time derivative appearing in the partial differential equation satisfied by $c(\boldsymbol{r}, t)$ rises by one. Indeed, eliminating the diffusion current $\boldsymbol{j}$ from the equation of continuity (3.5.13, 7.16) gives the partial differential equation

$$\left[\frac{\partial^2}{\partial t^2}+\frac{1}{\tau_{\mathrm{F}}}\left(\frac{\partial}{\partial t}-D\frac{\partial^2}{\partial \boldsymbol{r}^2}\right)\right]c(\boldsymbol{r}, t)=0\,, \tag{3.7.17}$$

which becomes the wave equation with the propagation velocity $\sqrt{D/\tau_{\mathrm{F}}}$ at high frequencies, because the $\partial/\partial t$ term can be neglected in comparison with the $\partial^2/\partial t^2$ term. In the case of heat conduction, this wave is observed as the second sound [3.12]. In the original Fick's law, as the concentration gradient changes, it induces change in the diffusion current at the same instant. Therefore, if the concentration is increased suddenly at a point in this system, then a diffusion current is induced without delay. This produces, in turn, the concentration gradient around the point so that the concentration ought to rise at all points of the system. In short, we should have an infinite propagation velocity. This irrationality has been remedied.

Let us determine the complex admittance corresponding to our new diffusion equation (3.7.17) by the method used in Sect. 3.5. Since the time differentiation has become second order, we need as the initial conditions not only the value of concentration, but also the value of its time derivative. The relaxation ought to satisfy an equation such as the second one of (3.7.13) because the translational motion of a molecule has inertia. Following (3.5.18), we set

$$\left[\frac{\partial c(\boldsymbol{r}, t)}{\partial t}\right]_{t=t_0}=0\,. \tag{3.7.18}$$

By taking the Fourier-Laplace transformation (3.5.16) of (3.7.17), we obtain, in place of (3.5.19),

$$\frac{1-\mathrm{i}\omega\tau_{\mathrm{F}}}{-\mathrm{i}\omega+D\boldsymbol{k}^2-\omega^2\tau_{\mathrm{F}}}=\int_0^\infty dt\int d\boldsymbol{r}\exp\left[-\mathrm{i}\left(\boldsymbol{k}\cdot\boldsymbol{r}-\omega t\right)\right]\Psi^{\mathrm{m}}(\boldsymbol{r}, t)\,. \tag{3.7.19}$$

The corresponding complex admittance is

$$\chi(\boldsymbol{k}, \omega) = \frac{\chi(\boldsymbol{k}, 0)}{1 - \mathrm{i}\omega\tau_\mathrm{D} - \omega^2 \tau_\mathrm{D}\tau_\mathrm{F}}. \tag{3.7.20}$$

The third term in the denominator is new with respect to (3.7.15). If the second term in the denominator may be neglected as compared to the third term at high frequencies $|\omega|\tau_\mathrm{F} \gg 1$, the complex admittance (3.7.20) has a pole representing the wave previously referred to and propagating with frequency

$$|\omega| = \frac{1}{\sqrt{\tau_\mathrm{D}\tau_\mathrm{F}}} = \sqrt{\frac{D}{\tau_\mathrm{F}}}\,|\boldsymbol{k}|. \tag{3.7.21}$$

When the second term cannot be neglected, this wave is damped.

Let us determine the parameter τ_F by using the sum rules. Substituting a complex number z such as $\mathrm{Im}\{z\} > 0$ in place of ω in (3.7.20) and expanding in powers of $1/z$ for a large $|z|$ yields the asymptotic expansion

$$\Xi(\boldsymbol{k}, z) = \left[\left(\frac{\mathrm{i}}{z}\right)^2 \frac{1}{\tau_\mathrm{D}\tau_\mathrm{F}} - \left(\frac{\mathrm{i}}{z}\right)^3 \frac{1}{\tau_\mathrm{D}\tau_\mathrm{F}^2} + \dots\right]\chi(\boldsymbol{k}, 0). \tag{3.7.22}$$

This equation corresponds to (3.7.9) generalized so as to include the wave vector $\boldsymbol{k}$. In accord with the first equation of (3.7.14), there is not first-order term with respect to i/z, so that the sum rule (3.7.2) ought to be satisfied. This is a matter of course, because we have chosen the assumption (3.7.18) in that way. However, this initial condition (3.7.18) could be introduced because the order of time differentiation was raised by one. The term of $(\mathrm{i}/z)^2$ can be made to agree with the corresponding term of (3.7.9), provided that τ_F satisfies

$$\frac{\chi(\boldsymbol{k}, 0)}{\tau_\mathrm{D}\tau_\mathrm{F}} = \int d\boldsymbol{r} \exp(-\mathrm{i}\boldsymbol{k}\cdot\boldsymbol{r}) \left[\frac{\partial \Phi(\boldsymbol{r}, t)}{\partial t}\right]_{t=0} \tag{3.7.23}$$

at least for such small wave numbers $|\boldsymbol{k}|$ that the validity of (3.7.16) is guaranteed. The right-hand side of (3.7.23) corresponds to the second equation of (3.7.14), and is calculable by means of the statistical mechanics of thermal equilibrium. Since by determining τ_F via (3.7.23) no undetermined parameters remain, the sum rule (3.7.8) is satisfied, but the sum rule (3.7.10) corresponding to the term of $(\mathrm{i}/z)^2$ will not in general be satisfied.

4. Statistical Mechanics of Linear Response

We have seen how linear irreversible processes can generally be described by the response function, the relaxation function, and especially the wave-vector-dependent complex admittance. The formulas by which these functions can be evaluated are derived from statistical mechanics, based on assumptions about the material structure. Making use of these formulas, we can connect these functions with new quantities, i.e., the correlation functions. The fluctuation-dissipation theorem developed from the Einstein relation and the Nyquist formula, as described in Sect. 1.6, will thus be generally proved. Statistical mechanics of equilibrium states, as discussed in [4.1], has offered us a means to understand static properties of materials by providing methods of calculating, e.g., the specific heat, the static dielectric constant and the static magnetic susceptibility. Were dynamical quantities such as the response function calculable, we would have at hand an even more powerful theory. This is the problem we are going to treat in this and subsequent chapter.

The first four sections discuss the reponse to mechanical external forces, and Sect. 4.5 gives an example of the internal field. Thermal internal forces are discussed in Sect. 4.6.

4.1 Static Response to External Force

Consider a particle system and denote by (p, q) the whole set of the momenta and coordinates of the particles. If the system is acted upon by an external force, the Hamiltonian of the system is a function of an external parameter x which represents the force. Then

$$\hat{A} = -\frac{\partial}{\partial x}\hat{\mathscr{H}}(p, q; x) \tag{4.1.1}$$

is the corresponding displacement. The circumflex ˆ indicates a dynamical variable belonging to the system, or in quantum-mechanical terminology, an operator. If the parameter x is changed by a small amount δx around its standard value x_0, the Hamiltonian can be assumed to be

$$\hat{\mathscr{H}}(p, q; x_0 + \delta x) = \hat{\mathscr{H}}(p, q) - \hat{A}(p, q)\,\delta x \tag{4.1.2}$$

to first order in δx. We shall write X for δx further below in this chapter and call it the external force. The parameter x_0 in the operator on the right-hand side of (4.1.2) is suppressed for simplicity. The force can be a function of time t, when the last term of (4.1.2) is the perturbation due to the force

$$\hat{\mathscr{H}}_{\mathrm{ext}}(t) = -\hat{A}(p, q)\, X(t)\,. \tag{4.1.3}$$

More generally, the force can be a vector or there may be more than one kind of force. Then (4.1.2) is generalized to

$$\hat{\mathscr{H}}_{\mathrm{t}}(p, q; X_1, X_2, \ldots) = \hat{\mathscr{H}}(p, q) - \sum_j A_j(p, q)\, X_j(t)\,. \tag{4.1.4}$$

For example, X can be an electric or magnetic field, when $\hat{A}$ is the electric or magnetic polarization. A force is called *mechanical* if its effect is represented by (4.1.4 or 2). Other kinds of forces exist which cannot be expressed in this manner. Inhomogeneities of temperature or chemical potential can be controlled externally and produce forces driving heat flow or material current, which can be called *thermal* internal forces. For the present we consider only mechanical forces, returning to the former further below in this chapter.

4.1.1 Static Admittance and the Canonical Correlation

Consider the static response of a system to a static force which is constant in time. If the system is in thermal equilibrium in contact with a heat reservoir of absolute temperature T, the statistical state of the system is described in terms of a canonical ensemble. The corresponding density matrix is given by

$$\hat{\varrho}_{\mathrm{eq}}(\delta x) = \exp\{\beta[F(\delta x) - \hat{\mathscr{H}} + A\,\delta x]\}\,, \tag{4.1.5}$$

where

$$\exp[-\beta F(\delta x)] = \mathrm{Tr}\,\{\exp[-\beta(\hat{\mathscr{H}} - \hat{A}\,\delta x)]\}\,.$$

Here $\beta = 1/kT$ and F is the Helmholtz free energy, and Tr denotes the diagonal sum of a matrix or the trace of an operator. In classical theory the trace operation goes over to the integration over the phase space. We are interested in a physical quantity $\hat{B}$ of the system. The equilibrium expectation of $\hat{B}$ is given by

$$\langle \hat{B} \rangle_{\delta x} = \mathrm{Tr}\,\{\hat{B}\,\hat{\varrho}_{\mathrm{eq}}(\delta x)\} \tag{4.1.6}$$

in the presence of the force δx and that in the absence of the force is

$$\langle \hat{B} \rangle \equiv \langle \hat{B} \rangle_0 = \mathrm{Tr}\,\{\hat{B}\,\varrho_{\mathrm{eq}}\} = \mathrm{Tr}\,\{\hat{B}\exp(-\beta\hat{\mathscr{H}})\}/\mathrm{Tr}\,\{\exp(-\beta\hat{\mathscr{H}})\}\,, \tag{4.1.7}$$

where

$$\hat{\varrho}_{eq} = \hat{\varrho}_{eq}(0)$$

for brevity, and the average $\langle\ \rangle$ hereafter includes $\hat{\varrho}_{eq}$ unless otherwise stated. The difference $\langle \hat{B} \rangle_{\delta x} - \langle \hat{B} \rangle_0$ is the static response of $\hat{B}$ to the force δx, which is conjugate to $\hat{A}$. The linear response defines the static isothermal susceptibility χ_{BA}^T by

$$\delta\langle \hat{B} \rangle = \langle \hat{B} \rangle_{\delta x} - \langle \hat{B} \rangle = \chi_{BA}^T\, \delta x\,. \tag{4.1.8}$$

To calculate χ_{BA}^T, there is a useful identity for an exponential function of two operators

$$\exp[\beta(\hat{a}+\hat{b})] = \exp(\beta\hat{a}) \left(1 + \int_0^\beta d\lambda \exp(-\lambda\hat{a})\, \hat{b} \exp[\lambda(\hat{a}+\hat{b})]\right). \tag{4.1.9}$$

This holds for arbitrary operators $\hat{a}$ and $\hat{b}$ which are not commutable and is easily confirmed by multiplying both sides from the left with $\exp(-\beta\hat{a})$ and differentiating with respect to β. To first order in $\hat{b}$, we may ignore $\hat{b}$ in the last exponential function, $\exp[\lambda(\hat{a}+\hat{b})]$. By making use of this identity, the density matrix (4.1.5) is expanded in δx:

$$\hat{\varrho}_{eq}(\delta x) = \hat{\varrho}_{eq} \left(1 + \int_0^\beta d\lambda \exp(\lambda\hat{\mathscr{H}})\, \Delta\hat{A} \exp(-\lambda\hat{\mathscr{H}})\, \delta x + \ldots\right), \tag{4.1.10}$$

where

$$\Delta\hat{A} = \hat{A} - \langle\hat{A}\rangle, \quad \langle\hat{A}\rangle = \mathrm{Tr}\{\hat{\varrho}_{eq}\hat{A}\}\,.$$

Thus

$$\chi_{BA}^T = \beta\langle \Delta\hat{B}; \Delta\hat{A}\rangle, \tag{4.1.11}$$

where we have introduced the *canonical correlation* of two quantities $\hat{a}$ and $\hat{b}$ by

$$\begin{aligned}\langle\hat{a};\hat{b}\rangle = \beta^{-1} \int_0^\beta d\lambda\, \mathrm{Tr}\{&\exp(-\beta\hat{\mathscr{H}}) \\ &\times \exp(\lambda\hat{\mathscr{H}})\, \hat{a} \exp(-\lambda\hat{\mathscr{H}})\, \hat{b}\}/\mathrm{Tr}\{\exp(-\beta\hat{\mathscr{H}})\}\end{aligned} \tag{4.1.12}$$

for the unperturbed Hamiltonian $\hat{\mathscr{H}}$. Note that the canonical correlation is symmetric, i.e.

$$\langle\hat{a};\hat{b}\rangle = \langle\hat{b};\hat{a}\rangle \tag{4.1.13}$$

and is positive, i.e.

$$\langle\hat{a};\hat{a}\rangle > 0 \tag{4.1.14}$$

if $\hat{a}$ is Hermitian. These are most easily seen by writing the explicit form of the canonical correlation taking the matrix representation diagonalizing the Hamiltonian $\hat{\mathscr{H}}$. Symmetry is not guaranteed if the correlation is defined simply by $\langle \hat{a}\hat{b} \rangle$.

In the classical limit $\hbar \to 0$, the trace operation is reduced to an integral over the phase space and the noncommutability of operators is no longer important. Thus, for (4.1.11), simply

$$\begin{aligned}\chi^T_{BA} &= \beta \langle \Delta B(p,q)\, \Delta A(p,q) \rangle \qquad (4.1.15)\\ &= \int d\Gamma \exp[-\beta \mathscr{H}(p,q)]\, \Delta B(p,q)\, \Delta A(p,q) / \int d\Gamma \exp[-\beta \mathscr{H}(p,q)],\end{aligned}$$

$d\Gamma = dp\, dq$ being the volume element of the phase space.

Equation (4.1.15) was first obtained by *Kirkwood* [4.2] for $B = A$. This shows that the static admittance χ^T_{AA}, for example, the electric or magnetic susceptibility, is determined by the thermal fluctuation of the displacement A in the *absence* of external force. This is a generalization of the Debye formula of the induced electric or magnetic polarization of noninteracting dipoles. Formula (4.1.11) is more general and applies to classical as well as quantal systems. It determines the response of the displacement under the quasi-static action of the external force in an isothermal condition.

Similar expressions can be obtained for adiabatic susceptibilities. If the parameter x changes by a small amount δx under an adiabatic condition, the temperature also changes. The corresponding change of $\beta = 1/kT$ is denoted by $\delta\beta$. The density matrix after the adiabatic change of x becomes

$$\begin{aligned}\hat{\varrho}' &= \exp[-(\beta + \delta\beta)(\hat{\mathscr{H}} - \hat{A}\,\delta x)] / \mathrm{Tr}\{\exp[-(\beta + \delta\beta)(\hat{\mathscr{H}} - \hat{A}\,\delta x)]\}\\ &= \hat{\varrho}_{\mathrm{eq}} \left(1 - \delta\beta\, \Delta\hat{\mathscr{H}} + \int_0^\beta d\lambda \exp(\lambda \hat{\mathscr{H}})\, \Delta\hat{A} \exp(-\lambda \hat{\mathscr{H}})\, \delta x + \dots \right), \qquad (4.1.16)\end{aligned}$$

where $\Delta\hat{\mathscr{H}} = \hat{\mathscr{H}} - \Delta\langle \hat{\mathscr{H}} \rangle$. The change $\delta\beta$ is related to δx by the adiabatic condition

$$\delta\langle \hat{\mathscr{H}} \rangle_{\mathrm{ad}} + \langle \hat{A} \rangle\, \delta x = 0\,, \qquad (4.1.17)$$

namely that the increase of the internal energy is equal to the work done by the external force. Now the energy change

$$\delta\langle \hat{\mathscr{H}} \rangle_{\mathrm{ad}} = \mathrm{Tr}\{\hat{\varrho}' (\hat{\mathscr{H}} - \hat{A}\,\delta x)\} - \mathrm{Tr}\{\hat{\varrho}_{\mathrm{eq}} \hat{\mathscr{H}}\}$$

is calculated using the expansion (4.1.16) to be

$$\delta\langle \hat{\mathscr{H}} \rangle_{\mathrm{ad}} = -\langle \hat{A} \rangle\, \delta x - \delta\beta \langle \Delta\hat{\mathscr{H}}\, \Delta\hat{\mathscr{H}} \rangle + \beta \langle \Delta\hat{\mathscr{H}}\, \Delta\hat{A} \rangle\, \delta x\,.$$

Therefore the adiabatic condition (4.1.17) yields

$$\delta\beta/\beta = \langle \Delta\hat{\mathscr{H}}\, \Delta\hat{A} \rangle\, \delta x / \langle \Delta\hat{\mathscr{H}}\, \Delta\hat{\mathscr{H}} \rangle. \qquad (4.1.18)$$

The change of the expectation value of $\hat{B}$ is found from (4.1.16) to be

$$\delta\langle\hat{B}\rangle_{\mathrm{ad}} = \beta\langle\Delta\hat{A};\Delta\hat{B}\rangle\,\delta x - \delta\beta\langle\Delta\hat{\mathscr{H}}\,\Delta\hat{B}\rangle .$$

Inserting (4.1.18) gives the expression for adiabatic susceptibility:

$$\chi_{BA}^{\mathrm{ad}} = \beta\left(\langle\Delta\hat{A};\Delta\hat{B}\rangle - \frac{\langle\Delta\hat{A}\,\Delta\hat{\mathscr{H}}\rangle\langle\Delta\hat{\mathscr{H}}\,\Delta\hat{B}\rangle}{\langle\Delta\hat{\mathscr{H}}\,\Delta\hat{\mathscr{H}}\rangle}\right). \tag{4.1.19}$$

In the above expressions, canonical correlations containing $\Delta\hat{\mathscr{H}}$ need not be distinguishd from ordinary ones since $\Delta\hat{\mathscr{H}}$ commutes with $\hat{\mathscr{H}}$. By virtue of the simple relation

$$\frac{\partial}{\partial\beta}\langle\hat{A}\rangle = \frac{\partial}{\partial\beta}\,\mathrm{Tr}\,\{\hat{\varrho}_{\mathrm{eq}}\bar{A}\} = -\langle\Delta\hat{\mathscr{H}}\,\Delta\hat{A}\rangle ,$$

which holds for an arbitrary quantity A, (4.1.19) is nothing but the thermodynamic equation

$$\chi_{BA}^{\mathrm{ad}} = \chi_{BA}^{T} - \frac{T}{C}\frac{\partial}{\partial T}\langle\hat{A}\rangle_x \frac{\partial}{\partial T}\langle\hat{B}\rangle_x , \tag{4.1.20}$$

where C is the heat capacity.

4.2 Dynamic Response to External Force

4.2.1 The Response Function and the Poisson Bracket

When the external force $X(t)$ in (4.1.3) changes with time, the corresponding change of the density matrix must be traced [4.3]. This is done by using the equation of motion

$$\frac{\partial}{\partial t}\hat{\varrho}(t) = \mathrm{i}\,[\mathscr{L} + \mathscr{L}_{\mathrm{ext}}(t)]\,\hat{\varrho}(t) , \tag{4.2.1}$$

where $\mathrm{i}\mathscr{L}$ and $\mathrm{i}\mathscr{L}_{\mathrm{ext}}(t)$ are the quantal Liouville operators introduced in Sect. 2.6:

$$\begin{aligned} \mathrm{i}\mathscr{L}\,\hat{\varrho} &= (\hat{\mathscr{H}},\varrho) \equiv [\hat{\mathscr{H}},\varrho]/\mathrm{i}\hbar \\ \mathrm{i}\mathscr{L}_{\mathrm{ext}}\,\varrho &= (\hat{\mathscr{H}}_{\mathrm{ext}},\varrho) \equiv [\hat{\mathscr{H}}_{\mathrm{ext}},\hat{\varrho}]/\mathrm{i}\hbar . \end{aligned} \tag{4.2.2}$$

Construction of a commutator is denoted by $[\hat{a},\hat{b}]$ and the corresponding quantal Poisson bracket by $(\hat{a},\hat{b})$, namely

$$(\hat{a},\hat{b}) \equiv \hat{a}\hat{b} - \hat{b}\hat{a} , \quad (\hat{a},\hat{b}) = [\hat{a},\hat{b}]/\mathrm{i}\hbar .$$

In the classical limit, the density matrix becomes the phase space distribution $f(p, q, t)$ which follows the equation of motion

$$\frac{\partial}{\partial t} f(p, q, t) = \mathrm{i}\,[\mathscr{L} + \mathscr{L}_{\mathrm{ext}}(t)]\, f(p, q, t)\,. \tag{4.2.3}$$

This is formally identical with (4.2.1). The difference is that the Liouville operators are defined by the corresponding classical Poisson bracket

$$\mathrm{i}\mathscr{L} f \equiv (\mathscr{H}, f) = \sum \left(\frac{\partial \mathscr{H}}{\partial q} \frac{\partial}{\partial p} - \frac{\partial \mathscr{H}}{\partial p} \frac{\partial}{\partial q} \right) f\,, \tag{4.2.4}$$

where the summation is taken over the whole set of canonical variables. Thus we can treat the quantal and classical systems in parallel.

Equation (4.2.1) is equivalent to the integral equation

$$\hat{\varrho}(t) = \exp[\mathrm{i}(t - t_0)\mathscr{L}]\, \hat{\varrho}(t_0) + \int_{t_0}^{t} dt' \exp[\mathrm{i}(t - t')\mathscr{L}]\, \mathrm{i}\mathscr{L}_{\mathrm{ext}}(t')\, \hat{\varrho}(t')\,, \tag{4.2.5}$$

as is easily seen by differentiating both sides with respect to t. Note that the exponential operator of the quantal Liouville operator in the above expression means

$$\exp(\mathrm{i}t\mathscr{L})\hat{g} = \exp(t\hat{\mathscr{H}}/\mathrm{i}\hbar)\, \hat{g}\, \exp(-t\hat{\mathscr{H}}/\mathrm{i}\hbar)\,. \tag{4.2.6}$$

Since we are interested in near-equilibrium states driven by external force, we may assume that the external perturbation started to work in the infinite past, i.e., $t_0 \to -\infty$, when the system was in equilibrium at a certain temperature. Thus we assume

$$\hat{\varrho}(-\infty) = \hat{\varrho}_{\mathrm{eq}} = \exp(-\beta\hat{\mathscr{H}})/\mathrm{Tr}\,\{\exp(-\beta\hat{\mathscr{H}})\}\,, \tag{4.2.7}$$

namely that the density matrix was initially canonical. Noticing that $\hat{\varrho}_{\mathrm{eq}}$ remains invariant with respect to the evolution (4.2.6), then (4.2.5) gives an approximate solution to first order in the perturbation $\mathrm{i}\mathscr{L}_{\mathrm{ext}}$. Thus

$$\hat{\varrho}(t) = \hat{\varrho}_{\mathrm{eq}} + \int_{-\infty}^{t} dt' \exp[(t - t')\hat{\mathscr{H}}/\mathrm{i}\hbar]\, (\hat{\mathscr{H}}_{\mathrm{ext}}(t'), \hat{\varrho}_{\mathrm{eq}})\, \exp[-(t - t')\hat{\mathscr{H}}/\mathrm{i}\hbar] + \dots \tag{4.2.8}$$

and, in the classical limit,

$$f(p, q, t) = f_{\mathrm{eq}}(p, q) + \int_{-\infty}^{t} dt' \exp[\mathrm{i}(t - t')\mathscr{L}]\, (\mathscr{H}_{\mathrm{ext}}(t'), f_{\mathrm{eq}}(p, q)) + \dots\,. \tag{4.2.9}$$

The statistical expectation of a quantity $\hat{B}$ is obtained from (4.2.8). Its deviation from the equilibrium value is given by

$$\begin{aligned}\delta\bar{B}(t) &= \mathrm{Tr}\,\{\hat{\varrho}(t)\,\hat{B}\} - \mathrm{Tr}\,\{\hat{\varrho}_{\mathrm{eq}}\,\hat{B}\}\\ &= \int_{-\infty}^{t} dt'\,\mathrm{Tr}\,\{(\hat{\mathscr{H}}_{\mathrm{ext}}(t'),\,\hat{\varrho}_{\mathrm{eq}})\,\Delta\hat{B}(t-t')\}\\ &= \int_{-\infty}^{t} dt'\,\mathrm{Tr}\,\{\hat{\varrho}_{\mathrm{eq}}\,(\Delta\hat{B}(t-t'),\hat{\mathscr{H}}_{\mathrm{ext}}(t'))\}\,,\end{aligned} \tag{4.2.10}$$

where

$$\begin{aligned}&\Delta\hat{B} = \hat{B} - \langle\hat{B}\rangle \quad \text{and}\\ &\Delta\hat{B}(t) = \exp(-t\hat{\mathscr{H}}/\mathrm{i}\hbar)\,\Delta\hat{B}\,\exp(t\hat{\mathscr{H}}/\mathrm{i}\hbar)\end{aligned} \tag{4.2.11}$$

is the Heisenberg operator obeying the equation of motion

$$\frac{d}{dt}\Delta B(t) = (\Delta\hat{B}(t),\hat{\mathscr{H}})\,. \tag{4.2.12}$$

Expression (4.2.10) is obtained by cyclically changing the order of operators inside Tr. Substituting (4.1.3) for $\hat{\mathscr{H}}_{\mathrm{ext}}(t)$ and noting that A in the Poisson bracket can be replaced by $\Delta\hat{A} = \hat{A} - \langle\hat{A}\rangle$, then

$$\delta\bar{B}(t) = \int_{-\infty}^{t} dt'\,\Phi_{BA}(t-t')\,X(t')\,, \tag{4.2.13}$$

where

$$\begin{aligned}\Phi_{BA}(t) &= \mathrm{Tr}\,\{(\hat{\varrho}_{\mathrm{eq}},\,\Delta\hat{A})\,\Delta\hat{B}(t)\}\\ &= \mathrm{Tr}\,\{\varrho_{\mathrm{eq}}\,(\Delta\hat{A},\,\Delta\hat{B}(t))\} = \langle(\Delta\hat{A},\,\Delta\hat{B}(t))\rangle\end{aligned} \tag{4.2.14}$$

is the response function expressing the response of B to the pulsed force $X(t) = \delta(t)$.

The same expression also holds for a classical case, the Poisson bracket being replaced by the classical one. The derivation is as follows. Using (4.2.9) we calculate the response function as

$$\begin{aligned}\Phi_{BA}(t) &= \iint dp\,dq\,\Delta B(p,q)\,\exp(\mathrm{i}t\mathscr{L})(-A(p,q),f_{\mathrm{eq}}(p,q))\\ &= \iint dp\,dq\,(f_{\mathrm{eq}}(p,q),A(p,q))\,\exp(-\mathrm{i}t\mathscr{L})\,\Delta B(p,q)\\ &= \iint dp\,dq\,(f_{\mathrm{eq}}(p,q),\Delta A(p,q))\,\Delta B(p_t,q_t)\\ &= \iint dp\,dq\,f_{\mathrm{eq}}(p,q)\,(\Delta A(p,q),\Delta B(p_t,q_t))\\ &= \langle(\Delta A,\,\Delta B(t))\rangle\,.\end{aligned} \tag{4.2.15}$$

To go to the second line from the first, partial integrations are carried out to make the Liouville operator operate on $\Delta B(p, q)$. The we use

$$\exp(-\mathrm{i}t\mathscr{L})\, G(p, q) = G(p_t, q_t)\,,$$

where (p_t, q_t) is the image of the initial phase point (p, q) determined by the Hamiltonian equation of motion (2.9.7),

$$\dot{p}_t = -\frac{\partial}{\partial q_t}\mathscr{H}(p_t, q_t)\,, \quad \dot{q}_t = \frac{\partial}{\partial p_t}\mathscr{H}(p_t, q_t)\,.$$

From the third to the fourth line, partial integration is again performed. The boundary values are assumed to vanish because the distribution function f should behave properly at the boundaries. Cyclic change of the order of operators inside the trace in quantal cases thus corresponds to partial integration for phase space integrals.

Thus the response function (4.2.14) is essentially the statistical expectation of the Poisson bracket of two quantities, namely $\hat{A}$, which is conjugate to the force, and $\hat{B}$, to be observed. It is important to note this physical meaning of the Poisson bracket. If the force has no influence on $\hat{B}$ at later times, $\hat{B}(t)$ commutes with $\hat{A}$.

For a periodic force, the response $\delta\bar{B}(t)$ is periodic. The complex admittance is obtained from (4.2.10) in the form

$$\chi_{BA}(t) = \int_0^\infty \Phi_{BA}(t)\, \mathrm{e}^{\mathrm{i}\omega t}\, dt\,. \qquad (4.2.16)$$

Equation (4.2.13) corresponds to (3.1.2), but it lacks the instantaneous response, showing that the system generally responds to the external force with some retardation. If the time precision of our observation is crude in a certain time scale, the corresponding part of the response may be regarded as almost instantaneous in the approximation and is included in χ_∞. It should be noted, however, that an instantaneous response can be present intrinsically if the observed quantity $\hat{B}$ explicitly depends on the force X. For a weak force X, B may be expanded as

$$\hat{B}(X) = \hat{B}(0) + \left(\frac{\partial \hat{B}}{\partial x}\right)_0 X + \dots\,.$$

The first term $\hat{B}(0)$ is treated as before, yielding only a retarded response, whereas the second term gives the instantaneous reponse

$$\chi_\infty X = \left\langle \frac{\partial \hat{B}}{\partial x} \right\rangle_0 X\,.$$

An example of this is the diamagnetic current responding to a time-dependent magnetic field.

4.2.2 Kubo Formula

Note the simple identity by *Kubo* [4.3]

$$(\exp[-\beta\hat{\mathscr{H}}], \hat{a}) = \exp(-\beta\hat{\mathscr{H}}) \int_0^\beta d\lambda\, \hat{a}(-\mathrm{i}\hbar\lambda)$$

$$= \exp(-\beta\hat{\mathscr{H}}) \int_0^\beta d\lambda \exp(\lambda\hat{\mathscr{H}})(\hat{a}, \hat{\mathscr{H}}) \exp(-\lambda\hat{\mathscr{H}}), \quad (4.2.17)$$

where $\hat{a}(-\mathrm{i}\hbar\lambda)$ is the Heisenberg operator (4.2.11) corresponding to $\hat{a}$ but with an imaginary time. Using this identity, (4.2.14) becomes

$$\Phi_{BA}(t) = \mathrm{Tr}\left\{\hat{\varrho}_{\mathrm{eq}} \int_0^\beta \Delta\hat{\dot{A}}(-\mathrm{i}\hbar\lambda)\, \Delta\hat{B}(t)\, d\lambda\right\}.$$

For later use, summarize here response function expressions:

$$\begin{aligned}\Phi_{BA}(t) &= \langle(\Delta\hat{A}, \Delta\hat{B}(t))\rangle \\ &= \beta\langle\Delta\hat{\dot{A}}; \Delta\hat{B}(t)\rangle \\ &= -\beta\langle\Delta\hat{A}; \Delta\hat{\dot{B}}(t)\rangle. \end{aligned} \quad (4.2.18)$$

Note the stationarity of a canonical correlation:

$$\langle\hat{a}(t_1); \hat{b}(t_2)\rangle = \langle\hat{a}(t_1+\tau); \hat{b}(t_2+\tau)\rangle.$$

This is seen to be true because the expectation is defined in terms of the equilibrium density matrix $\hat{\varrho}_{\mathrm{eq}}$. Since the right-hand side is independent of τ,

$$\langle\hat{\dot{a}}(t_1); \hat{b}(t_2)\rangle + \langle\hat{a}(t_1); \hat{\dot{b}}(t_2)\rangle = 0. \quad (4.2.19)$$

In the classical limit, (4.2.17) changes to

$$(\exp[-\beta\mathscr{H}], a) = \beta\exp(-\beta\mathscr{H})(a, \mathscr{H}) = \beta\exp(-\beta\mathscr{H})\,\dot{a}$$

and (4.2.18) to

$$\Phi_{BA}(t) = \beta\langle\Delta\dot{A}\,\Delta B(t)\rangle = -\beta\langle\Delta A(0)\Delta\dot{B}(t)\rangle. \quad (4.2.20)$$

The canonical correlation in the classical limit is simply the canonical average of the product of two quantities.

Let us consider the case where $\hat{B}$ is the current $\hat{J}_\mu$ corresponding to the displacement $\hat{A}_\mu$ appearing in the perturbing term in (4.1.4). It is defined by an equation of the same form as (3.2.22) in phenomenological theory:

$$\hat{J}_\mu(t) = \frac{d}{dt}\hat{A}_\mu(t) \equiv [\hat{A}_\mu(t), \hat{\mathscr{H}}]/\mathrm{i}\hbar. \quad (4.2.21)$$

Here, obviously the equilibrium average vanishes

$$\langle\hat{J}_\mu\rangle = 0. \quad (4.2.22)$$

The response function is obtained from (4.2.18)

$$\Phi_{\mu\nu}(t) = \beta \langle \hat{J}_\nu, \hat{J}_\mu(t) \rangle . \tag{4.2.23}$$

The complex conductivity is thus given by

$$\sigma_{\mu\nu}(\omega) = \beta \int_0^\infty dt\, e^{i\omega t} \langle \hat{J}_\nu ; \hat{J}_\mu(t) \rangle . \tag{4.2.24}$$

This is often referred to as the *Kubo formula of conductivity.*

The simplest example is the complex mobility of a Brownian particle

$$\mu(\omega) = \beta \int_0^\infty dt\, e^{i\omega t} \langle \hat{u}(0) ; \hat{u}(t) \rangle , \tag{4.2.25}$$

where $\hat{u}$ is the velocity of the particle. In the classical limit this reduces to

$$\mu(\omega) = \beta \int_0^\infty dt\, e^{i\omega t} \langle u(0)\, u(t) \rangle$$

which was called the first fluctuation-dissipation theorem in Chap. 2. Thus the present treatment provides us with the proof of the theorem. The treatment in Chap. 2 applies to quantal cases as well, simply by reinterpreting correlation functions as canonical correlations.

Next let us consider the case where B is the displacement A itself. For a pulsed force $X_\nu = \delta(t)$ conjugate to $\hat{A}_\nu$, the response of $\hat{A}_\mu$ at later times is given by

$$\Phi_{\mu\nu}(t) = \langle ((\hat{A}_\nu(0), \hat{A}_\mu(t))) \rangle , \tag{4.2.26}$$

which is valid for an arbitrary equilibrium ensemble. For the canonical ensemble, we can use (4.2.18) to rewrite the above expression as

$$\Phi_{\mu\nu}(t) = -\beta \frac{d}{dt} \langle \Delta\hat{A}_\nu(0) ; \Delta\hat{A}_\mu(t) \rangle .$$

Substituting this into the definition (3.1.11) of the relaxation function yields

$$\Psi_{\mu\nu}(t) = \beta \langle \Delta\hat{A}_\nu(0) ; \Delta\hat{A}_\mu(t) \rangle - C , \tag{4.2.27}$$

where the constant C denotes the limiting value of the first term on the right-hand side at $t \to \infty$, namely

$$\begin{aligned} C &= \lim_{t\to\infty} \beta \langle \Delta\hat{A}(0) ; \Delta\hat{A}_\mu(t) \rangle \\ &= \beta \lim_{\varepsilon\to 0} \varepsilon \int_0^\infty e^{-\varepsilon t} \langle \Delta\hat{A}_\nu(0) ; \Delta\hat{A}_\mu(t) \rangle\, dt . \end{aligned} \tag{4.2.28}$$

Now note that

$$\lim_{\varepsilon \to 0} \varepsilon \int_0^\infty \exp(-\varepsilon t + \mathrm{i} v t)\, dt = 1 \quad \text{if } v = 0$$
$$= 0 \quad \text{otherwise}.$$

Therefore, only the zero-frequency part of $\Delta \hat{A}_\mu(t)$ remains in (4.2.28). Superscript 0 denotes the zero-frequency part of a Heisenberg operator. Then we find that

$$C = \beta \langle \Delta \hat{A}_v^0; \Delta \hat{A}_\mu^0 \rangle = \beta \langle \Delta \hat{A}_v^0 \, \Delta A_\mu^0 \rangle$$

since the canonical correlation is reduced to the ordinary correlation when the operators inside commute with the Hamiltonian. Therefore the relaxation function (4.2.27) becomes

$$\begin{aligned} \Psi_{\mu v}(t) &= \beta \{\langle \Delta \hat{A}_v(0); \Delta \hat{A}_\mu(t) \rangle - \langle \Delta \hat{A}_\mu^0 \, \Delta \hat{A}_v^0 \rangle\} \\ &= \beta \langle \Delta \hat{A}_v'; \Delta \hat{A}_\mu'(t) \rangle, \end{aligned} \tag{4.2.29}$$

where the prime denotes the nonzero-frequency part of the operator,

$$\hat{A}' = \hat{A} - \hat{A}^0.$$

The relaxation function is expressed in terms of the canonical correlation of the nonzero-frequency parts of the relevant operators.

The complex susceptibility (4.2.16) can be expressed in terms of the relaxation function. With the use of

$$\Phi_{\mu v}(t) = -\frac{d}{dt} \Psi_{\mu v}(t), \quad \Psi_{\mu v}(\infty) = 0,$$

we get

$$\chi_{\mu v}(\omega) = \chi_{\mu v}^{\text{iso}} + \mathrm{i}\omega \int_0^\infty \mathrm{e}^{\mathrm{i}\omega t} \Psi_{\mu v}(t)\, dt, \tag{4.2.30}$$

where

$$\chi_{\mu v}^{\text{iso}} = \chi_{\mu v}(0) = \Psi_{\mu v}(0) = \beta(\langle \Delta \hat{A}_\mu; \Delta \hat{A}_v \rangle - \langle \Delta A_\mu \, \Delta A_v \rangle) \tag{4.2.31}$$

is called the *isolated susceptibility* since the system is isolated from the outer world throughout its evolution under the influence of the applied force, as is implied by (4.2.1) on which our whole derivation is based. A magnetic system exemplifies this formalism, when the canonical correlation of the magnetic moments gives the dynamic susceptibilities.

The isolated susceptibility is generally not equal to the isothermal susceptibility (4.1.11) nor to the adiabatic susceptibility (4.1.19). Inequalities [4.4]

$$\chi_{\mu\mu}^T > \chi_{\mu\mu}^{\text{ad}} > \chi_{\mu\mu}^{\text{iso}} \tag{4.2.32}$$

exist between these three susceptibilities. The first inequality is seen from (4.1.11, 19). The second is seen to be true in the following way.

The unperturbed motion governed by the Hamiltonian may have a certain number of constants of motion. The whole set of dynamical quantities corresponding to these constants of motion is denoted by $\{H_j\}$, which can be chosen to satisfy the condition

$$\langle \hat{H}_j \hat{H}_k \rangle = \langle \hat{H}_j^2 \rangle \, \delta_{jk}$$

by Schmidt's procedure of orthogonalization. Then the diagonal part of an operator $\hat{a}$ is a linear combination of H_j's

$$\hat{a}^{\mathrm{d}} = \sum_j \hat{H}_j \langle \hat{H}_j \hat{a} \rangle / \langle \hat{H}_j^2 \rangle \, .$$

Using such expressions for $\Delta\hat{A}_\mu$ and $\Delta\hat{A}_\nu$ in (4.2.31) gives

$$\chi_{\mu\nu}^{\mathrm{iso}} = \chi_{\mu\nu}^{T} - \beta \sum_j \langle \hat{H}_j \, \Delta\hat{A}_\mu \rangle \langle \hat{H}_j \, \Delta\hat{A}_\nu \rangle / \langle \hat{H}_j^2 \rangle . \tag{4.2.33}$$

One $\hat{H}_j$, say $\hat{H}_0$, may be taken as $\Delta\hat{\mathscr{H}}$ which appeared in (4.1.19). Therefore

$$\chi_{\mu\nu}^{\mathrm{iso}} = \chi_{\mu\nu}^{\mathrm{ad}} - \beta \sum{}' \langle \hat{H}_j \, \Delta\hat{A}_\mu \rangle \langle \hat{H}_j \, \Delta\hat{A}_\nu \rangle / \langle \hat{H}_j^2 \rangle , \tag{4.2.34}$$

where the summation over the constants of motion excludes $\Delta\hat{H}_0$. This proves (4.2.32) for $\mu = \nu$. More generally, the tensors

$$\chi_{\mu\nu}^{T} - \chi_{\mu\nu}^{\mathrm{ad}}, \; \chi_{\mu\nu}^{T} - \chi_{\mu\nu}^{\mathrm{iso}}$$

and

$$\chi_{\mu\nu}^{\mathrm{ad}} - \chi_{\mu\nu}^{\mathrm{iso}}$$

are all positive definite.

Expression (4.2.30) is a generalization of the *Kirkwood* formula (4.1.15) to quantal and dynamic cases [4.2.3]. It provides us with a starting point to calculate dynamic susceptibilities on the bases of microscopic theory. As seen in the above, the zero-frequency limit of the calculated susceptibility does not necessarily coincide with a thermodynamic susceptibility. The question depends on what we consider as the system. If the dynamic structure of the system is such that the Hamiltonian $\hat{\mathscr{H}}$ of the system allows no constant of motion other than $\mathscr{H}$ itself, the calculated susceptibility reduces to the adiabatic one as the frequency of the applied force approaches zero. If, furthermore, the relevant quantities $\hat{A}$ are related only to a very small fraction of the degrees of freedom of the system, the difference between isothermal and adiabatic susceptibility is negligible. This is true, for instance, if the heat bath is included in the system, when the heat capacity C becomes practically infinite in (4.1.20). The interaction between the subsystem with which we are primarily concerned and the bath may be treated stochastically, Sect. 2.7.

4.2.3 Initial Values of the Response Function and Its Derivatives

As mentioned in Sect. 3.7, we have to calculate the value of the response function or its time derivatives exactly at the point $t = 0$ in order to use the moment sum rules. Since the response function is generally given by (4.2.18), we need only evaluate the expectation values of the commutator or multiple commutators at the thermal equilibrium state such that

$$\Phi_{\mu\nu}(0) = \langle(\hat{A}_\nu, \hat{B}_\mu)\rangle$$
$$\left[\frac{d^n}{dt^n}\Phi_{\mu\nu}(t)\right]_{t=0} = \langle(\hat{\mathscr{H}}, (\hat{\mathscr{H}}, \dots (\hat{\mathscr{H}}, \hat{A}_\nu) \dots), \hat{B}_\mu)\rangle . \tag{4.2.35}$$

If $\hat{B}_\mu$ commutes with the displacement $\hat{A}_\nu$, then

$$\Phi_{\mu\nu}(0) = 0 , \tag{4.2.36}$$

as assumed in (3.7.14). An obvious example is when $\hat{B}_\mu$ itself is equal to the displacement $\hat{A}_\mu$. If $\hat{B}_\mu$ is the current defined by (4.2.21) $\hat{J}_\mu$, $\Phi_{\mu\nu}(0)$ can have a nonzero value, but its first derivative vanishes. Examples for calculating (4.2.35) are given in Sect. 4.5.

4.3 Symmetry and the Dispersion Relations

The formulas for the response function, the relaxation function, and the complex admittance derived in the previous section can be used to prove various points. In this section, we examine symmetry properties of correlation functions as preliminaries to the following section.

4.3.1 Spectral Function and Its Symmetry

To elucidate the discussions on symmetry and other properties, it is convenient to introduce the spectral function. This function is the Fourier transform of a quantity similar to the correlation function and is introduced by generalizing the fact that the Fourier transform of the correlation function gives the spectral intensity by virtue of the Wiener-Khintchine theorem described in Sect. 1.3.

Consider two dynamical quantities $\hat{F}(p, q)$ and $\hat{G}(p, q)$ for a given system whose Hamiltonian is $\hat{\mathscr{H}}$. We have seen in the previous sections that the canonical correlation plays a fundamental role in linear-response theory. There are, however, a great variety of definitions of correlation functions, depending on the purposes for their use. Green's functions used in many-body theories are, in fact, such correlation functions.

Let us consider the simplest form of a correlation function

$$\langle \hat{F}(t)\, \hat{G}(t') \rangle = \mathrm{Tr}\,\{\hat{\varrho}_{\mathrm{eq}}\, \hat{F}(t)\, \hat{G}(t')\}\,, \tag{4.3.1}$$

where $\hat{F}(t)$ is the Heisenberg operator defined by (4.2.11) and

$$\varrho_{\mathrm{eq}} = \exp(-\beta \hat{\mathscr{H}})/\mathrm{Tr}\,\{\exp(-\beta \hat{\mathscr{H}})\}$$

is the equilibrium density matrix. By virtue of this definition, we obviously have the stationary property

$$\langle \hat{F}(t)\, \hat{G}(t') \rangle = \langle \hat{F}(t-t')\, \hat{G}(0) \rangle\,. \tag{4.3.2}$$

The spectral function for this correlation function is defined by

$$J_{FG}(\omega) = \int_{-\infty}^{\infty} \mathrm{e}^{\mathrm{i}\omega t} \langle \hat{F}(t)\, \hat{G}(0) \rangle\, dt\,. \tag{4.3.3}$$

For convenience we introduce the function

$$j_{FG}(E, E') = \mathrm{Tr}\,\{\delta(E - \hat{\mathscr{H}})\, \hat{F}\, \delta(E' - \hat{\mathscr{H}})\, \hat{G}\} \tag{4.3.4}$$

to write the spectral function as

$$J_{FG}(\omega) = \iint dE\, dE'\, \varrho_{\mathrm{eq}}(E)\, 2\pi\hbar\, \delta(E + \hbar\omega - E')\, j_{FG}(E, E')\,, \tag{4.3.5}$$

where

$$\varrho_{\mathrm{eq}}(E) = \mathrm{e}^{-\beta E}/\mathrm{Tr}\,\{\exp(-\beta \hat{\mathscr{H}})\}\,.$$

Then the correlation function is expressed by

$$\langle \hat{F}(t)\, \hat{G}(t') \rangle = \frac{1}{2\pi} \int_{-\infty}^{\infty} d\omega\, \exp[-\mathrm{i}\omega(t-t')]\, J_{FG}(\omega)\,. \tag{4.3.6}$$

The function $j_{FG}(E, E')$ has the following properties:

1) $$j_{GF}(E', E) = j_{FG}(E, E') \tag{4.3.7}$$

obvious from the definition by cyclic change of operators inside Tr;

2) $$[j_{FG}(E, E')]^* = j_{F^+G^+}(E', E) = j_{G^+F^+}(E, E')\,, \tag{4.3.8}$$

where * denotes the complex conjugate of the inside of the bracket and the dagger the Hermitian conjugate, seen by taking the complex conjugate of (4.3.4);

3) $$\begin{aligned} j_{FG}(E, E'; \boldsymbol{B}) &= \varepsilon_F\, \varepsilon_G\, j_{GF}(E, E', -\boldsymbol{B}) \\ &= \varepsilon_F\, \varepsilon_G\, j_{FG}(E', E, -\boldsymbol{B})\,, \end{aligned} \tag{4.3.9}$$

where $\boldsymbol{B}$ is the magnetic field applied to the system and $\hat{F}$ and $\hat{G}$ are Hermitian operators having time-reversal symmetry such that

$$\hat{F}(-p, q, -\boldsymbol{\sigma}) = \varepsilon_F \hat{F}(p, q, \boldsymbol{\sigma}), \tag{4.3.10}$$

with $\varepsilon_F = \pm 1$ and $p, q,$ and $\boldsymbol{\sigma}$ representing the whole sets of canonical momenta, coordinates and spin variables. This is concerned with time-reversal symmetry. Since some misunderstanding seems to exist about the nature of this symmetry, we shall discuss it a little more in the nonrelativistic Pauli approximation below.

The time-reversal operation means the transformation

$$t \to -t, \; p \to -p, \; \boldsymbol{\sigma} \to -\boldsymbol{\sigma}, \quad \text{and} \quad \boldsymbol{B} \to -\boldsymbol{B}. \tag{4.3.11}$$

A dynamical system is said to possess *time-reversal symmetry* if its Hamiltonian is invariant with regard to time reversal:

$$\hat{\mathscr{H}}(p, q, \boldsymbol{\sigma}, \boldsymbol{B}) = \hat{\mathscr{H}}(-p, q, -\boldsymbol{\sigma}, -\boldsymbol{B}). \tag{4.3.12}$$

The Schrödinger equation is written as

$$\mathrm{i}\hbar \frac{\partial}{\partial t} \varphi = \hat{\mathscr{H}}(p, q, \boldsymbol{\sigma}, \boldsymbol{B}) \varphi, \tag{4.3.13}$$

where the wave function φ is a vector consisting of 2^N components if the system has N Fermi particles, say electrons. Corresponding to a given state $\varphi(q, t, \boldsymbol{B})$, we define its reciprocal state φ by

$$\varphi'(q, t, \boldsymbol{B}) = S\bar{\varphi}(q, -t, -\boldsymbol{B}), \tag{4.3.14}$$

where $\bar{\varphi}$ is the complex conjugate of φ and

$$S = S^{-1} = \prod \sigma_y,$$

the product being taken over all spins. Here we use the customary representation of spin matrices

$$\sigma_x = \begin{pmatrix} 0 & 1 \\ 1 & 0 \end{pmatrix}, \quad \sigma_y = \begin{pmatrix} 0 & -\mathrm{i} \\ \mathrm{i} & 0 \end{pmatrix}, \quad \sigma_2 = \begin{pmatrix} 1 & 0 \\ 0 & -1 \end{pmatrix}$$

and note that

$$\bar{\boldsymbol{\sigma}} = -S\boldsymbol{\sigma}S.$$

Transforming (4.3.13) by (4.3.11) and using the above relation, clearly

$$\mathrm{i}\hbar \frac{\partial}{\partial t} \varphi'(q, t, \boldsymbol{B}) = \hat{\mathscr{H}}(-p, q, -\boldsymbol{\sigma}, -\boldsymbol{B}) \varphi'(q, t, \boldsymbol{B})$$

or, under the time-reversal symmetry (4.3.12)

$$\mathrm{i}\hbar \frac{\partial}{\partial t} \varphi'(q, t, \boldsymbol{B}) = \mathscr{H}(p, q, \sigma, \boldsymbol{B})\, \varphi'(q, t, \boldsymbol{B}). \tag{4.3.15}$$

Consequently, if $\varphi(q, t, B)$ is a realizable state of the system, then its reciprocal state $\varphi'(q, t, B)$ is also realizable. In particular, if $E_n(\boldsymbol{B})$ is an eigenvalue belonging to the eigenfunction $\phi_n(q, \boldsymbol{B})$, namely if

$$\varphi_n(q, t, \boldsymbol{B}) = \phi_n(q, \boldsymbol{B}) \exp[-\mathrm{i}\, t E_n(\boldsymbol{B})/\hbar],$$

$$\hat{\mathscr{H}} \phi_n(q, \boldsymbol{B}) \doteq E_n(\boldsymbol{B})\, \phi_n(q, \boldsymbol{B}),$$

then $E_n(-\boldsymbol{B})$ is also an eigenvalue of $\hat{\mathscr{H}}(p, q, \sigma, \boldsymbol{B})$ belonging to

$$\phi_n' = S \bar{\phi}_n(q, -\boldsymbol{B}).$$

Here $E_n(-\boldsymbol{B})$ is not necessarily equal to $E_n(\boldsymbol{B})$, but the set of eigenvalues and eigenfunctions $E_n(-\boldsymbol{B})$, $\phi_n'(q, \boldsymbol{B})$ is identical with the set of $E_n(\boldsymbol{B})$, $\phi_n(q, \boldsymbol{B})$. Applying this symmetry to (4.3.4),

$$\begin{aligned} j_{FG}(E, E'; \boldsymbol{B}) &= \sum_m \sum_n \delta(E - E_m(\boldsymbol{B}))\, (m\,|\hat{F}|\, n)\, \delta(E' - E_n(\boldsymbol{B}))\, (n\,|\hat{G}|\, m) \\ &= \sum_m \sum_n \delta(E - E_m(-\boldsymbol{B}))\, (m'\,|\hat{F}|\, n')\, \delta(E' - E_n(-\boldsymbol{B}))\, (n'\,|\hat{G}|\, m') \\ &= \sum_m \sum_n \delta(E - E_m(-\boldsymbol{B}))\, (m\,|\hat{G}'|\, n)\, \delta(E' - E_n(-\boldsymbol{B}))\, (n\,|\hat{F}'|\, m) \end{aligned}$$

where we define the primed operators by

$$\begin{aligned} (m\,|\hat{F}'|\, n) &= (n'\,|\hat{F}|\, m') \\ &= \int dq\, \phi_n(q)\, SF\left(\frac{\hbar}{\mathrm{i}} \frac{\partial}{\partial q}, q, \sigma\right) S\bar{\phi}_m(q) \\ &= \int dq\, \bar{\phi}_m(q)\, F\left(-\frac{\hbar}{\mathrm{i}} \frac{\partial}{\partial q}, q, -\sigma\right) \phi_n(q). \end{aligned}$$

Therefore $\hat{F}' = \hat{F}(-p, q, -\boldsymbol{\sigma})$ if $\hat{F}$ is Hermitian. In some cases, we may be concerned with quantities which explicitly contain $\boldsymbol{B}$. A simple example is the current

$$\hat{J}_\mu(p, q, \boldsymbol{B}) = \sum \frac{1}{m}\left(\hat{p}_\mu - \frac{e}{c} \hat{A}_\mu\right),$$

for which

$$J_\mu'(-p, q, -\boldsymbol{B}) = \sum \frac{1}{m}\left(-\hat{p}_\mu - \frac{e}{c} \hat{A}_\mu\right) = -\hat{J}_\mu(p, q, -\boldsymbol{B}),$$

where $\hat{A}$ is the vector potential from which the magnetic field $\boldsymbol{B}$ is derived.

4.3.2 Symmetry in the Current Response

Let us examine the response function (4.2.23). Introducing the spectral function (4.3.4) for $\hat{J}_\mu$ and $\hat{J}_\nu$,

$$J^C_{\nu\mu}(\omega) = \int\int_{-\infty}^{\infty} dE\, dE'\, \varrho_{eq}(E)\, 2\pi\hbar\,\delta(E + \hbar\omega - E')\, j^C_{\nu\mu}(E, E')$$

$$j^C_{\nu\mu}(E, E') = \mathrm{Tr}\,\{\delta(E - \hat{\mathscr{H}})\,\hat{J}_\nu\,\delta(E' - \hat{\mathscr{H}})\,\hat{J}_\mu\}\,, \tag{4.3.16}$$

we obtain

$$\begin{aligned}\Phi_{\mu\nu}(t) &= \frac{1}{2\pi}\int_{-\infty}^{\infty} d\omega\, e^{i\omega t}\int_0^\beta d\lambda\, e^{-\lambda\hbar\omega}\, J^C_{\nu\mu}(\omega)\\ &= \frac{1}{2\pi}\int_{-\infty}^{\infty} d\omega\, e^{i\omega t}\,\frac{1 - \exp(-\beta\hbar\omega)}{\hbar\omega}\, J^C_{\nu\mu}(\omega)\\ &= \int\int_{-\infty}^{\infty} dE\, dE'\,\frac{\varrho_{eq}(E') - \varrho_{eq}(E)}{E - E'}\,\exp[-i(E - E')\,t/\hbar]\, j^C_{\nu\mu}(E, E').\end{aligned} \tag{4.3.17}$$

This expression is convenient for discussing symmetry and other properties.

First, let us ascertain that the response function is real. Since the current operator $\hat{J}_\mu$ and $\hat{J}_\nu$ are self-adjoint, we obtain from (4.3.8)

$$[j^C_{\nu\mu}(E, E')]^* = j^C_{\nu\mu}(E', E)\,.$$

Interchanging the integration variables E and E' in the equation complex conjugate of (4.3.17) gives the same equation as (4.3.17) by virtue of the relation just obtained, and hence

$$\Phi^*_{\mu\nu}(t) = \Phi_{\mu\nu}(t)\ :\ \text{real}. \tag{4.3.18}$$

Next, let us define the response function for $t < 0$ by (4.2.23) or (4.3.17), although essentially it is not needed except for $t \geqq 0$ because of the causality. Then, by using (4.3.7), we can prove similarly to the above discussion that

$$\Phi_{\mu\nu}(-t) = \Phi_{\nu\mu}(t)\,. \tag{4.3.19}$$

We introduce the symmetric and antisymmetric parts of the tensor by

$$\Phi^s_{\mu\nu} = \tfrac{1}{2}(\Phi_{\mu\nu} + \Phi_{\nu\mu})\,,\quad \Phi^a_{\mu\nu} = \tfrac{1}{2}(\Phi_{\mu\nu} - \Phi_{\nu\mu})\,.$$

They satisfy the relations:

$$\Phi^s_{\mu\nu}(-t) = \Phi^s_{\mu\nu}(t)\,,\quad \Phi^a_{\mu\nu}(-t) = -\Phi^a_{\mu\nu}(t)\,. \tag{4.3.20}$$

We thus find that the symmetric part should be an even function of time while the antisymmetric part must be an odd function.

Lastly, through (4.3.9) the magnetic field dependence is obtained

$$\Phi_{\mu\nu}(t;-\boldsymbol{B}) = \varepsilon_\mu \varepsilon_\nu \Phi^*_{\mu\nu}(-t;\boldsymbol{B}), \tag{4.3.21}$$

or by making use of (4.3.18, 19)

$$\Phi_{\mu\nu}(t;-\boldsymbol{B}) = \varepsilon_\mu \varepsilon_\nu \Phi_{\nu\mu}(t;\boldsymbol{B}). \tag{4.3.22}$$

This relation requires that if the currents $\hat{J}_\mu$ and $\hat{J}_\nu$ are of the same kind and hence if $\varepsilon_\mu \varepsilon_\nu = 1$, the symmetric part of the response function should be an even function of the external magnetic field $\boldsymbol{B}$, while the antisymmetric part should be an odd function, or more generally that

$$\begin{aligned} \Phi^{\mathrm{s}}_{\mu\nu}(t;-\boldsymbol{B}) &= \varepsilon_\mu \varepsilon_\nu \Phi^{\mathrm{s}}_{\mu\nu}(t;\boldsymbol{B}), \\ \Phi^{\mathrm{a}}_{\mu\nu}(t;-\boldsymbol{B}) &= -\varepsilon_\mu \varepsilon_\nu \Phi^{\mathrm{a}}_{\mu\nu}(t;\boldsymbol{B}). \end{aligned} \tag{4.3.23}$$

Since the complex admittance is more frequently used than response function, let us examine it similarly. First, the angular frequency dependence is at once obvious from (4.3.18):

$$\chi'_{\mu\nu}(-\omega) = \chi'_{\mu\nu}(\omega), \quad \chi''_{\mu\nu}(-\omega) = -\chi''_{\mu\nu}(\omega). \tag{4.3.24}$$

The magnetic field dependence is obtained by using (4.3.22 or 23) in (3.2.5) as

$$\begin{aligned} \chi_{\mu\nu}(\omega;-\boldsymbol{B}) &= \varepsilon_\mu \varepsilon_\nu \chi_{\nu\mu}(\omega;\boldsymbol{B}), \\ \chi^{\mathrm{s}}_{\mu\nu}(\omega;-\boldsymbol{B}) &= \varepsilon_\mu \varepsilon_\nu \chi^{\mathrm{s}}_{\mu\nu}(\omega;\boldsymbol{B}), \\ \chi^{\mathrm{a}}_{\mu\nu}(\omega;-\boldsymbol{B}) &= -\varepsilon_\mu \varepsilon_\nu \chi^{\mathrm{a}}_{\mu\nu}(\omega;\boldsymbol{B}). \end{aligned} \tag{4.3.25}$$

Since the kinetic coefficient is the static admittance of the current, i.e., the limit of (3.2.5) as $\omega \to 0$, [as is obvious from (3.1.5)], these relations give the general form of the *Onsager* reciprocity theorem [4.5], which is the basis of irreversible thermodynamics.

4.3.3 Symmetry in the Displacement Response

The relaxation function was given in (4.2.29). In the same way as for (4.3.16), let us introduce the spectral function

$$\begin{aligned} J^{\mathrm{D}}_{\nu\mu}(\omega) &= \int\int_{-\infty}^{\infty} dE\, dE'\, \varrho_{\mathrm{eq}}(E)\, 2\pi\hbar\,\delta(E-E'+\hbar\omega)\, j^{\mathrm{D}}_{\nu\mu}(E,E') \\ j^{\mathrm{D}}_{\nu\mu}(E,E') &= \mathrm{Tr}\,\{\delta(E-\mathscr{H})\,\Delta A'_\nu\,\delta(E'-\mathscr{H})\,\Delta A'_\mu\}. \end{aligned} \tag{4.3.26}$$

Analogously to (4.3.14)

$$\Psi_{\mu\nu}(t) = \int\int_{-\infty}^{\infty} dE\, dE' \frac{\varrho_{\text{eq}}(E) - \varrho_{\text{eq}}(E')}{E' - E} \exp[\mathrm{i}(E' - E)\, t/\hbar]\, j^{\mathrm{D}}_{\nu\mu}(E, E') . \tag{4.3.27}$$

Therefore (4.3.18–23) are valid as

$$\begin{aligned}
&\Psi_{\mu\nu}(t)\text{: real,} \quad \Psi_{\mu\nu}(-t) = \Psi_{\nu\mu}(t) , \\
&\Psi_{\mu\nu}(t; -\boldsymbol{B}) = \varepsilon_\mu\, \varepsilon_\nu\, \Psi_{\nu\mu}(t; \boldsymbol{B}) , \\
&\Psi^{\mathrm{s}}_{\mu\nu}(-t) = \Psi^{\mathrm{s}}_{\mu\nu}(t), \quad \Psi^{\mathrm{a}}_{\mu\nu}(-t) = -\Psi^{\mathrm{a}}_{\mu\nu}(t) , \\
&\Psi^{\mathrm{s}}_{\mu\nu}(t; -\boldsymbol{B}) = \varepsilon_\mu\, \varepsilon_\nu\, \Psi^{\mathrm{s}}_{\mu\nu}(t; \boldsymbol{B}) , \\
&\Psi^{\mathrm{a}}_{\mu\nu}(t; -\boldsymbol{B}) = -\varepsilon_\mu\, \varepsilon_\nu\, \Psi^{\mathrm{a}}_{\mu\nu}(t; \boldsymbol{B}) .
\end{aligned} \tag{4.3.28}$$

The complex admittance is given by (3.2.6), and its real and imaginary parts become

$$\begin{aligned}
&\chi'_{\mu\nu}(\omega) = \Psi_{\mu\nu}(0) - \omega \int_0^\infty dt\, \Psi_{\mu\nu}(t) \sin \omega t , \\
&\chi'_{\mu\nu}(\omega) = \omega \int_0^\infty dt\, \Psi_{\mu\nu}(t) \cos \omega t
\end{aligned} \tag{4.3.29}$$

because the relaxation function is real. It is obvious that the angular frequency dependence is in accordance with (4.3.24) and the magnetic field dependence with (4.3.25).

Observing the property (4.3.21) of the response function for the current and that (4.3.28) of the relaxation function for the displacement, we infer that the problem of the magnetic field dependence is closely connected with time-reversal symmetry. *Onsager* [4.5] ascribed the validity of the reciprocity theorem to the invariance of dynamics against time reversal.

4.3.4 Proof of the Dispersion Relations

Although we have given the proof based on the theory of complex function in Sect. 3.6, we shall describe here a proof based on the Fourier transformation partly to give some background to the next section. The response function can be Fourier-analyzed because it has just been extended into negative times:

$$\Phi_{\mu\nu}(t) = \frac{1}{2\pi} \int_{-\infty}^{\infty} \Phi_{\mu\nu}[\omega]\, \mathrm{e}^{-\mathrm{i}\omega t}\, d\omega . \tag{4.3.30}$$

Substituting this expression into the defining equation of the complex admittance (3.2.5), and carrying out the time integral by using

$$\lim_{\theta \to \infty} \int_0^\theta dt\, e^{i\omega t} = \lim_{\varepsilon \to 0+} \int_0^\infty dt \exp(-\varepsilon t + i\omega t)$$

$$= \lim_{\varepsilon \to 0+} \frac{i}{\omega + i\varepsilon}$$

$$= \pi\delta(\omega) + i\frac{\mathscr{P}}{\omega}, \qquad (4.3.31)$$

we obtain

$$\chi_{\mu\nu}(\omega) = \lim_{\varepsilon \to 0} \int_{-\infty}^{\infty} \frac{d\omega'}{2\pi} \frac{1}{\omega' - \omega - i\varepsilon} \Phi_{\mu\nu}[\omega']$$

$$= \frac{1}{2}\Phi_{\mu\nu}[\omega] - \frac{i}{2\pi}\int_{-\infty}^{\infty} d\omega' \frac{\mathscr{P}}{\omega' - \omega}\Phi_{\mu\nu}[\omega']. \qquad (4.3.32)$$

On the other hand, according to (4.3.18, 19)

$$\Phi^*_{\mu\nu}[\omega] = \Phi_{\nu\mu}[\omega] = \Phi_{\mu\nu}[-\omega]$$

and hence

$$\Phi^{s}_{\mu\nu}[\omega] = \Phi^{s}_{\mu\nu}[-\omega]: \ \text{real},$$
$$\Phi^{a}_{\mu\nu}[\omega] = -\Phi^{a}_{\mu\nu}[-\omega]: \ \text{pure imaginary}. \qquad (4.3.33)$$

With these equations, we can derive from (4.3.32)

$$\chi'^{s}_{\mu\nu}(\omega) = \tfrac{1}{2}\Phi^{s}_{\mu\nu}[\omega],$$

$$\chi''^{s}_{\mu\nu}(\omega) = -\frac{1}{2\pi}\int_{-\infty}^{\infty} d\omega' \frac{\mathscr{P}}{\omega' - \omega}\Phi^{s}_{\mu\nu}[\omega'],$$

$$\chi'^{a}_{\mu\nu}(\omega) = -\frac{i}{2\pi}\int_{-\infty}^{\infty} d\omega' \frac{\mathscr{P}}{\omega' - \omega}\Phi^{a}_{\mu\nu}[\omega'], \qquad (4.3.34)$$

$$\chi''^{a}_{\mu\nu}(\omega) = -\frac{i}{2}\Phi^{a}_{\mu\nu}[\omega].$$

The first and fourth equations are written in the forms

$$\chi'^{s}_{\mu\nu}(\omega) = \frac{1}{2}\int_{-\infty}^{\infty} dt\, \Phi^{s}_{\mu\nu}(t)\cos\omega t$$

and

$$\chi''^{a}_{\mu\nu}(\omega) = \frac{1}{2}\int_{-\infty}^{\infty} dt\, \Phi^{a}_{\mu\nu}(t)\sin\omega t.$$

Eliminating $\Phi^{s}_{\mu\nu}[\omega]$ and $\Phi^{a}_{\mu\nu}[\omega]$ from (4.3.34) results in the symmetric part of the second equation of the dispersion relation (3.6.8) and the antisymmetric part of the first equation, respectively. By taking the inverse Hilbert transformation of these results, we obtain the symmetric part of the first equation and the antisymmetric part of the second equation, respectively.

For the relaxation function of the displacement, by making the Fourier analysis

$$\Psi_{\mu\nu}(t) = \frac{1}{2\pi}\int_{-\infty}^{\infty} d\omega\, \mathrm{e}^{-\mathrm{i}\omega t}\, \Psi_{\mu\nu}[\omega], \tag{4.3.35}$$

we can write the complex admittance (3.2.6) in the form

$$\begin{aligned}\chi_{\mu\nu}(\omega) &= \lim_{\varepsilon\to 0}\frac{1}{2\pi}\int_{-\infty}^{\infty} d\omega' \frac{\omega' - \mathrm{i}\varepsilon}{\omega' - \omega - \mathrm{i}\varepsilon}\Psi_{\mu\nu}[\omega'] \\ &= \frac{1}{2\pi}\int_{-\infty}^{\infty} d\omega' \frac{\mathscr{P}}{\omega' - \omega}\omega' \Psi_{\mu\nu}[\omega'] + \frac{\mathrm{i}\omega}{2}\Psi_{\mu\nu}[\omega].\end{aligned} \tag{4.3.36}$$

Since we can prove

$$\begin{aligned}&\Psi^{s}_{\mu\nu}[\omega] = \Psi^{s}_{\mu\nu}[-\omega]\text{: real,} \\ &\Psi^{a}_{\mu\nu}[\omega] = -\Psi^{a}_{\mu\nu}[-\omega]\text{: pure imaginary,}\end{aligned} \tag{4.3.37}$$

similarly as for (4.3.33), (4.3.36) yields

$$\begin{aligned}\chi'^{s}_{\mu\nu}(\omega) &= \frac{1}{2\pi}\int_{-\infty}^{\infty} d\omega' \frac{\mathscr{P}}{\omega' - \omega}\omega' \Psi^{s}_{\mu\nu}[\omega'], \\ \chi''^{s}_{\mu\nu}(\omega) &= \frac{1}{2}\omega \Psi^{s}_{\mu\nu}[\omega], \\ \chi'^{a}_{\mu\nu}(\omega) &= \frac{1}{2}\omega \Psi^{a}_{\mu\nu}[\omega], \\ \chi''^{a}_{\mu\nu}(\omega) &= \frac{1}{2\pi\mathrm{i}}\int_{-\infty}^{\infty} d\omega' \frac{\mathscr{P}}{\omega' - \omega}\omega' \Psi^{a}_{\mu\nu}[\omega'],\end{aligned} \tag{4.3.38}$$

from which, by eliminating $\Psi^{s}_{\mu\nu}[\omega]$ and $\Psi^{a}_{\mu\nu}[\omega]$, we arrive at the dispersion relation (3.6.8). Note that the second and third equations are written in the form

$$\chi''^{s}_{\mu\nu}(\omega) = \frac{1}{2}\omega\int_{-\infty}^{\infty} dt\, \Psi^{s}_{\mu\nu}(t)\cos\omega t$$

and

$$\chi'^{a}_{\mu\nu}(\omega) = -\frac{1}{2}\omega\int_{-\infty}^{\infty} dt\, \Psi^{a}_{\mu\nu}(t)\sin\omega t\,.$$

4.4 Fluctuation and Dissipation Theorem

4.4.1 Symmetrized Correlation

In classical statistical mechanics the correlation function is the expectation value of the product of two dynamical variables at different times. In quantum mechanics, two dynamical variables at different times generally do not commute with each other, even if they represent the same physical quantity and the order in the product becomes important. This is made clearer by the spectral function described at the beginning of the previous section. According to (4.3.5, 7)

$$J_{G,F}(\omega) = \int\int_{-\infty}^{\infty} dE\, dE'\, \varrho_{\text{eq}}(E)\, 2\pi\hbar\delta(E + \hbar\omega - E')\, j_{F,G}(E', E)$$

and, by interchanging E and E',

$$J_{GF}(\omega) = \mathrm{e}^{\beta\hbar\omega} J_{FG}(\omega)\,. \tag{4.4.1}$$

Thus the extra factor $\exp(\beta\hbar\omega)$ appears in the integrand of the expectation value for the product with the interchanged order

$$\begin{aligned}\langle \hat{G}(t')\,\hat{F}(t)\rangle &= \int_{-\infty}^{\infty} \frac{d\omega}{2\pi} \exp[\mathrm{i}\omega(t-t')] \exp(\beta\hbar\omega)\, J_{F,G}(-\omega) \\ &= \int_{-\infty}^{\infty} \frac{d\omega}{2\pi} \exp[-\mathrm{i}\omega(t-t')] \exp(-\beta\hbar\omega)\, J_{F,G}(\omega)\,.\end{aligned} \tag{4.4.2}$$

This factor may be approximated by 1 when $\hbar\omega \ll kT$, and hence only when the contributions from the region $\hbar\omega \gtrsim kT$ may be neglected in (4.4.2) does the order of the product become insignificant. This is the condition of use of the correlation function of classical statistical mechanics.

In quantum-statistical mechanics, as is always the case when proceeding from classical to quantum theory, the kind of quantity we should take as the *correlation* function is not unique but depends on the method of measurement. In Sect. 4.1 we introduced the canonical correlation. As another possibility, let us consider the expectation value of the symmetrized product and call it the *symmetrized correlation.* The Fourier transform of the expectation value of the symmetrized product is given by (4.3.6) and (4.4.2) as

$$\left\langle \frac{1}{2}\{\hat{F}(t)\,\hat{G}(t') + \hat{G}(t')\,\hat{F}(t)\}\right\rangle = \int_{-\infty}^{\infty} \frac{d\omega}{2\pi} \times \exp[-\mathrm{i}\omega(t-t')]\, \frac{1+\exp(-\beta\hbar\omega)}{2}\, J_{F,G}(\omega)\,. \tag{4.4.3}$$

This symmetrized correlation not only satisfies the symmetry corresponding to (4.1.13) but also conserves the property of the classical correlation function in that it reduces to $\langle \hat{F}^2 \rangle$ when $\hat{F} = \hat{G}$ and $t = t'$.

As background for Sect. 4.4.2, let us Fourier-analyze the commutator

$$\left\langle \frac{1}{\mathrm{i}\hbar} [\hat{F}(t), \hat{G}(t')] \right\rangle = \int_{-\infty}^{\infty} \frac{d\omega}{2\pi} \exp[-\mathrm{i}\omega(t-t')] \Lambda_{F,G}(\omega), \tag{4.4.4}$$

where the spectral function is given by

$$\Lambda_{F,G}(\omega) = \frac{1 - \exp(-\beta\hbar\omega)}{\mathrm{i}\hbar} J_{F,G}(\omega). \tag{4.4.5}$$

The convenience of the spectral function is that it makes it easy to relate various kinds of expectation values with one another. For example, substituting $J_{FG}(\omega)$ obtained from (4.4.5) into (4.4.3) gives

$$\begin{aligned} &\langle \tfrac{1}{2} [\hat{F}(t)\hat{G}(t') + \hat{G}(t')\hat{F}(t)] \rangle \\ &= \int_{-\infty}^{\infty} \frac{d\omega}{2\pi} \exp[-\mathrm{i}\omega(t-t')] \frac{\mathrm{i}}{\omega} E_\beta(\hbar\omega) \Lambda_{F,G}(\omega) \end{aligned} \tag{4.4.6}$$

and further, substituting into this the inverse Fourier transformation of (4.4.4), we can express the expectation value of the symmetrized products in terms of that of the commutator. In (4.4.6), the function

$$E_\beta(\hbar\omega) = \frac{\hbar\omega}{2} + \frac{\hbar\omega}{\exp(\beta\hbar\omega) - 1} = \frac{\hbar\omega}{2} \coth\left(\frac{\beta\hbar\omega}{2}\right) \tag{4.4.7}$$

agrees with the average energy of a harmonic oscillator with an angular frequency ω, but (4.4.6) itself is not directly concerned with either the harmonic oscillator or the Bose statistics. It is generated only from the ratio of the expectation values of the symmetrized product and the commutator.

4.4.2 The Equivalence Between the Symmetrized Correlation Function and the Response or the Relaxation Function

The symmetrized correlation function has been introduced above. Now we shall show that it contains knowledge equivalent to that contained in the response or relaxation function or the complex admittance.

For the response of the current to the external force, we need only to relate the symmetrized correlation of the current

$$\begin{aligned} C_{\mu\nu}(t-t') &= \left\langle \frac{1}{2} \{\hat{J}_\mu(t)\hat{J}_\nu(t') + \hat{J}_\nu(t')\hat{J}_\mu(t)\} \right\rangle \\ &= \int_{-\infty}^{\infty} \frac{d\omega}{2\pi} \exp[-\mathrm{i}\omega(t-t')] C_{\mu\nu}[\omega] \end{aligned} \tag{4.4.8}$$

to the response function (4.2.23). The response function has been rewritten in the form of (4.3.17). In the same way we can rewrite (4.4.8) as

$$C_{\mu\nu}(t) = \int_{-\infty}^{\infty} \frac{d\omega}{2\pi} e^{i\omega t} \frac{1+\exp(-\beta\hbar\omega)}{2} J_{\nu\mu}^{C}(\omega)$$
$$= \int\int_{-\infty}^{\infty} dE\, dE' \frac{\varrho_{\text{eq}}(E)+\varrho_{\text{eq}}(E')}{2} \exp[-i(E-E')t/\hbar]\, j_{\nu\mu}^{C}(E,E') \tag{4.4.9}$$

by virtue of (4.4.3). Comparing this with the expression for the response function (4.3.17), we see that the difference between the two expressions is the appearance of the arithmetic mean of $\varrho_{\text{eq}}(E)$ in (4.4.9) in place of the difference quotient $[\varrho_{\text{eq}}(E') - \varrho_{\text{eq}}(E)]/(E-E')$ in (4.3.17). As stated in (4.4.3, 4), this difference is due to the difference between the spectral function of the commutator and that of the symmetrized product. We now consider the Fourier component $C_{\mu\nu}[\omega]$ of (4.4.8):

$$C_{\mu\nu}[\omega] = \int\int_{-\infty}^{\infty} dE\, dE' \frac{\varrho_{\text{eq}}(E)+\varrho_{\text{eq}}(E')}{2} 2\pi\hbar\,\delta(E-\hbar\omega-E')\, j_{\nu\mu}^{C}(E,E') \tag{4.4.10}$$

and use the identity

$$\frac{1}{2}[\varrho_{\text{eq}}(E)+\varrho_{\text{eq}}(E')] = E_{\beta}(E-E')\frac{\varrho_{\text{eq}}(E')-\varrho_{\text{eq}}(E)}{E-E'} \tag{4.4.11}$$

to substitute into (4.4.10). We may replace $E_{\beta}(E-E')$ with $E_{\beta}(\hbar\omega)$ because of the factor $\delta(E-E'-\hbar\omega)$ in (4.4.10) and take it out of the integral sign. Then as the integral we obtain the Fourier transform (4.3.30) of the response function (4.3.17), thus proving that

$$C_{\mu\nu}[\omega] = E_{\beta}(\hbar\omega)\,\Phi_{\mu\nu}[\omega]\,. \tag{4.4.12}$$

We can proceed in the same way for the response of the displacement to the external force. This time we need only to connect the symmetrized correlation function of the displacement

$$D_{\mu\nu}(t-t') = \langle \tfrac{1}{2}(\Delta\hat{A}'_{\mu}(t)\,\Delta\hat{A}'_{\nu}(t') + \Delta\hat{A}'_{\nu}(t')\,\Delta\hat{A}'_{\mu}(t)]\rangle$$
$$= \frac{1}{2\pi}\int_{-\infty}^{\infty} d\omega \exp[-i\omega(t-t')]\, D_{\mu\nu}[\omega] \tag{4.4.13}$$

with the relaxation function (4.2.29). It will at once be obvious that an equation of the same form as (4.4.12)

$$D_{\mu\nu}[\omega] = E_{\beta}(\hbar\omega)\,\Psi_{\mu\nu}[\omega] \tag{4.4.14}$$

is valid between the Fourier transform (4.3.35) of the relaxation function (4.3.27) and that of the symmetrized correlation (4.4.13).

4.4.3 Fluctuation-Dissipation Theorem

The symmetrized correlation of the current or the displacement has been shown to contain information equivalent to the corresponding response function or the relaxation function, and therefore becomes equivalent to the complex admittance according to the statement in Sect. 3.2. To show this explicitly, we need only use the relation connecting the complex admittance to the Fourier transform of the response function (4.3.34) or to that of the relaxation function (4.3.38). The equivalence of the canonical correlation and the complex admittance is proved by means of the Kubo formulas described in Sect. 4.2. That part of the interrelation thus obtained between the symmetrized correlation and the complex admittance which is concerned with energy dissipation has long been known as an example of the Nyquist theorem connecting the thermal noise of the electric current to the electric resistance.

Regarding the response of the current, the real symmetric and the imaginary antisymmetric parts of the complex admittance (3.2.25) are related to the energy dissipation, so that we need to use only the first and fourth equations of (4.3.34). For the symmetrized correlation function, the symmetry is the same as that of the response function by virtue of (4.4.12), and the relations of the same form as (4.3.33)

$$\begin{aligned} C^{\mathrm{s}}_{\mu\nu}[\omega] &= C^{\mathrm{s}}_{\mu\nu}[-\omega]\text{: real,} \\ C^{\mathrm{a}}_{\mu\nu}[\omega] &= -C^{\mathrm{a}}_{\mu\nu}[-\omega]\text{: pure imaginary} \end{aligned} \tag{4.4.15}$$

hold. Thus we obtain at once

$$\begin{aligned} C^{\mathrm{s}}_{\mu\nu}[\omega] &= 2E_\beta(\hbar\omega)\,\chi'^{\mathrm{s}}_{\mu\nu}[\omega]\,, \\ C^{\mathrm{a}}_{\mu\nu}[\omega] &= 2\mathrm{i}\,E_\beta(\hbar\omega)\,\chi''^{\mathrm{a}}_{\mu\nu}[\omega]\,, \end{aligned} \tag{4.4.16}$$

or, by inverse Fourier transformation,

$$\begin{aligned} C^{\mathrm{s}}_{\mu\nu}(t) &= \frac{2}{\pi}\int_0^\infty d\omega\, E_\beta(\hbar\omega)\,\chi'^{\mathrm{s}}_{\mu\nu}(\omega)\cos\omega t\,, \\ C^{\mathrm{a}}_{\mu\nu}(t) &= \frac{2}{\pi}\int_0^\infty d\omega\, E_\beta(\hbar\omega)\,\chi''^{\mathrm{a}}_{\mu\nu}(\omega)\sin\omega t\,. \end{aligned} \tag{4.4.17}$$

The first of these relations is the generalization of the Nyquist theorem, and contains the factor $E_\beta(\hbar\omega)$ inferred by *Nyquist* [4.6].

Regarding the response of the displacement, the quantities connected with the energy dissipation are the imaginary symmetric and the real antisymmetric parts of the complex admittance. By making use of the second and third equations of (4.3.38), then similarly

$$\begin{aligned} D^{\mathrm{s}}_{\mu\nu}[\omega] &= 2E_\beta(\hbar\omega)\,\chi''^{\mathrm{s}}_{\mu\nu}(\omega)/\omega\,, \\ D^{\mathrm{a}}_{\mu\nu}[\omega] &= 2E_\beta(\hbar\omega)\,\chi'^{\mathrm{a}}_{\mu\nu}(\omega)/\mathrm{i}\,\omega \end{aligned} \tag{4.4.18}$$

and

$$D^{\mathrm{s}}_{\mu\nu}(t) = \frac{2}{\pi}\int_0^\infty d\omega\, E_\beta(\hbar\omega)\,\frac{\chi''^{\mathrm{s}}_{\mu\nu}(\omega)}{\omega}\cos\omega t\,,$$
$$D^{\mathrm{a}}_{\mu\nu}(t) = -\frac{2}{\pi}\int_0^\infty d\omega\, E_\beta(\hbar\omega)\,\frac{\chi'^{\mathrm{a}}(\omega)}{\omega}\sin\omega t\,. \tag{4.4.19}$$

Since the correlation function serves as a measure of the thermal fluctuations, (4.4.16, 17 or 18, 19) are the predominant relations connecting the energy dissipation in irreversible processes to the thermal fluctuation in equilibrium. They are called the fluctuation-dissipation theorem. *Nyquist* [4.6] first gave an example and provided a proof based on the second law of thermodynamics. As discussed in Sect. 4.6, *Onsager* [4.5] postulated this for his proof of the reciprocity law of kinetic coefficients. Generalizations have been made by many authors, among whom we refer here to the works by *Callen* and *Welton* [4.7], *Takahasi* [4.8] and *Kubo* [4.3]. From the current point of view, the theorem is more general than the name would indicate, representing a very general relationship between response, relaxation and fluctuations.

If we represent thermal fluctuation by the canonical correlation, we obtain the fluctuation-dissipation theorem in a form in which kT appears in place of the factor $E_\beta(\hbar\omega)$ in (4.4.17, 19). Taking this point of view, we are led to a wider view of the fluctuation-dissipation theorem, holding that the propagating behavior of the effects of the external force is closely connected to the fluctuation behavior of the thermal equilibrium state, irrespective of whether this behavior is irreversible or not. This is nothing but the viewpoint mentioned in Sect. 4.2.

4.5 Density Response, Conduction and Diffusion

As an example of application of the theory of linear response, let us treat here the problem of density response and related subjects for a system of interacting particles which is supposed to be a fluid with no long-range order [4.9]. We discuss responses of the system to an externally applied potential observed in the changes of density and current. A complexity arising here is the distinction between external and internal fields. This is a good example for applying the first and second fluctuation-dissipation theorems.

4.5.1 Density and Current in Response to the External Field

For simplicity, we consider a system consisting of a single kind of particles. We assume there is a background consisting of other particles or a fictitious medium, for example, a counter-balancing uniform positive background as one considers a jellium model of electrons or a solvent in which the particles exist as solute molecules. This means that momentum is *not conserved,* and accordingly the sound wave modes are ignored. The particle number density at a given position is defined by (to simplify notation, henceforth omit the circumflex denoting a physical quantity)

$$n(\boldsymbol{r}, t) = \sum_l \delta(\boldsymbol{r} - \boldsymbol{r}_l(t)) = L^{-3} \sum_{\boldsymbol{k}} n_{\boldsymbol{k}}(t) \exp(\mathrm{i}\,\boldsymbol{k} \cdot \boldsymbol{r}) \tag{4.5.1}$$

with

$$n_{\boldsymbol{k}}(t) = \int n(\boldsymbol{r}, t) \exp(-\mathrm{i}\,\boldsymbol{k} \cdot \boldsymbol{r})\, d\boldsymbol{r} = \sum_l \exp[-\mathrm{i}\,\boldsymbol{k} \cdot \boldsymbol{r}_l(t)], \tag{4.5.2}$$

where L^3 is the whole volume of the system which contains N particles and $\sum$ means sum over possible discrete values of $\boldsymbol{k}$ allowed by the imposed boundary condition. In the limit of $L \to \infty$, (4.5.1) reduces to

$$n(\boldsymbol{r}, t) = (2\pi)^{-3} \int d\boldsymbol{k}\, n_{\boldsymbol{k}}(t) \exp(\mathrm{i}\,\boldsymbol{k} \cdot \boldsymbol{r}). \tag{4.5.3}$$

Similarly the current density is defined by

$$\begin{aligned} \boldsymbol{j}(\boldsymbol{r}, t) &= \sum_l \left[\frac{1}{m} \boldsymbol{p}_l\, \delta(\boldsymbol{r} - \boldsymbol{r}_l(t))\right] \\ &= L^{-3} \sum_{\boldsymbol{k}} \boldsymbol{j}_{\boldsymbol{k}}(t) \exp(\mathrm{i}\,\boldsymbol{k} \cdot \boldsymbol{r}) \\ &= (2\pi)^{-3} \int d\boldsymbol{k}\, \boldsymbol{j}_{\boldsymbol{k}}(t) \exp(\mathrm{i}\,\boldsymbol{k} \cdot \boldsymbol{r}) \end{aligned} \tag{4.5.4}$$

with

$$\boldsymbol{j}_{\boldsymbol{k}}(t) = \sum_l \left[\frac{1}{m} \boldsymbol{p}_l \exp(-\mathrm{i}\,\boldsymbol{k} \cdot \boldsymbol{r}_l)\right], \tag{4.5.5}$$

where the curly bracket denotes a symmetrized product of the operators inside. Note

$$\frac{\partial}{\partial t} n(\boldsymbol{r}, t) + \nabla \boldsymbol{j}(\boldsymbol{r}, t) = 0$$

or

$$\dot{n}_{\boldsymbol{k}}(t) + \mathrm{i}\,\boldsymbol{k}\,\boldsymbol{j}_{\boldsymbol{k}} = 0 \quad \text{and} \tag{4.5.6}$$

$$(\boldsymbol{j}_{-\boldsymbol{k}}, n_{\boldsymbol{k}})\, L^{-3} = (n/m)\, \mathrm{i}\,\boldsymbol{k}. \tag{4.5.7}$$

When an external potential $\varphi^{\mathrm{e}}(\boldsymbol{r}, t)$ is imposed, the perturbation is represented by the Hamiltonian

$$\begin{aligned} \mathscr{H}_{\mathrm{ext}}(t) &= \sum_l \varphi^{\mathrm{e}}(\boldsymbol{r}_l(t), t) \\ &= \int n(\boldsymbol{r}, t)\, \varphi^{\mathrm{e}}(\boldsymbol{r}, t)\, d\boldsymbol{r} \\ &= \int d\boldsymbol{k}\, n_{-\boldsymbol{k}}(t)\, \varphi^{\mathrm{e}}(\boldsymbol{k}, t), \end{aligned} \tag{4.5.8}$$

where the Fourier component of the potential is introduced by

$$\varphi^{e}(\boldsymbol{k}, t) = (2\pi)^{-3} \int d\boldsymbol{r}\, \varphi^{e}(\boldsymbol{r}, t) \exp(-\mathrm{i}\boldsymbol{k} \cdot \boldsymbol{r}) .$$

If the potential is purely periodic in space with the wave vector $\boldsymbol{k}$, its Fourier components are

$$\varphi^{e}(\boldsymbol{k}', t) = \mathrm{Re}\,\{\varphi^{e}_{\boldsymbol{k}}(t)\, \delta(\boldsymbol{k}' - \boldsymbol{k})\} \sim (L/2\pi)^3\, \mathrm{Re}\,\{\varphi^{e}_{\boldsymbol{k}}(t)\, \delta_{\boldsymbol{k},\boldsymbol{k}'}\} \tag{4.5.9}$$

and

$$\mathscr{H}_{\mathrm{ext}}(t) = \mathrm{Re}\,\{n_{-\boldsymbol{k}}(t)\, \varphi^{e}_{\boldsymbol{k}}(t)\} . \tag{4.5.10}$$

The linear response of the system is expressed by

$$\overline{n(\boldsymbol{r}, t)} = -\int_{-\infty}^{t} dt' \int d\boldsymbol{r}'\, \chi(\boldsymbol{r} - \boldsymbol{r}', t - t')\, \varphi^{e}(\boldsymbol{r}', t') \tag{4.5.11}$$

for the density, where χ is the *nonlocal density susceptibility*, and

$$\overline{\boldsymbol{j}(\boldsymbol{r}, t)} = -\int_{-\infty}^{t} dt' \int d\boldsymbol{r}\, \mu^{e}(\boldsymbol{r} - \boldsymbol{r}', t - t')\, \nabla \varphi^{e}(\boldsymbol{r}', t') \tag{4.5.12}$$

for the current density, where μ^{e} is the *external nonlocal mobility* tensor. We have assumed the system to be uniform with translational invariance.

The linear-response theory states that

$$\chi(\boldsymbol{r} - \boldsymbol{r}'\, t - t') = \langle (n(\boldsymbol{r}', t'), n(\boldsymbol{r}, t)) \rangle \tag{4.5.13}$$

and

$$\mu^{e}(\boldsymbol{r} - \boldsymbol{r}', t - t') = \beta \langle \boldsymbol{j}(\boldsymbol{r}', t'); \boldsymbol{j}(\boldsymbol{r}, t) \rangle . \tag{4.5.14}$$

The former is a direct consequence of (4.2.14). To derive the latter we express the current response to the potential by (4.2.18) and perform partial integration using (4.5.6).

If the potential is purely periodic in space and time

$$\varphi^{e}_{\boldsymbol{k}}(t) = \mathrm{Re}\,\{\varphi^{e}_{\boldsymbol{k}} \exp(\mathrm{i}\boldsymbol{k} \cdot \boldsymbol{r} - \mathrm{i}\omega t)\} , \tag{4.5.15}$$

the linear responses are

$$\overline{n_{\boldsymbol{k}}(t)} = -L^3\, \mathrm{Re}\,\{\chi[\boldsymbol{k}, \omega]\, \varphi^{e}_{\boldsymbol{k}}\, \mathrm{e}^{-\mathrm{i}\omega t}\} , \tag{4.5.16}$$

$$\overline{\boldsymbol{j}_{\boldsymbol{k}}(t)} = -L^3\, \mathrm{Re}\,\{\mu^{e}[\boldsymbol{k}, \omega]\, \mathrm{i}\boldsymbol{k}\, \varphi^{e}_{\boldsymbol{k}}\, \mathrm{e}^{-\mathrm{i}\omega t}\} , \tag{4.5.17}$$

where

$$\chi[\boldsymbol{k}, \omega] = \int_{0}^{\infty} dt\, \mathrm{e}^{\mathrm{i}\omega t} \lim_{L \to \infty} \langle (n_{-\boldsymbol{k}}(0), n_{\boldsymbol{k}}(t)) \rangle\, L^{-3} , \tag{4.5.18}$$

$$\mu^{e}[\boldsymbol{k}, \omega] = \beta \int_{0}^{\infty} dt\, \mathrm{e}^{\mathrm{i}\omega t} \lim_{L \to \infty} \langle \boldsymbol{j}_{-\boldsymbol{k}}(0); \boldsymbol{j}_{\boldsymbol{k}}(t) \rangle\, L^{-3} , \tag{4.5.19}$$

where $\chi[\boldsymbol{k}, \omega]$ is the *density susceptibility* (admittance) and $\mu^{\mathrm{e}}[\boldsymbol{k}, \omega]$ is the *external mobility*. These can be also obtained from (4.5.11, 12) by Fourier transformation taking account of (4.5.9). They are related to each other by

$$\mathrm{i}\omega\chi[\boldsymbol{k}, \omega] + \boldsymbol{k} \cdot \mu^{\mathrm{e}}[\boldsymbol{k}, \omega] \cdot \boldsymbol{k} = 0\,. \tag{4.5.20}$$

4.5.2 Relaxation of the Density Response and the Density Fluctuation

When the potential is static and has been applied from the infinite past to the present, the induced density response is

$$\bar{n}_{\boldsymbol{k}} = L^3\,\chi[\boldsymbol{k}, 0]\,\varphi^{\mathrm{e}}_{\boldsymbol{k}}\,,$$

where $\chi[\boldsymbol{k}, 0]$ is the static susceptibility

$$\chi[\boldsymbol{k}, 0] = -\,L^{-3}\frac{\partial \bar{n}_{\boldsymbol{k}}}{\partial \varphi^{\mathrm{e}}_{\boldsymbol{k}}} = \beta\langle n_{-\boldsymbol{k}}; n_{\boldsymbol{k}}\rangle\,L^{-3}\,, \tag{4.5.21}$$

see (4.1.11). In the limit of vanishing $\boldsymbol{k}$, this becomes a thermodynamic quantity:

$$\lim_{k\to 0}\chi[\boldsymbol{k}, 0] = \partial n/\partial\zeta = n^2\varkappa\,, \tag{4.5.22}$$

where n is the equilibrium density, ζ the chemical potential and $\varkappa$ is the isothermal compressibility. The first equality in (4.5.22) is obtained by identifying the local change of $-\,\varphi^{\mathrm{e}}$ with that of the chemical potential. The second equality follows from the Gibbs-Duhem relation.

If the potential φ^{e} is removed at $t = 0$ after it has been applied from the infinite past, then the induced density change relaxes to zero. The relaxation is desribed by

$$\overline{n_{\boldsymbol{k}}(t)}/\bar{n}_{\boldsymbol{k}} = \Psi_{\boldsymbol{k}}(t)\,,$$

where the relaxation function is given by (4.2.29):

$$\Psi_{\boldsymbol{k}}(t) = \langle n_{-\boldsymbol{k}}(0); n_{\boldsymbol{k}}(t)\rangle/\langle n_{-\boldsymbol{k}}; n_{\boldsymbol{k}}\rangle\,. \tag{4.5.23}$$

We introduce the Fourier-Laplace transform of the canonical correlation of the density fluctuation,

$$\Lambda[\boldsymbol{k}, \omega] = \int_0^\infty dt\,\mathrm{e}^{\mathrm{i}\omega t}\lim_{L\to\infty}\langle n_{-\boldsymbol{k}}(0); n_{\boldsymbol{k}}(t)\rangle\,L^{-3}\,. \tag{4.5.24}$$

Then (4.2.30) is written as

$$\chi[\boldsymbol{k}, \omega] = \chi[\boldsymbol{k}, 0] + \mathrm{i}\omega\beta\Lambda[\boldsymbol{k}, \omega]\,. \tag{4.5.25}$$

We now write $\Lambda[\boldsymbol{k}, \omega]$, (4.5.24), in the form

$$\begin{aligned} \Lambda[\boldsymbol{k}, \omega] &= \frac{\langle n_{-\boldsymbol{k}}; n_{\boldsymbol{k}}\rangle}{-\mathrm{i}\omega + \gamma_{\boldsymbol{k}}[\omega]} \\ &= \frac{\chi[\boldsymbol{k}, 0]}{\beta(-\mathrm{i}\omega + \gamma_{\boldsymbol{k}}[\omega])}\,. \end{aligned} \tag{4.5.26}$$

As shown in Sects. 1.6 and 2.9, (4.5.26) implies that the continuity equation (4.5.6) is regarded as a Langevin equation for the density fluctuation in equilibrium with *no external potential,*

$$\dot{n}_{\boldsymbol{k}}(t) = -\mathrm{i}\boldsymbol{k}\cdot\boldsymbol{j}^0_{\boldsymbol{k}}(t) - \mathrm{i}\boldsymbol{k}\cdot\boldsymbol{j}'_{\boldsymbol{k}}(t)\,, \tag{4.5.27}$$

where the systematic current $\boldsymbol{j}^0_{\boldsymbol{k}}(t)$ defined by

$$-\mathrm{i}\boldsymbol{k}\cdot\boldsymbol{j}^0_{\boldsymbol{k}}(t) = -\int_{-\infty}^{t} \gamma_{\boldsymbol{k}}(t-t')\, n_{\boldsymbol{k}}(t')\, dt \tag{4.5.28}$$

and

$$\boldsymbol{j}'_{\boldsymbol{k}}(t) = \boldsymbol{j}_{\boldsymbol{k}}(t) - \boldsymbol{j}^0_{\boldsymbol{k}}(t) \tag{4.5.29}$$

is the random current. The systematic current is the part of the total current $\boldsymbol{j}_{\boldsymbol{k}}(t)$ induced by the density change. Equation (4.5.27) corresponds to (1.6.7, 20 and 2.9.20), so that through the second fluctuation-dissipation theorem the function $\gamma_{\boldsymbol{k}}[\omega]$ is determined by the time correlation of the random current. Thus we may write it as

$$\gamma_{\boldsymbol{k}}[\omega] = \frac{\boldsymbol{k}\cdot\mu^*[\boldsymbol{k}, \omega]\,\boldsymbol{k}}{\chi[\boldsymbol{k}, 0]} = \boldsymbol{k}\cdot D[\boldsymbol{k}, \omega]\cdot\boldsymbol{k}\,, \tag{4.5.30}$$

where the *local mobility* tensor is defined by

$$\mu^*[\boldsymbol{k}, \omega] = \beta\int_0^\infty dt\, e^{\mathrm{i}\omega t} \lim_{L\to\infty} \langle \boldsymbol{j}'_{-\boldsymbol{k}}(0); \boldsymbol{j}'_{\boldsymbol{k}}(t)\rangle / L^{-3} \tag{4.5.31}$$

and the diffusion constant D by

$$D[\boldsymbol{k}, \omega] = \mu^*[\boldsymbol{k}, \omega]/\chi[\boldsymbol{k}, 0]\,. \tag{4.5.32}$$

The induced current is then expressed as

$$\boldsymbol{j}^0(\boldsymbol{r}, t) = -\int_{-\infty}^{t} dt' \int d\boldsymbol{r}'\, D(\boldsymbol{r}-\boldsymbol{r}', t-t')\, \nabla n(\boldsymbol{r}, t)\,, \tag{4.5.33}$$

where $D(\boldsymbol{r}-\boldsymbol{r}', t-t')$ is the corresponding *nonlocal retarded diffusion* coefficient.

Mori's method (Sect. 2.9) produces an explicit form for the random current $\boldsymbol{j}'_{\boldsymbol{k}}$. The projection (2.9.8) is now defined by

$$\mathscr{P} g = n_{\boldsymbol{k}} \langle n_{-\boldsymbol{k}}; g\rangle / \langle n_{-\boldsymbol{k}}; n_{\boldsymbol{k}}\rangle \tag{4.5.34}$$

for an arbitrary operator g. Corresponding to (2.9.21) we introduce

$$j'_k(t) = \exp(-\mathrm{i}t\mathcal{P}'\mathcal{L})\, j_k \tag{4.5.35}$$

and replace the lower limit of the integration (4.5.28) by $t = 0$.

4.5.3 Shielding of the External Potential

Equations (2.9.33, 34) can be used to relate $\mu^*[k, \omega]$ and $\mu^{\mathrm{e}}[k, \omega]$ to obtain

$$\mu^*[\boldsymbol{k}, \omega] = \frac{\mu^{\mathrm{e}}[\boldsymbol{k}, \omega]}{1 + \boldsymbol{k}\cdot\mu^{\mathrm{e}}[\boldsymbol{k}, \omega]\cdot\boldsymbol{k}(\mathrm{i}\omega\chi[\boldsymbol{k}, 0])^{-1}} \tag{4.5.36}$$

or

$$\begin{aligned}\mu^{\mathrm{e}}[\boldsymbol{k}, \omega] &= \frac{\mu^*[\boldsymbol{k}, \omega]}{1 - \boldsymbol{k}\cdot\mu^*[\boldsymbol{k}, \omega]\cdot\boldsymbol{k}(\mathrm{i}\omega\chi[\boldsymbol{k}, 0])^{-1}} \\ &= \frac{\mu^*[\boldsymbol{k}, \omega]}{\varepsilon^*[\boldsymbol{k}, \omega]}\,.\end{aligned} \tag{4.5.37}$$

In the last equation we introduced the shielding factor $\varepsilon^*[k, \omega]$ by the definition

$$\varepsilon^*[\boldsymbol{k}, \omega] = 1 - \frac{\boldsymbol{k}\cdot\mu^*[\boldsymbol{k}, \omega]\cdot\boldsymbol{k}}{\mathrm{i}\omega\chi[\boldsymbol{k}, 0]}\,, \tag{4.5.38}$$

which can be written as

$$\frac{1}{\varepsilon^*[\boldsymbol{k}, \omega]} = 1 + \frac{\boldsymbol{k}\cdot\mu^{\mathrm{e}}[\boldsymbol{k}, \omega]\cdot\boldsymbol{k}}{\mathrm{i}\omega\chi[\boldsymbol{k}, 0]} = 1 - \frac{\chi[\boldsymbol{k}, \omega]}{\chi[\boldsymbol{k}, 0]} \tag{4.5.39}$$

by virtue of (4.5.37, 20).

The current response (4.5.17) is written as

$$\overline{j_k(t)} = -\operatorname{Re}\{\mu^*[k, \omega]\,\mathrm{i}\boldsymbol{k}\varphi^*_{\boldsymbol{k}}\,\mathrm{e}^{-\mathrm{i}\omega t}\}\,, \tag{4.5.40}$$

where the local field potential $\varphi^*_{\boldsymbol{k}}(t)$ is introduced by the definition

$$\varphi^*_{\boldsymbol{k}}(t) = \varphi^{\mathrm{e}}_{\boldsymbol{k}}(t)/\varepsilon^*[\boldsymbol{k}, \omega]\,, \tag{4.5.41}$$

which can be written, with (4.5.39, 16), as

$$\varphi^*_{\boldsymbol{k}}(t) = \varphi^{\mathrm{e}}_{\boldsymbol{k}}(t) - \frac{\chi[\boldsymbol{k}, \omega]}{\chi[\boldsymbol{k}, 0]}\varphi^{\mathrm{e}}_{\boldsymbol{k}}(t) = \varphi^{\mathrm{e}}_{\boldsymbol{k}}(t) + \frac{\overline{n_{\boldsymbol{k}}(t)}}{\chi[\boldsymbol{k}, 0]}\,. \tag{4.5.42}$$

This shows that the local field results from the external field and the effective field induced by the density change. The latter shields the applied field. For small nonzero $\boldsymbol{k}$'s (4.5.42) may be approximated by

$$\varphi^*(\boldsymbol{r}, t) = \varphi^{\mathrm{e}}(\boldsymbol{r}, t) + \frac{\partial \zeta}{\partial n} n(\boldsymbol{r}, t),$$

which means that the local field correction is the chemical potential change induced by the density change.

The current may be considered to be driven either by the external field with external mobility $\mu^{\mathrm{e}}[\boldsymbol{k}, \omega]$ or by the local field with local mobility $\mu^*[\boldsymbol{k}, \omega]$. Or the induced current is regarded as being composed of the current driven by the gradient of the external potential with the local mobility μ^* and the diffusion current driven by the gradient of the induced density change with the diffusion constant D,

$$\overline{\boldsymbol{j}_k(t)} = -\operatorname{Re}\{\mu^*[\boldsymbol{k}, \omega]\,\mathrm{i}\boldsymbol{k}\,\varphi_k^{\mathrm{e}}(t)\} - D[\boldsymbol{k}, \omega]\,\mathrm{i}\boldsymbol{k}\,\overline{n_k(t)}\}. \tag{4.5.43}$$

As long as the particle interaction is short ranged, it can generally be proved that the static susceptibility $\chi[\boldsymbol{k}, 0]$ is nonsingular at $\boldsymbol{k}=0$, so that the local field correction vanishes in the long-wavelength limit $\boldsymbol{k} \to 0$, i.e.,

$$\lim_{\boldsymbol{k}\to 0} \varepsilon^*[\boldsymbol{k}, \omega] = 1 \quad \text{for a finite } \omega, \tag{4.5.44}$$

if the particles are not bound. The mobilities should remain finite and are equal in this limit, i.e.,

$$\mu^{\mathrm{e}}[0, \omega] = \mu^*[0, \omega]. \tag{4.5.45}$$

On the other hand, if ω is made to vanish first, keeping $\boldsymbol{k}$ finite, then from (4.5.39)

$$\lim_{\omega\to 0} \varepsilon^*[\boldsymbol{k}, \omega] = \infty \quad \text{for a finite } \boldsymbol{k}, \tag{4.5.46}$$

which means that the external potential is completely shielded, namely that $\varphi^* = 0$ in the static limit. No current is present in equilibrium under the action of an external potential with a finite wavelength. Accordingly

$$\mu^{\mathrm{e}}[\boldsymbol{k}, 0] = 0$$

or more precisely

$$\boldsymbol{k}\cdot\mu^{\mathrm{e}}[\boldsymbol{k}, \omega]\,\boldsymbol{k} \sim -\chi[\boldsymbol{k}, 0]\,\mathrm{i}\omega \quad \text{as} \quad \omega \sim 0. \tag{4.5.47}$$

This is a special case of the general theorem (2.9.37) since the current is derived from the density and the density fluctuation is bounded. On the other hand, the local mobility remains finite in the static limit.

Equation (4.5.32) gives the *generalized Einstein relation*

$$D = D[0,0] = \mu\,[0,0]/\frac{\partial n}{\partial \zeta}\,. \tag{4.5.48}$$

The classical factor $1/kT$ is replaced here by $\partial n/\partial \zeta$, because the current is driven by the gradient of the chemical potential rather than that of the density.

4.5.4 Resistivity Formula

We show here that (4.5.17) for the current is equivalent to

$$\frac{d}{dt}\overline{j_k(t)} = -\frac{n}{m}\mathrm{i}\boldsymbol{k}\,\varphi_k^{\mathrm{e}}(t) - \int_{-\infty}^{t} \varrho_k(t-t')\cdot\overline{j_k(t')}\,dt' \tag{4.5.49}$$

or in Fourier transforms

$$-\mathrm{i}\omega\,\overline{j_k[\omega]} = -\frac{N}{m}\mathrm{i}\boldsymbol{k}\,\varphi_k^{\mathrm{e}} - \varrho\,[\boldsymbol{k},\omega]\cdot\overline{j_k[\omega]} \tag{4.5.50}$$

for a harmonic driving field. This represent the balance of forces acting on the particles.

For this purpose, we carry out partial integration in (4.5.19) and obtain

$$\mu^{\mathrm{e}}[\boldsymbol{k},\omega] = -\frac{1}{\mathrm{i}\omega}\left[\frac{n}{m} + \beta\int_0^\infty dt\,\mathrm{e}^{\mathrm{i}\omega t}\lim_{L\to\infty}\langle \boldsymbol{j}_{-k}(0);\dot{\boldsymbol{j}}_k(t)\rangle\,L^{-3}\right],$$

with

$$\beta\langle \boldsymbol{j}_{-k};\boldsymbol{j}_k\rangle = N/m\,,$$

which is seen easily from (4.5.6) and (4.2.18). Using the above expression of $\mu^{\mathrm{e}}[\boldsymbol{k},\omega]$, we transform (4.5.17) as

$$-\mathrm{i}\omega\,\overline{j_k[\omega]}\left[1+\frac{m\beta}{N}\int_0^\infty dt\,\mathrm{e}^{\mathrm{i}\omega t}\langle \dot{\boldsymbol{j}}_{-k};\dot{\boldsymbol{j}}_k(t)\rangle - \frac{m\beta}{N\mathrm{i}\omega}\int_0^\infty dt\,\mathrm{e}^{\mathrm{i}\omega t}\langle \boldsymbol{j}_{-k};\dot{\boldsymbol{j}}_k(t)\rangle\right]$$

$$= -\left[\frac{N}{m}1 + \beta\int_0^\infty dt\,\mathrm{e}^{\mathrm{i}\omega t}\langle \boldsymbol{j}_{-k};\dot{\boldsymbol{j}}_k(t)\rangle\right]\mathrm{i}\boldsymbol{k}\,\varphi_k^{\mathrm{e}}\,,$$

noting that

$$\int_0^\infty dt\,\mathrm{e}^{\mathrm{i}\omega t}\langle \boldsymbol{j}_{-k};\dot{\boldsymbol{j}}_k(t)\rangle = \frac{1}{\mathrm{i}\omega}\int_0^\infty dt\,\mathrm{e}^{\mathrm{i}\omega t}\langle \dot{\boldsymbol{j}}_{-k};\dot{\boldsymbol{j}}_k(t)\rangle\,.$$

This yields (4.5.50) with the *resistivity* given by

$$\varrho[\boldsymbol{k},\omega]=\frac{\int_0^\infty dt\,\exp(\mathrm{i}\omega t)\,\langle\dot{\boldsymbol{j}}_{-\boldsymbol{k}};\dot{\boldsymbol{j}}_{\boldsymbol{k}}(t)\rangle/\langle\boldsymbol{j}_{-\boldsymbol{k}};\boldsymbol{j}_{\boldsymbol{k}}\rangle}{1+\int_0^\infty dt\,\exp(\mathrm{i}\omega t)\,\langle\boldsymbol{j}_{-\boldsymbol{k}};\dot{\boldsymbol{j}}_{\boldsymbol{k}}(t)\rangle/\langle\boldsymbol{j}_{-\boldsymbol{k}};\boldsymbol{j}_{\boldsymbol{k}}\rangle}$$

or

$$\varrho[\boldsymbol{k},\omega]=\frac{\int_0^\infty dt\,\exp(\mathrm{i}\omega t)\,\langle\dot{\boldsymbol{j}}_{-\boldsymbol{k}};\dot{\boldsymbol{j}}_{\boldsymbol{k}}(t)\rangle/\langle\boldsymbol{j}_{-\boldsymbol{k}};\boldsymbol{j}_{\boldsymbol{k}}\rangle}{1+\frac{1}{\mathrm{i}\omega}\int_0^\infty dt\,\exp(\mathrm{i}\omega t)\,\langle\dot{\boldsymbol{j}}_{-\boldsymbol{k}};\dot{\boldsymbol{j}}_{\boldsymbol{k}}(t)\rangle/\langle\boldsymbol{j}_{-\boldsymbol{k}};\boldsymbol{j}_{\boldsymbol{k}}\rangle}\,, \tag{4.5.51}$$

in terms of which the external mobility is expressed as

$$\mu^{\mathrm{e}}[\boldsymbol{k},\omega]=\frac{n}{m}\,\frac{1}{-\mathrm{i}\omega+\varrho[\boldsymbol{k},\omega]}\,.$$

Here we are essentially repeating the treatment given in Sect. 2.9. Equation (4.5.51) is an example of the general equation (2.9.34) (note that the sign of i is reversed here). It gives the resistivity in terms of the force-force correlation since $m\dot{\boldsymbol{j}}_{-k}$ and $m\dot{\boldsymbol{j}}_k$ are the density of force coming from particle interactions, and it is, in fact, equivalent to the conductivity formula (4.5.19). There has been unfortunate confusion in the literature regarding these formulas [4.10]. Implications of the resistivity formula were briefly discussed in Sect. 2.9.

4.5.5 Dielectric Shielding and Electric Conductivity

For charged particles, the function $\chi[\boldsymbol{k},0]$ is singular at $\boldsymbol{k}=0$ in such a way that

$$\chi[\boldsymbol{k},0]^{-1}\sim 4\pi\mathrm{e}^2/\boldsymbol{k}^2 \tag{4.5.52}$$

is the dominant term for small k's, e being the charge on each particle. This comes from the Coulomb interaction energy which contributes a part of the free energy change when a density fluctuation is induced with the wave number $\boldsymbol{k}$. Treating only $\boldsymbol{k}$'s much smaller than the inverse of the Debye shielding length, only this term of $\chi[\boldsymbol{k},0]$ may define the local field

$$\varphi_k(t)=\varphi_k^{\mathrm{e}}(t)+\frac{4\pi\mathrm{e}^2}{\boldsymbol{k}^2}\,\overline{n_k(t)}\,L^{-3}\,. \tag{4.5.53}$$

Then the shielding factor becomes equal to the usual dielectric constant $\varepsilon[\boldsymbol{k},\omega]$. The external field derived from φ^{e} corresponds to the electric displacement $\boldsymbol{D}$, whereas the local field is the electric field $\boldsymbol{E}$, namely

$$\boldsymbol{E}[\boldsymbol{k},\omega]=\boldsymbol{D}[\boldsymbol{k},\omega]/\varepsilon[\boldsymbol{k},\omega]\,.$$

In the static limit

$$4\pi e^2 \bar{n}_k / L^3 + k^2 \varphi_k^e = 0$$

from (4.5.16), which means that the induced space charge exactly cancels out the external charges producing the external potential. This is true only for small $\boldsymbol{k}$.

The conductivity is customarily defined by

$$\sigma[\boldsymbol{k}, \omega] = e^2 \mu^* [\boldsymbol{k}, \omega],$$

which is distinguished from the external conductivity $\sigma^e [\boldsymbol{k}, \omega]$. Corresponding to (4.5.38, 39), there are two expressions for the dielectric constant,

$$\varepsilon[\boldsymbol{k}, \omega] = 1 - \frac{4\pi}{i\omega k^2} \boldsymbol{k} \cdot \sigma[\boldsymbol{k}, \omega] \cdot \boldsymbol{k} \tag{4.5.54}$$

and

$$\frac{1}{\varepsilon[\boldsymbol{k}, \omega]} = 1 + \frac{4\pi}{i\omega k^2} \boldsymbol{k} \cdot \sigma^e [\boldsymbol{k}, \omega] \cdot \boldsymbol{k}, \tag{4.5.55}$$

where

$$\sigma[\boldsymbol{k}, \omega] = e^2 \beta \int_0^\infty dt\, e^{i\omega t} \lim_{L \to \infty} \langle \boldsymbol{j}'_{-\boldsymbol{k}}(0); \boldsymbol{j}'_{\boldsymbol{k}}(t) \rangle L^{-3} \tag{4.5.56}$$

and

$$\sigma^e [\boldsymbol{k}, \omega] = e^2 \beta \int_0^\infty dt\, e^{i\omega t} \lim_{L \to \infty} \langle \boldsymbol{j}_{-\boldsymbol{k}}(0); \boldsymbol{j}_{\boldsymbol{k}}(t) \rangle L^{-3} \tag{4.5.57}$$

are the local and external conductivities. They are mutually related by

$$\sigma^e [\boldsymbol{k}, \omega] = \frac{\sigma[\boldsymbol{k}, \omega]}{\varepsilon[\boldsymbol{k}, \omega]} = \frac{\sigma[\boldsymbol{k}, \omega]}{1 - 4\pi\sigma[\boldsymbol{k}, \omega]/i\omega}$$

and

$$\sigma[\boldsymbol{k}, \omega] = \frac{\sigma^e [\boldsymbol{k}, \omega]}{1 + 4\pi\sigma^e [\boldsymbol{k}, \omega]/i\omega}$$

if the system is isotropic. Equation (4.5.55) inserted into (4.5.57) is the *Noziéres-Pines formula* [4.11]. It is commonly expressed in terms of the density Green's function as given by (4.5.18), which replaces $\boldsymbol{k} \cdot \sigma^e \cdot \boldsymbol{k}/i\omega$. We shall give an example of a calculation in Sect. 5.2.2.

Rigorous calculation of electric conductivities is not simple because of the complication introduced by the presence of both long-ranged Coulomb interactions and short-ranged scattering potentials. A common approximation is to use (4.5.56) regarding the random current to be produced by charge carriers scattered by properly shielded scatterers. For example, an effective Hamiltonian can describe the electron-phonon interaction when treating the electronic conductivity of a metal, so that direct electron-electron interactions do not have to be included.

4.5.6 Kramers-Kronig Relations and the Sum Rules

The dispersion relations (3.6.8) yield

$$\mathrm{Re}\left\{\frac{1}{\varepsilon[\boldsymbol{k},\omega]}-1\right\}=\frac{1}{\pi}\int_0^{\infty}d\omega'\,\frac{\mathcal{P}}{\omega'-\omega}\,\mathrm{Im}\left\{\frac{1}{\varepsilon[\boldsymbol{k},\omega']}\right\},$$

$$\mathrm{Im}\left\{\frac{1}{\varepsilon[\boldsymbol{k},\omega]}\right\}=-\frac{1}{\pi}\int_0^{\infty}d\omega'\,\frac{\mathcal{P}}{\omega'-\omega}\,\mathrm{Re}\left\{\frac{1}{\varepsilon[\boldsymbol{k},\omega']}-1\right\} \qquad (4.5.58)$$

for the complex admittance (4.5.55), while for (4.5.54)

$$\begin{aligned}\varepsilon'[\boldsymbol{k},\omega]-1&=\frac{1}{\pi}\int_{-\infty}^{\infty}d\omega'\,\frac{\mathcal{P}}{\omega'-\omega}\,\varepsilon''[\boldsymbol{k},\omega'],\\ \varepsilon''[\boldsymbol{k},\omega]&=-\frac{1}{\pi}\int_{-\infty}^{\infty}d\omega'\,\frac{\mathcal{P}}{\omega'-\omega}\,(\varepsilon'[\boldsymbol{k},\omega']-1)\,.\end{aligned} \qquad (4.5.59)$$

The latter are the expressions given by *Kramers* and *Kronig* [4.12].

A simple application of the moment sum rule (3.7.2, 8) can be made for the dielectric function (4.5.55). Using (4.5.6), we write (4.5.57) as

$$\begin{aligned}\boldsymbol{k}\cdot\sigma^{\mathrm{e}}[\boldsymbol{k},\omega]\cdot\boldsymbol{k}&=e^2\beta\int_0^{\infty}dt\,\mathrm{e}^{\mathrm{i}\omega t}\lim_{L\to\infty}\langle\dot{n}_{-\boldsymbol{k}}(0);\dot{n}_{\boldsymbol{k}}(t)\rangle L^{-3}\\ &=-\mathrm{i}\omega e^2\beta\int_0^{\infty}dt\,\mathrm{e}^{\mathrm{i}\omega t}\lim\langle\dot{n}_{-\boldsymbol{k}}(0);n_{\boldsymbol{k}}(t)\rangle L^{-3}\end{aligned} \qquad (4.5.60)$$

after partial integration, noticing that

$$\langle\dot{n}_{-\boldsymbol{k}}(0);n_{\boldsymbol{k}}(0)\rangle=\langle(n_{-\boldsymbol{k}},n_{\boldsymbol{k}})\rangle=0\,. \qquad (4.5.61)$$

Using the third expression of (4.5.60)

$$\int_{-\infty}^{\infty}d\omega\left(\frac{1}{\varepsilon[\boldsymbol{k},\omega']}-1\right)=0 \qquad (4.5.62)$$

by virtue of (4.5.61). Using the second expression of (4.5.60) and

$$\begin{aligned}\beta\langle\dot{n}_{-\boldsymbol{k}};\dot{n}_{\boldsymbol{k}}\rangle L^{-3}&=\beta\,\mathrm{i}\,\boldsymbol{k}\cdot\langle\boldsymbol{j}_{-\boldsymbol{k}};\dot{n}_{\boldsymbol{k}}\rangle L^{-3}\\ &=\mathrm{i}\,\boldsymbol{k}\cdot\langle(\boldsymbol{j}_{-\boldsymbol{k}},n_{\boldsymbol{k}})\rangle L^{-3}=n\boldsymbol{k}^2/m\end{aligned}$$

or

$$\beta\langle\boldsymbol{j}_{-\boldsymbol{k}};\boldsymbol{j}_{\boldsymbol{k}}\rangle L^{-3}=n/m\,,$$

we obtain

$$\int_{-\infty}^{\infty} d\omega \left(\frac{1}{\varepsilon[\boldsymbol{k}, \omega]} - 1 \right) \omega = - \mathrm{i} \pi \omega_\mathrm{p}^2 , \tag{4.5.63}$$

where $\omega_\mathrm{p} = (4\pi n e^2/m)^{1/2}$ denotes the plasma frequency.

Generalizing the *oscillator strength* defined in the case of the dielectric function as almost equal to 1, we define it as

$$f[\boldsymbol{k}, \omega] = \frac{m\omega}{2\pi^2 e^2} \operatorname{Im} \left\{ 1 - \frac{1}{\varepsilon[\boldsymbol{k}, \omega]} \right\} . \tag{4.5.64}$$

According to (4.3.24), this is an even function with frequency ω. The imaginary part of the second of the sum rules (4.5.63) can be written as

$$\int_0^{\infty} d\omega f[\boldsymbol{k}, \omega] = n = N/V \tag{4.5.65}$$

and the first of the dispersion relations (4.5.58) becomes

$$\operatorname{Re} \left\{ \frac{1}{\varepsilon[\boldsymbol{k}, \omega]} - 1 \right\} = - \frac{4\pi e^2}{m} \int_0^{\infty} d\omega' \frac{\mathscr{P}}{\omega'^2 - \omega^2} f[\boldsymbol{k}, \omega'] . \tag{4.5.66}$$

Taking (4.5.65) into account, the last equation shows that we may interpret $f[\boldsymbol{k}, \omega]$ as the number per unit volume of oscillators with an eigenfrequency ω. In the language of the classical electron theory, the equation of motion for a "dispersion electron" having the eigenfrequency ω' may be written with respect to its component in the direction of the electric field as

$$\frac{d^2}{dt^2} x(t) = - \omega'^2 x(t) - \frac{e}{m} D(t) ,$$

provided that the eigenfrequency ω' is determined by considering the effects of interactions with the plasma. If we take an external field of the form $D(t) = D_0 \exp(-\mathrm{i}\omega t + \varepsilon t)$ $(\varepsilon \to +0)$, the contribution of that electron to the electric polarization is given by

$$- e x(t) = \frac{e^2/m}{\omega'^2 - (\omega + \mathrm{i}\varepsilon)^2} D(t) .$$

If $f[\boldsymbol{k}, \omega'] \, d\omega'$ such electrons exist per unit of volume, the electronic displacement becomes

$$D(t) = E(t) + 4\pi \int_0^{\infty} d\omega' f[\boldsymbol{k}, \omega'] \frac{e^2/m}{\omega'^2 - (\omega + \mathrm{i}\varepsilon)^2} D(t) ,$$

and hence the complex dielectric function is given by

$$\frac{1}{\varepsilon[\boldsymbol{k},\omega]}-1=-\int_0^\infty d\omega f[\boldsymbol{k},\omega']\frac{4\pi e^2}{m}\frac{1}{\omega'^2-(\omega+\mathrm{i}\varepsilon)^2}. \tag{4.5.67}$$

Taking the real parts of both sides of this equation yields (4.5.66). In this sense $f[\boldsymbol{k},\omega]$ is also called the number density of dispersion electrons.

4.6 Response to Thermal Internal Forces

As mentioned in Sect. 4.1, the internal thermal forces within a fluid are produced by the presence of a velocity gradient, a concentration gradient, or a temperature gradient. These parameters are, by themselves, statistical in nature, unlike those which characterize external mechanical disturbances. They characterize deformations of the distribution function or the density matrix, or the states of the reservoirs in contact with the system under consideration. Therefore, the treatment developed in Sect. 4.2 does not seem to be directly applicable to nonequilibrium states produced by such thermal forces, in order to obtain explicit formulas expressing the responses to thermal forces. It is generally accepted, however, that such formulas exist and are of the same form as those for responses to mechanical disturbances. Namely, they are given in terms of canonical correlation functions or generally some sort of Green's functions of the relevant physical quantities.

A simple example has already been given in previous sections. We have seen that a particle current is produced by the presence of an external field $\varphi_{\boldsymbol{k}}^{\mathrm{e}}$. The induced current consists of two parts, one directly driven by the external field and the other caused by the nonuniformity of particle density. The latter current is an indirect consequence of the external field. But it will be agreed that the same current is produced by the same nonuniformity of density caused by some other means. This is a diffusion process and (4.5.32) with $\mu^*[\boldsymbol{k},\omega]$ (4.5.31) gives the generalized diffusion coefficient. In the limit, $\boldsymbol{k}\to 0$ and $\omega\to 0$ in this order, the diffusion constant is given by (4.5.48)

$$D=(\partial n/\partial\zeta)^{-1}\beta\lim_{\omega\to 0}\int_0^\infty dt\, \mathrm{e}^{\mathrm{i}\omega t}\lim_{\substack{\boldsymbol{k}\to 0\\ L\to\infty}}\langle j_{-\boldsymbol{k}}(0);j_{\boldsymbol{k}}(t)\rangle L^{-3}, \tag{4.6.1}$$

which is the celebrated *Einstein relation.*

Similar treatments are possible for other kinds of nonequilibrium transport processes. A mechanical disturbance generally causes various kinds of responses; flows of momentum, energy and material particles are induced by the disturbance. Some flows are considered as direct responses and others are indirectly driven by thermal forces produced by the disturbance. Thus, in principle, we should be able to identify each kinetic coefficient by analyzing the expression for mechanical admittance.

We shall not pursue this line of approach any further. Instead we consider in the following different approaches which directly treat nonequilibrium states with thermal forces present. This aids in better understanding the nature of the problem of nonequilibrium statistical mechanics.

4.6.1 Onsager's Postulate

In his celebrated paper of 1931 [4.5], *Onsager* introduced a postulate upon which he based his proof of the reciprocity law of kinetic coefficients. The postulate asserts that the average regression of thermal fluctuations in equilibrium follows the macroscopic law of relaxation. Since the microscopic law of fluctuation is dynamical, the time-reversal symmetry of dynamics naturally reflects the symmetry of the kinetic coefficients, for example, the tensor of heat conductivity.

Consider, for example, the temperature distribution in a bulk system in thermal equilibrium, characterized by a uniform temperature T_0. Temperature fluctuates in space and time so that the deviation $\Delta T(\boldsymbol{r}, t)$ from T_0 is considered as a random field. We select samples from the ensemble which have the same initial values and construct the conditional average $\overline{\Delta T(\boldsymbol{r}, t)}$. Then *Onsager's* postulate means that it follows

$$c\varrho \frac{\partial}{\partial t} \overline{\Delta T(\boldsymbol{r}, t)} = \operatorname{div} \{\varkappa \operatorname{grad} \overline{\Delta T(\boldsymbol{r}, t)}\}, \tag{4.6.2}$$

where c is the specific heat, ϱ the density, and $\varkappa$ is the heat conductivity tensor. Instead of fixing the initial values, we can take the average over their equilibrium distribution to obtain the correlation function of temperature fluctuation,

$$\langle \Delta T(\boldsymbol{r}, t)\, \Delta T(\boldsymbol{r}', t') \rangle .$$

As a function of $\boldsymbol{r}$ and t, this obeys (4.6.2). More conveniently we introduce the Fourier components

$$A(t) = L^{-3/2} \int \Delta T(\boldsymbol{r}, t) \exp(-\mathrm{i}\boldsymbol{k} \cdot \boldsymbol{r})\, d\boldsymbol{r},$$

where L^3 is the volume of the system. Here the A_k's are random variables and their correlation functions decay in time as

$$\langle A_k(t)\, A_{-k}(0) \rangle = \langle A_k A_{-k} \rangle \exp(-\alpha_k t), \tag{4.6.3}$$

where

$$\alpha_k = a\boldsymbol{k}^2$$

and $a = \varkappa / c\varrho$ is the thermal diffusivity if the system is isotropic.

4.6.2 Fluctuation of Macrovariables as Brownian Motion

Like the above example of temperature fluctuation, macrovariables such as particle density, energy density, or momentum density fluctuate in thermal equilibrium. These densities are local quantities defined for an appropriate size of spatial element smaller than the bulk system and yet larger than the microscopic size containing a large number of particles. Thus the wave numbers of their Fourier components are much smaller than the inverse of interparticle distances and yet larger than $1/L$.

Einstein first noticed in 1906 [4.13] that such fluctuations of macrovariables are, like the motion of small particles in fluids, a sort of Brownian motion. If a Fourier component of such a macrovariable is A_k, it follows the Langevin equation,

$$\dot{A}_k(t) = -\alpha_k A_k(t) + R_k(t), \tag{4.6.4}$$

where $R_k(t)$ is the random force causing random motion of A_k. As an ideal Brownian motion, it is a Gaussian white noise. We have emphasized in Chaps. 1 and 2, particularly in Sect. 2.2, that the process of A_k is then described by the Fokker-Planck equation (2.2.31)

$$\frac{\partial}{\partial t} P(A_k, t) = \frac{\partial}{\partial A_k} \alpha_k A_k P(A_k, t) + D \frac{\partial^2}{\partial A_k \partial A_{-k}} P(A_k, t), \tag{4.6.5}$$

where

$$D = \int_0^\infty \langle R_{-k}(t_0) R_k(t_0 + t) \rangle dt.$$

The equilibrium distribution $P_e(A)$ is given by

$$P_e(A_k) = \text{const} \cdot \exp(-\alpha_k |A_k|^2 / 2D).$$

Therefore

$$\alpha_k = D / \langle |A_k|^2 \rangle = \frac{1}{\langle |A_k|^2 \rangle} \int_0^\infty \langle R_{-k}(t_0) R_k(t_0 + t) \rangle dt, \tag{4.6.6}$$

as shown in Chap. 1. We go a step further and write it as

$$\alpha_k = \frac{1}{\langle |A_k|^2 \rangle} \int_0^{\tau_p} \langle \dot{A}_{-k}(t_0) \dot{A}_k(t_0 + t) \rangle dt, \tag{4.6.7}$$

where the upper limit of integration is cut off at a certain time τ_p satisfying the condition

$$\tau_c \ll \tau_p \ll \tau_r = 1/\alpha_k. \tag{4.6.8}$$

The cut-off is necessary because the time integral ought to vanish if it is extended to infinity as noted in Sect. 2.9. In (4.6.8) τ_c is the correlation time of the correlation function inside the integral (4.6.7) and τ_r is the relaxation time of A_k. We assume here that the characteristic times τ_c and τ_r greatly differ in magnitude, which makes it possible to choose τ_p to satisfy (4.6.8).

The difference between $\dot{A}_k(t)$ and $R_k(t)$ is $-\alpha_k A_k(t)$, which changes slowly with the time scale of τ_r. On the other hand, the random noise $R_k(t)$ changes very rapidly, the time scale of which is idealized here to zero. Therefore, the cross-correlation of $\dot{A}_k(t)$ and $R_k(t)$ vanishes while the time integration of the term $\alpha_k^2 \langle A_k(t_0) A_{-k}(t_0+t)\rangle / \langle |A_k|^2 \rangle$ is $\alpha_k^2 \tau_p$, which is negligibly small in comparison to the integral (4.6.6) if the cut-off limit τ_p satisfies the condition (4.6.8). Chapter 2 shows that the random force $R(t)$ is not a simple object so that it is advantageous to express α_k in terms of $\dot{A}_k(t)$ instead of $R_k(t)$. As already noted in Sect. 2.9, (4.6.7) was first obtained by *Kirkwood* [4.14]. Now the important point is that if the decay rate γ_k of average regression behaves like $a\boldsymbol{k}^2$, then we can push τ_r to infinity and accordingly τ_p is also made infinite, so (4.6.7) becomes

$$a = \int_0^\infty dt \lim_{k \to 0} \langle \dot{A}_{-k}(t_0) \dot{A}_k(t_0 + t) \rangle / k^2 , \tag{4.6.9}$$

which is the exact expression of the kinetic coefficient appearing in the macroscopic law of relaxation

$$\dot{A}_k(t) = -a k^2 A_k(t) \tag{4.6.10}$$

if we accept Onsager's postulate. An example of the above is the diffusion constant which was already given by (4.6.1). If A_k represents a Fourier component of the particle density, $\dot{A}_k$ is equal to $(-\mathrm{i}\boldsymbol{k}\cdot \boldsymbol{j}_k)$ through (4.5.6). Then (4.6.9) is nothing but (4.6.1).

It is important to note that this is more clearly understood using (2.9.34) in which $\gamma[\omega]$ and $\gamma_t[\omega]$ are both proportional to $\boldsymbol{k}^2$ for small $\boldsymbol{k}$'s. Therefore

$$\lim_{\omega \to 0} \lim_{k \to 0} \gamma[\omega]/\boldsymbol{k}^2 = \lim_{\omega \to 0} \lim_{k \to 0} \frac{(\mathrm{i}\,\omega/\boldsymbol{k}^2)\,\gamma_t[\omega]/\boldsymbol{k}^2}{(\mathrm{i}\,\omega/\boldsymbol{k}^2) + \gamma_t[\omega]/\boldsymbol{k}^2} = \lim_{\omega \to 0} \lim_{k \to 0} \gamma_t[\omega]/\boldsymbol{k}^2 .$$

by making $\boldsymbol{k}$ approach zero first. The last expression is identical with that on the right-hand side of (4.6.9).

It should be possible to derive (4.6.5) from the dynamic Liouville equation by projecting the phase-space distribution onto the reduced distribution describing the evolution of the set of macrovariables $A_1, A_2, \dots$ if we consider more than one such variable. *Green* [4.15] was the first who showed how this can be achieved. He derived under a few assumptions a Fokker-Planck equation generalizing (4.6.5), giving an expression for the kinetic coefficient in terms of the relevant correlation function. This is considered as a special case of a more general expression of dynamic admittance considered by *Kubo* et al. [4.3, 19]. Hence such an expression for the kinetic coefficients is often called the Green-Kubo formula.

Green's method was reformulated by several authors. In particular, *Zwanzig* applied a more systematic method of projection operators to the Liouville equation [4.16]. An important generalization was the extension to nonlinear cases by *Mori* and co-workers [4.17]. Nonlinear coupling between long-wavelength fluctuations becomes essential for transport phenomena near critical points [4.18]. This is, however, beyond the scope of the present book, so we shall not enter into this topic.

4.6.3 A General Formulation of Onsager's Postulate

Let us treat the problem more along the lines of *Onsager's* original approach [4.19]. We consider a set of macrovariables (gross variables) $\{A_1, A_2, \dots\}$ which we assume to be all zero in equilibrium. For convenience, we write them as a column vector

$$\boldsymbol{A} = \begin{pmatrix} A_1 \\ A_2 \\ \cdot \\ \cdot \end{pmatrix}$$

and as a row vector

$$\boldsymbol{A}^+ = (A_1, A_2, \dots) .$$

If the macrovariables are forced to deviate from zero, the entropy decreases. The statistical tendency to make entropy increase is the source of thermal forces. Thus

$$X_l = \partial S / \partial A_l \tag{4.6.11}$$

is considered as the thermal force conjugate to the variable A_l.

The macroscopic law of irreversible relaxation of the A's is expressed by a set of differential equations (as in the previous section we omit circumflexes for dynamical quantities)

$$\frac{d}{dt}\boldsymbol{A}(t) = \mathrm{i}\Omega \boldsymbol{A} + G\boldsymbol{X} , \tag{4.6.12}$$

where G is the matrix of kinetic coefficients for the irreversible process and $\mathrm{i}\Omega$ is a matrix representing dynamic coupling of the variables, for example that between a coordinate and its conjugate momentum. Its elements are of the form, see (2.9.48),

$$\Omega_{lm} = \mathrm{i}\omega_{lm} , \quad \omega_{lm} = -\omega_{ml} .$$

Switching to the microscopic viewpoint, the above equation becomes

$$\overline{A(t+\Delta t|A')}-A'=(\mathrm{i}\Omega A'+GX)\,\Delta t\,. \tag{4.6.13}$$

Here, Δt is microscopically finite but is very short in comparison with the macroscopic relaxation time like τ_r in (4.6.8). We denote by $A(t+\Delta t|A')$ the value of A realized at time $t+\Delta t$ when A is specified as A' at t. It must show fluctuations so that its conditional expectation is identified with the macrosopic value.

Now we observe spontaneous thermal fluctuations in equilibrium. The values of A realized as fluctuations are much smaller than those realized in a macroscopic relaxation process. But following Onsager, we assume that the fluctuations follow the same law of relaxation (4.6.13) on the average, leading to the decay law of the correlation function

$$\langle A(t+\Delta t)A^+(t)\rangle-\langle A(t)A^+(t)\rangle=(\mathrm{i}\Omega\langle AA^+\rangle-k_\mathrm{B}G)\,\Delta t\,, \tag{4.6.14}$$

where k_B is the Boltzmann constant. To see this, we multiply (4.6.13) by $A'W(A')$, where W is the probability distribution function for A' related to the entropy by $S=k_\mathrm{B}\log W$, and integrate over A'. The second term on the right-hand side is

$$\begin{aligned}&\int\sum_m G_{lm}\frac{\partial S}{\partial A'_m}A'_n W(A')\,dA'\\&\quad=k_\mathrm{B}\int\sum_m G_{lm}A'_n\frac{\partial}{\partial A'_m}W\,dA'=-k_\mathrm{B}\sum_m G_{lm}\,\delta_{mn}\,.\end{aligned}$$

So far we considered A's as classical c numbers. If they are treated as quantal quantities, we expect the correlation functions to be replaced by the corresponding canonical correlations. In the following we show this and derive the explicit forms of $\mathrm{i}\Omega$ and G,

$$\langle A(t+\Delta t);A^+(t)\rangle-\langle A(t);A^+(t)\rangle=(\mathrm{i}\Omega\langle A;A^+\rangle-kG)\,\Delta t\,, \tag{4.6.15}$$

where

$$\mathrm{i}\Omega=\langle\dot{A};A^+\rangle\langle A;A^+\rangle^{-1}\quad\text{and} \tag{4.6.16}$$

$$G=\frac{1}{k_\mathrm{B}}\int_0^{\tau_\mathrm{p}}\langle\dot{A}(\tau);\dot{A}^+(0)\rangle\,d\tau\,. \tag{4.6.17}$$

For this purpose let us first consider the entropy $S(A')$ as a function of the specified values of the macrovariables $A'_1, A'_2, \dots$. The entropy S is defined in terms of the density matrix by

$$S=-k_\mathrm{B}\,\mathrm{Tr}\,\{\varrho\log\varrho\}\,.$$

This is maximized under the subsidiary conditions

$$\begin{aligned} \mathrm{Tr}\,\{\varrho \mathscr{H}\} &= \langle \mathscr{H} \rangle = E\,, \\ \mathrm{Tr}\,\{\varrho A_l\} &= \langle A_l \rangle = A_l' \end{aligned} \tag{4.6.18}$$

to obtain the *local equilibrium* density matrix

$$\varrho = \varrho_{\mathrm{loc}} = \exp\left[\Phi(0) + \Delta\Phi(A') - \beta\mathscr{H} + \sum_l \xi_l A_l\right], \tag{4.6.19}$$

where β and ξ are the undetermined multipliers subject to (4.6.18). The second condition of (4.6.18) gives

$$-\,\partial\Delta\Phi/\partial\xi_l = A_l' \quad \text{or}$$

$$\sum_m \langle A_l; A_m \rangle\, \xi_m = A_l'$$

because by expanding (4.6.19) in ξ

$$\Delta\Phi = -\tfrac{1}{2}\,\boldsymbol{\xi}^+ \cdot \langle \boldsymbol{A}; \boldsymbol{A}^+ \rangle \cdot \boldsymbol{\xi}\,,$$

using the vector notations for $\boldsymbol{A}$ and $\boldsymbol{\xi}$. Accordingly the entropy is

$$S(\boldsymbol{A}') = S(0) - \tfrac{1}{2}\,k_{\mathrm{B}}\,\boldsymbol{A}'^+ \cdot \langle \boldsymbol{A}; \boldsymbol{A}^+ \rangle^{-1} \cdot \boldsymbol{A}'\,,$$

where $\langle \boldsymbol{A}; \boldsymbol{A}^+ \rangle^{-1}$ is the inverse of the matrix of canonical correlations $\langle \boldsymbol{A}; \boldsymbol{A}^+ \rangle$. The thermal force (4.6.11) is then

$$X_l = \partial S/\partial A_l' = -\,k_{\mathrm{B}}\,\xi_l = -\,k_{\mathrm{B}} \sum_m \langle \boldsymbol{A}; \boldsymbol{A}^+ \rangle_{lm}^{-1}\, A_m'\,. \tag{4.6.20}$$

If the statistical ensemble at time t is given by (4.6.19), then the expectation of A_l at $t+\Delta t$ is

$$\overline{A_l(t+\Delta t\,|\,\boldsymbol{A}')} = \mathrm{Tr}\,\{\varrho_{\mathrm{loc}}\, A_l(\Delta t)\} = \sum_m \langle A_l(\Delta t); A_m \rangle\, \xi_m\,,$$

where $\boldsymbol{A}(\Delta t)$ is the Heisenberg operator. Thus

$$\overline{A_l(t+\Delta t\,|\,\boldsymbol{A}')} - A_l' = \sum_m \left[\langle A_l(\Delta t); A_m \rangle - \langle A_l; A_m \rangle\right] \xi_m\,. \tag{4.6.21}$$

Now we note the identity

$$\langle A_l(\Delta t); A_m(0) \rangle - \langle A_l; A_m \rangle = \Delta t\, \langle \dot{A}_l; A_m \rangle - \int_0^{\Delta t} (\Delta t - \tau)\, \langle \dot{A}_l(\tau); \dot{A}_m(0) \rangle\, d\tau\,. \tag{4.6.22}$$

Inserting this into (4.6.21) and using (4.6.20) yields (4.6.16, 17). For the latter we need the condition (4.6.8). If τ_{p} can be made infinite by taking the macroscopic limit, this gives

$$G_{lm} = k_B^{-1} \int_0^\infty dt \lim_{\text{macro}} \langle \dot{A}_l(t); \dot{A}_m(0) \rangle . \tag{4.6.23}$$

Equation (4.6.13) corresponds to the Langevin equation (2.9.46) with

$$\Gamma(t-s) = G\delta(t-s) .$$

If condition (4.6.8) cannot be met, the macroscopic relaxation is not a type of exponential decay, or we may say it is nonlocal in time as is represented by the retardation kernel Γ.

4.6.4 Nonequilibrium Density Matrix

If a macrovariable A_l is a density, it is related to a current $\boldsymbol{j}_l$ by

$$\dot{A}_l + \operatorname{div} \boldsymbol{j}_l = 0 . \tag{4.6.24}$$

We consider here a stationary state with nonuniform densities $\boldsymbol{A}$ and associated currents $\boldsymbol{j}$ maintained by driving thermal forces, say temperature or chemical potential gradients. Such a nonuniformity may be imposed on the system by bringing it into contact with suitably prepared reservoirs as long as the spatial variation is very much slower in comparison with the range of interparticle forces. We allow for the nonuniform states by taking the thermodynamic limit and do not treat explicitly the interaction between the system and the reservoirs. Then we interpret (4.6.12) as the stationary currents induced by the imposed values of the A's and X's. Our problem is now to find an appropriate expression of the density matrix ϱ_{st} representing the system in such a stationary state.

First, note that the local equilibrium density matrix (4.6.19) does not give a dissipative flow, because

$$\begin{aligned} \operatorname{Tr}\{\varrho_{loc}\dot{A}_l\} &= \operatorname{Tr}\left\{\varrho\left(1+\beta^{-1}\int_0^\beta d\lambda \exp(\lambda\mathcal{H})\sum \xi_m A_m \exp(-\lambda\mathcal{H}) + \dots\right)\dot{A}_l\right\} \\ &= \sum_m \xi_m \langle A_m; \dot{A}_l\rangle + \dots \end{aligned} \tag{4.6.25}$$

to first order in ξ. This corresponds to the term $\mathrm{i}\Omega \boldsymbol{A}$ in (4.6.12). Let us now try to find ϱ_{st} which is close to ϱ_{loc}. This would satisfy the stationary condition

$$\partial\varrho_{st}/\partial t = \mathrm{i}\mathcal{L}\varrho_{st} = 0 , \tag{4.6.26}$$

where $\mathrm{i}\mathcal{L}$ is the quantal Liouville operator (4.2.2). Assume

$$\varrho_{st} = \varrho_{loc} + \Delta\varrho_{st} \tag{4.6.27}$$

and determine $\Delta \varrho$ from

$$i\mathscr{L}\,\Delta\varrho_{st} + i\mathscr{L}\,\varrho_{loc} = 0\,. \tag{4.6.28}$$

We expand ϱ_{loc} as

$$\varrho_{loc} = \varrho_e + \Delta\varrho_{loc} + \dots\,,$$

where ϱ_e is the equilibrium density matrix and

$$\Delta\varrho_{loc} = \varrho_e\,\beta^{-1}\int_0^\beta d\lambda\,\exp(\lambda\mathscr{H})\sum \xi_m A_m \exp(-\lambda\mathscr{H}) \tag{4.6.29}$$

to the linear approximation in ξ. The formal solution of (4.6.28) is given by

$$\Delta\varrho_{st} = \lim_{\varepsilon\to 0}\int_{-\infty}^{0} dt\,\exp[(\varepsilon - i\mathscr{L})\,t]\,i\mathscr{L}\,\Delta\varrho_{loc} \tag{4.6.30}$$

since $i\mathscr{L}\varrho_e = 0$. Inserting (4.6.29) into (4.6.30) and noticing that $\exp(-i\mathscr{L}t)\,\boldsymbol{A} = \boldsymbol{A}(t)$ is a Heisenberg operator, then

$$\Delta\varrho_{st} = -\,\varrho_e\,\beta^{-1}\lim_{\varepsilon\to 0}\int_{-\infty}^{0} dt\,e^{\varepsilon t}\int_0^\beta d\lambda\,\exp(\lambda\mathscr{H})\sum \xi_m \dot{A}_m(t)\,\exp(-\lambda\mathscr{H})\,, \tag{4.6.31}$$

where $\xi_m = -X_m/k_\beta$. Therefore

$$\bar{\dot{A}}_l = \mathrm{Tr}\,\{\varrho_{st}\dot{A}_l\} = \sum_m (i\Omega)_{lm} A'_m + \sum_m G_{lm} X_m \tag{4.6.32}$$

with

$$\begin{aligned} G_{lm} &= \lim_{\varepsilon\to 0} k_\beta^{-1}\int_{-\infty}^{0} dt\,e^{\varepsilon t}\,\langle \dot{A}_m(t);\dot{A}_l\rangle \\ &= \lim_{\varepsilon\to 0} k_\beta^{-1}\int_0^\infty dt\,e^{-\varepsilon t}\,\langle \dot{A}_m;\dot{A}_l(t)\rangle \end{aligned} \tag{4.6.33}$$

in accordance with (4.6.23).

The above derivation is, however, somewhat tricky. One may wonder if the only result from this calculation would be zero, because (4.6.26) would end up with $\Delta\varrho_{st} = -\,\Delta\varrho_{loc}$ so that ϱ_{st} is simply equal to ϱ_e: namely, the only stationary solution of (4.6.26) is uniform equilibrium. This is connected with the fact that the time integral of a current-current correlation would vanish if the integral is carried to infinity, (2.9.37). The key to solving the puzzle is to realize that the local equilibrium acts as a sort of a boundary condition, implicitly assuming the thermodynamic limit when deriving (4.6.31, 33). A similar problem is present in the case of the classical kinetic theory. When a Boltzmann equation is solved for a stationary flow by the Enskog-Chapman method, the thermodynamic limit is implicitly implied.

Let us examine this point a little closer following *Zubarev* [4.20]. The presence of reservoirs or the environment drives the system under observation to a nonequilibrium state close to local equilibrium. This effect may be represented by a source term added to the Liouville equation for the density matrix of the system. Here we have two choices. One is to assume that

$$\frac{\partial}{\partial t}\varrho(t) - \mathrm{i}\mathscr{L}\varrho(t) = -\varepsilon[\varrho(t) - \varrho_{\mathrm{loc}}(t)], \tag{4.6.34}$$

with $\varepsilon = 0+$ for the density matrix. The other choice is to write

$$\frac{\partial}{\partial t}\eta(t) - \mathrm{i}\mathscr{L}\eta(t) = -\varepsilon[\eta(t) - \eta_{\mathrm{loc}}(t)] \tag{4.6.35}$$

for

$$\eta(t) = \log\varrho(t).$$

The latter choice is possible since $\eta(t)$ follows the same Liouville equation in the absence of the source term. This is easily proved by noticing that

$$\dot{\varrho}(t) = \mathrm{e}^{\eta}\int_0^1 d\lambda\, \mathrm{e}^{-\lambda\eta}\,\dot{\eta}(t)\,\mathrm{e}^{\lambda\eta}$$

and using the identity

$$\int_0^1 d\lambda\, \mathrm{e}^{-\lambda\eta}[\mathscr{H}, \eta]\,\mathrm{e}^{\lambda\eta} = \mathrm{e}^{-\eta}\mathscr{H}\mathrm{e}^{\eta} - \mathscr{H}.$$

Here we generally allow the state of the environment to change in time as represented by (4.6.19) with time-dependent ξ. The right-hand side of (4.6.34 or 35) means relaxation of the density matrix toward the corresponding local equilibrium at each instant, caused by the interaction of the system with the environment. But neither the interaction details nor the relaxation rate is relevant, because after all we let ε go to zero. The mere existence of relaxation is important.

We consider the formal solution of (4.6.34),

$$\varrho(t) = \lim_{\varepsilon\to 0+}\varepsilon\int_{-\infty}^{t} dt' \exp[-(\varepsilon - \mathrm{i}\mathscr{L})(t-t')]\,\varrho_{\mathrm{loc}}(t'). \tag{4.6.36}$$

Using (4.6.19) we write this as

$$\begin{aligned}\varrho(t) = \lim_{\varepsilon\to 0+}\int_{-\infty}^{t} \varepsilon\, dt' \exp[&\varepsilon(t'-t) + \Phi(A'(t')) - \beta\mathscr{H}\\ &+ \textstyle\sum \xi_m(t')\,A_m(t'-t)].\end{aligned} \tag{4.6.37}$$

Similarly

$$\eta(t) = \lim_{\varepsilon \to 0+} \int_{-\infty}^{t} \varepsilon \, dt' \, e^{\varepsilon(t'-t)} [\Phi(A'(t')) - \beta \mathscr{H} + \sum \xi_m(t') A_m(t'-t)] . \tag{4.6.38}$$

Now we consider the stationary case and introduce

$$\tilde{A}_m = \lim_{\varepsilon \to 0+} \varepsilon \int_{-\infty}^{0} dt \, e^{\varepsilon t} A_m(t) = A_m - \lim_{\varepsilon \to 0} \int_{-\infty}^{0} dt \, e^{\varepsilon t} \dot{A}_m(t) \tag{4.6.39}$$

and call this the *invariant (zero-frequency) part* of A_m, (4.2.2). This is equal to the long-time average of A_m,

$$\bar{A}_m^T = \lim_{T \to \infty} T^{-1} \int_{-T}^{0} dt' A_m(t') .$$

Then (4.6.38) takes the simple form

$$\eta_{\rm st} = \Phi(A') - \beta \mathscr{H} + \sum_m \xi_m \tilde{A}_m \tag{4.6.40}$$

and the corresponding nonequilibrium stationary density matrix is given by

$$\varrho_{\rm st} = \exp \eta_{\rm st} . \tag{4.6.41}$$

When expanded in ξ's, the first-order term is equal to $\Delta \varrho_{\rm loc} + \Delta \varrho_{\rm st}$ (4.6.29, 31). This elucidates the meaning of the previous procedure deriving (4.6.33). On the other hand, the density matrix (4.6.37) is not expressed in terms of $\tilde{A}_m$, because it contains invariant parts of operator products which are not equal to products of the invariant parts. In this sense our second choice is simpler and more acceptable. It should be noted, however, that for nonstationary states, neither (4.6.35) nor (4.6.37) can claim such general validity as claimed by the theory for mechanical disturbance, because the source terms there do not represent the physical situation with such generality.

Now the question arises if such an invariant part really does exist. If the quantity A is a sort of density of a conserved quantity, its invariant part may most probably be only the zero wave-number component, namely the integration over the whole volume of the system. This is trivial and does not lead us to the theory of nonuniform states. Therefore we have to consider a system with an infinite extension keeping the wave numbers finite. So we are concerned with three limit processes in the following order: first the thermodynamic limit, second the wave number $\boldsymbol{k} \to 0$, and third $\varepsilon \to 0$. In these limits $\tilde{A}$ (4.6.39) has the meaning of a quasi-invariant or local integral of motion.

Let us consider a few examples of the foregoing treatment. If a nonuniform temperature distribution is caused by some means, for instance by bringing the system in contact with heat reservoirs at different temperatures, the local equilibrium density matrix is represented by

$$\varrho_{\text{loc}} = \exp\{\Phi[\beta(\boldsymbol{r})] - \int \beta(\boldsymbol{r})\, \varepsilon(\boldsymbol{r})\, d\boldsymbol{r}\}\,, \tag{4.6.42}$$

where $\Phi[\beta(\boldsymbol{r})]$ is for normalization and is a functional of the function $\beta(\boldsymbol{r})$, $\varepsilon(\boldsymbol{r})$ is the energy density, and $\beta(\boldsymbol{r}) = 1/kT(\boldsymbol{r})$ corresponds to ξ in (4.6.19). The energy density is given by

$$\varepsilon(\boldsymbol{r}) = \sum_j \left[\frac{1}{2m}\boldsymbol{p}_j^2 + \frac{1}{2}\sum_l \phi(\boldsymbol{r}_j - \boldsymbol{r}_l)\right] \delta(\boldsymbol{r} - \boldsymbol{r}_l) \tag{4.6.43}$$

for a classical particle system with pair interactions. We introduce the Fourier components

$$\begin{aligned}\varepsilon_{\boldsymbol{k}} &= \int \varepsilon(\boldsymbol{r}) \exp(-\mathrm{i}\boldsymbol{k}\cdot\boldsymbol{r})\, d\boldsymbol{r}\\ &= \sum_j \left[\frac{1}{2m}\boldsymbol{p}_j^2 + \frac{1}{2}\sum_l \phi(\boldsymbol{r}_j - \boldsymbol{r}_l)\right] \exp(-\mathrm{i}\boldsymbol{k}\cdot\boldsymbol{r}_l)\end{aligned} \tag{4.6.44}$$

and

$$\beta_{\boldsymbol{k}} = L^{-3} \int \beta(\boldsymbol{r}) \exp(-\mathrm{i}\boldsymbol{k}\cdot\boldsymbol{r})\, d\boldsymbol{r} \tag{4.6.45}$$

and write

$$\int \beta(\boldsymbol{r})\, \varepsilon(\boldsymbol{r})\, d\boldsymbol{r} = \sum_{\boldsymbol{k}} \beta_{-\boldsymbol{k}}\, \varepsilon_{\boldsymbol{k}}\,.$$

Note now that

$$\lim_{\boldsymbol{k}\to 0} \langle \varepsilon_{\boldsymbol{k}}; \varepsilon_{-\boldsymbol{k}} \rangle = L^3 c\, \varrho\, k_{\mathrm{B}} T^2\,. \tag{4.6.46}$$

When the cause of temperature nonuniformity is removed, each Fourier component $\varepsilon_{\boldsymbol{k}}$ as a macrovariable relaxes to zero. From (4.6.12, 20, 23) follows

$$\dot{\varepsilon}_{\boldsymbol{k}} = -\int_0^\infty dt\, \langle \dot{\varepsilon}_{\boldsymbol{k}}(t); \dot{\varepsilon}_{-\boldsymbol{k}}(0) \rangle \langle \varepsilon_{\boldsymbol{k}}; \varepsilon_{-\boldsymbol{k}} \rangle^{-1} \varepsilon_{\boldsymbol{k}} \tag{4.6.47}$$

as $\boldsymbol{k} \to 0$. This yields the relaxation

$$\dot{\varepsilon}_{\boldsymbol{k}} = -\,\varepsilon_{\boldsymbol{k}}/\tau_{\mathrm{r}}$$

with the relaxation rate

$$1/\tau_{\mathrm{r}} = \varkappa \boldsymbol{k}^2/c\varrho$$

in accord with (4.6.3). Here $\varkappa$ is the thermal conductivity tensor given by

$$\varkappa = \lim_{\boldsymbol{k}\to 0} \frac{1}{k_{\mathrm{B}} T^2} \int_0^\infty dt\, \langle j_{\boldsymbol{k}}^{\varepsilon}(t); j_{-\boldsymbol{k}}^{\varepsilon}(0) \rangle\, L^{-3}\,, \tag{4.6.48}$$

where the energy current density $\boldsymbol{j}_{\boldsymbol{k}}^{\varepsilon}$ is defined by

$$\dot{\varepsilon}_k = -\mathrm{i}\,\boldsymbol{k}\cdot\boldsymbol{j}_k^{\varepsilon},$$

$$\boldsymbol{j}_k^{\varepsilon} = \sum_j \left\{\left[\frac{1}{2m}\boldsymbol{p}_j^2 + \frac{1}{2}\sum_l \phi(r_{jl})\right]\frac{\boldsymbol{p}_j}{m} + \frac{1}{4}\sum_l \frac{1}{m}(\boldsymbol{p}_j+\boldsymbol{p}_l)(\boldsymbol{F}_{jl}\cdot\boldsymbol{r}_{jl})\right\}\exp(-\mathrm{i}\,\boldsymbol{k}\cdot\boldsymbol{r}_j) \tag{4.6.49}$$

with

$$\boldsymbol{F}_{jl} = -\partial\phi(\boldsymbol{r}_{jl})/\partial\boldsymbol{r}_{jl}, \quad \boldsymbol{r}_{jl} = \boldsymbol{r}_j - \boldsymbol{r}_l.$$

Equation (4.6.49) is valid for $\boldsymbol{k}$'s small in comparison with the force range of interactions.

The same result is obtained by using (4.6.11), which now reads

$$X_{-\boldsymbol{k}} = \frac{\partial S}{\partial \varepsilon_{\boldsymbol{k}}} = k_{\mathrm{B}}\,\beta_{-\boldsymbol{k}} \tag{4.6.50}$$

and

$$\lim_{\boldsymbol{k}\to 0} d\varepsilon_{\boldsymbol{k}}/dt = \frac{dE}{dT}\frac{\partial T}{\partial t} = L^3 c\,\frac{\partial T}{\partial t}$$

for (4.6.12).

If the nonuniformity of temperature is maintained constant by keeping the reservoirs in contact, the stationary density matrix ϱ_{st} (4.6.27) yields the heat flow, calculated with (4.6.31). The result is

$$\boldsymbol{j}_k^{\varepsilon} = -\varkappa\,\mathrm{grad}\,T \quad (\boldsymbol{k}\to 0),$$

with $\varkappa$ defined by (4.6.48).

When a field of nonuniform mass flow is present in a fluid, the local equilibrium density matrix has the form

$$\varrho_{\mathrm{loc}} = \exp\{\Phi[\boldsymbol{v}(\boldsymbol{r})] - \beta(\mathscr{H} - \int \boldsymbol{v}(\boldsymbol{r})\cdot\boldsymbol{j}^m(\boldsymbol{r})\,d\boldsymbol{r})\}, \tag{4.6.51}$$

where $\boldsymbol{j}^m(\boldsymbol{r})$ is the operator of particle current density, $\boldsymbol{v}(\boldsymbol{r})$ the velocity field and $\Phi[\boldsymbol{v}(\boldsymbol{r})]$ is a functional of $\boldsymbol{v}(\boldsymbol{r})$. In terms of the Fourier components

$$\boldsymbol{j}_k^m = \sum_j \frac{\boldsymbol{p}_j}{m}\exp(-\mathrm{i}\,\boldsymbol{k}\cdot\boldsymbol{r}_j),$$

$$\boldsymbol{v}_k = L^{-3}\int \boldsymbol{v}(\boldsymbol{r})\exp(-\mathrm{i}\,\boldsymbol{k}\cdot\boldsymbol{r})\,d\boldsymbol{r},$$

then

$$\int v(\boldsymbol{r})\cdot\boldsymbol{j}^m(\boldsymbol{r})\,d\boldsymbol{r} = \sum_k \boldsymbol{v}_{-\boldsymbol{k}}\cdot\boldsymbol{j}_k^m.$$

Thus, if we take $\boldsymbol{j}_{\boldsymbol{k}}^{m}$ as A_l in (4.6.19), $\beta \boldsymbol{v}_{-\boldsymbol{k}}$ corresponds to ξ_l. We assume a shear flow, $\boldsymbol{v}=(v_x(\boldsymbol{r}), 0, 0)$, $\partial v_x/\partial x = 0$, $\partial v_x/\partial y \neq 0$, $\partial v_x/\partial z = 0$, and consider only $j_{\boldsymbol{k}x}^{m}$ with $\boldsymbol{k} \parallel y$. Then (4.6.12) is written as

$$\frac{d}{dt} j_{\boldsymbol{k}x}^{m} = -\boldsymbol{k}^2 \eta v_{\boldsymbol{k}x} \tag{4.6.52}$$

with (4.6.20, 23). The viscosity coefficient η is thus expressed by

$$\eta = \beta \int_0^\infty dt \, \langle R_{xy}(t); R_{xy}(0) \rangle \, L^{-3} \tag{4.6.53}$$

with

$$R_{xy}(t) = \lim_{k \to 0} k^{-1} j_{\boldsymbol{k}x}^{m}(t) = \sum_j \left(\frac{1}{m} p_{jx} p_{jy} - \frac{1}{2} \sum x_{jl} F_{jly} \right). \tag{4.6.54}$$

There are other methods of deriving the expression of nonequilibrium density matrix with flow. *Mac Lennan* [4.21] derived an equation of motion for η treating the problem in classical mechanics. He showed how the effect of the environment is expressed as nonconservative forces and obtained essentially the same result as (4.6.38). *Robertson* [4.22] elaborated a method to determine the nonequilibrium density matrix as a functional of the time-dependent macroparameters. Admittedly each of these methods contains delicate points with some ambiguities, but all agree in the linear regimes. So we can safely say that the theory has been well founded as far as the linear responses are concerned. On the other hand, beyond the linear regimes, general theories are still far from established.

4.7 Some Remarks on the Linear-Response Theory

4.7.1 The Kinetic Method versus the Linear-Response Theory

Statistical mechanics is indeed a very difficult theory. Treating a dynamical system with an enormously large number of degrees of freedom, it must transform the dynamic evolution into a stochastic one by some sort of coarse graining. This we call in the following *stochastization* (we have not seen this word in the literature but use it here simply because it is the best we can find). A fundamental problem of statistical mechanics is to clarify the logic of this stochastization with a satisfactory degree of mathematical rigor. But this is a task we have not been able to achieve. The whole framework of equilibrium statistical mechanics is constructed on the basis of the principle of equal weights of microscopic states. Although the ergodic theories have progressed in recent years, as shown in Chap. 5 of [4.1], we still lack a rigorous proof of the equal weight principle. Despite this fact, no one doubts the validity of equilibrium statistical mechanics.

The foundation of nonequilibrium statistical mechanics is perhaps far more difficult to establish than that of equilibrium statistical mechanics if we want such a level of rigor as is aimed at by the ergodic theories. In the first place, the concept of nonequilibrium is too wide to be unified under a few principles. Therefore we must modestly confine ourselves to a limited category of nonequilibrium states if we want a general method of approach. We discuss two such categories.

One is the *kinetic method.* This assumes that stochastization can be achieved, for example, by means of a Boltzmann-type equation or a Markovian equation for an appropriate distribution function. This method can be applied only if the system considered has a sufficiently amenable structure and if we limit ourselves to a certain class of physical properties corresponding to the crudeness required for stochastization. Thus the Boltzmann-Bloch-type transport equation of electrons in a metal or a semiconductor cannot be used when the mean-free paths are too short or the frequency of the applied field is too high. Within its own limitation of validity, such a kinetic method is very powerful. It can also be used for nonlinear problems such as nonohmic conduction.

Another category of nonequilibrium phenomena which allow a general formulation is the *near-equilibrium processes.* We have seen in this chapter how the linear-response theory relates the nonequilibrium properties of a system directly to fluctuations in equilibrium and thus has almost the same validity as equilibrium statistical mechanics. The important point is that the linearization is made *before* stochastization. So the applicability of this approach does not depend on whether a stochastization is possible or not.

If a stochastization is possible, the relevant correlation function of fluctuation can be calculated by transforming the time evolution operator in the expression into an appropriate stochastic operator. The result is, of course, the same as that obtained by the kinetic method, namely by first setting up the transport equation and secondly linearizing it with respect to the driving field. If a stochastization is not possible or is unnecessary, the linear response is very simple. For example, if it is applied to an atom or a simple molecule exposed to radiation, the optical response is expressed by the Heisenberg-Kramers dispersion formula [4.23]. For a more complex system containing an internal relaxation mechanism, the linear response is essentially a generalization of this formula. In general cases, some sort of stochastization is made somewhere in the calculation of the correlation function or the Green's function as a computational approximation. However, the nature of the stochastization is not always very clear. Only in simple cases is it Markovian. In general it is neither Markovian nor Gaussian. (We treat the Green's function method in Chap. 5.) Thus we may say that the kinetic method and the linear-response theory neither exclude nor are incompatible with each other: the domains of their validity overlap each other. Where both are valid, the results are the same. The kinetic method has an advantage in that it applies beyond the linear regime. On the other hand, the linear-response theory applies even when the kinetic method is useless. A well-known example is transport phenomena in a highly disordered system. We cannot write down a transport equation for electrons in an amorphous semiconductor but the general expression of conductivity (4.2.24) here comes to rescue, giving the starting point.

Confining courselves to near-equilibrium cases, we may go beyond the linear response by taking second-order terms in the external perturbation. This is useful in some cases, for example, in treating second-order optical responses. Nonlinear admittance functions are then expressed in terms of higher-order correlation functions. In other cases, however, the nonlinear response theory is not so successful because then the generation of heat has to be included. This can be done, but so far no general method has been developed.

4.7.2 Van Kampen's Objection

Van Kampen [4.24] once made a severe criticism of the linear-response theory. He objected to the use of simplistic perturbational calculation by which a response such as (4.2.10) was obtained. As shown in [Ref. 4.1, Chap. 5], dynamical trajectories in phase space are essentially unstable and are very sensitive to perturbation. Therefore a perturbational calculation of microscopic dynamics has only an extremely small range of validity. On the other hand, the linear response actually observed in macroscopic systems has a physically significant range of validity. Thus he claims that *microscopic linearity* and *macroscopic linearity* are totally different. The latter can be understood only by a kinetic approach such as the use of a Boltzmann-Bloch equation of electrons in an applied external field. He also points out that the linear-response theory introduces implicitly a sort of randomization by linearization, namely, independently superposing the effects of external perturbation at each instant. He asserts that this is a simulation of stochastization but is not proper.

Replying to this criticism, we first point out that the linear-response theory does not use a perturbational calculation for phase trajectories. Instead, it uses it for a limited class of phase space distribution functions which are sufficiently smooth since we are considering near equilibrium states. It is true that the characteristics of the Liouville equation are the phase trajectories and so the Liouville equation and the Hamilton equation of motion are equivalent in this sense. However, the limitation of the class of distribution functions on which the Liouville operator operates makes them nonequivalent. Unfortunately it is not easy to formulate the mathematical condition for this statement. But we realize that the instabilities of trajectoris are in fact the cause of mixing, favoring rather than unfavoring the stability of distribution functions.

Secondly, as was already discussed, the difference between the kinetic method and the linear-response theory is the interchange of the order of stochastization and linearization. Although it is difficult to prove it rigorously, this interchange is legitimate. The linearity considered in the linear-response theory is indeed macroscopic and not microscopic when it is applied to a macroscopic system. Stochastization, if it is ever to be done, is made at the second stage in the linear-response theory. This comment is related to the question of irreversibility. The linear-response formula applies to reversible as well as to irreversible processes. If the system is simple enough, no irreversibility is observed. If the system is sufficiently complex, some feature of irreversibility appears and is represented by some sort of stochastization. Thermodynamic limits often play an essential role.

Incidentally, there is another difficulty in the use of the perturbation method in a many-particle system. The perturbation term is $O(N)$ if the number of particles is N and so it is not small. This difficulty is present even in the equilibrium theory when we made a simple expansion such as that in (4.1.10). If we want to circumvent this difficulty, we could use reduced distribution functions or reduced density matrices or their cumulants. The naive use of perturbation can be justified in this way. It should be noted that the relaxation function (4.2.27, 29) is just a slight generalization of (4.1.11).

4.7.3 Spurious Singularities at the Zero Value of the External Field

The simplest model of electronic conduction is given by the kinetic equation

$$\frac{\partial}{\partial t} f(p_x, t) = -eE \frac{\partial}{\partial p_x} f(p_x, t) - \frac{1}{\tau}[f(p_x, t) - f_e(p_x)], \tag{4.7.1}$$

where $f_e(p_x)$ is the equilibrium distribution function of the momentum in zero field. The simple assumption for scattering is called a *strong collision* model, which means that electrons undergo collisions at the averaged rate $1/\tau$ and after a single collision the distribution reaches equilibrium. This is a great simplification of the collision kernel by assuming their eigenvalues are all degenerate, except the zero eigenvalue. Because of this simplification, the steady solution of $\partial f/\partial t = 0$

$$-eE\frac{\partial}{\partial p_x} f - \frac{1}{\tau}(f - f_e) = 0$$

is easily obtained as

$$f(p_x) = \int_{-\infty}^{p_x} \exp[-(p_x - p')/\xi] f_e(p')\, dp'/\xi, \tag{4.7.2}$$

where

$$\xi = eE\tau.$$

Using the solution (4.7.2), the average current is found to be proportional to the field E and the electric conductivity is the familiar one

$$\sigma = e^2 n\tau/m\,.$$

The same answer is obtained by setting $f = f_e + g$ and ignoring g in the drift term. This is the usual derivation of conductivity from a kinetic equation.

Bakshi and *Gross* [4.25] noted that this is a fortituous coincidence because $E=0$ is an essential singularity of the solution (4.7.2). This may make one uneasy about the legitimacy of the linear-response theory. However, this is a spurious singularity originating from the oversimplification of the collision term to that of strong collision. One can convince oneself at once that *no* such singularity occurs if the collision term is more realistic, for example, the momentum diffusion (2.2.31) or a collision integral with nonsingular kernel. Mathematically, it is well known that a perturbation in the term with the highest order in a differential equation causes such a singularity. In the present case, the strong collision model is to be blamed for the singularity. Thus this example should not be taken as any failing of the linear-response theory. In physics, idealization of physical models often leads to paradoxes, many examples of which are known in hydrodynamics.

4.7.4 Singularities at $k = 0$, $\omega = 0$

The behavior of a response function or an admittance in the neighborhood of $\boldsymbol{k}=0$ and $\omega=0$ is a direct reflection of macroscopic laws such as that of hydrodynamics [4.26]. Analyzing the calculated admittance function or a Green's function, the macroscopic dynamical laws are discovered. We must keep in mind that the well-known hydrodynamics laws are based on a few conservation laws and some empirical facts. There is no obvious way to generalize hydrodynamics to apply to phenomena with shorther wavelengths and higher frequencies. In this respect a generalized hydrodynamics is an outstanding problem.

As was pointed out repeatedly in this chapter, the order of limit for $\boldsymbol{k}\to 0$ and $\omega \to 0$ is important; $\boldsymbol{k}$ is first and ω second. Here we propose a conjecture that the point $\boldsymbol{k}=0$, $\omega=0$ may well be a singularity of an admittance function. To see the nature of the problem, let us consider the function $[\mathrm{i}\omega(\mathrm{i}\omega + 2D\boldsymbol{k}^2)]^{-1/2}$. In the above-mentioned limit

$$[\mathrm{i}\omega(\mathrm{i}\omega + 2D\boldsymbol{k}^2)]^{-1/2} \sim (\mathrm{i}\omega + D\boldsymbol{k}^2)^{-1} \tag{4.7.3}$$

as the asymptotic form of the function. This means that the relevant field quantities, say ϕ, obey the macroscopic law

$$\frac{\partial}{\partial t}\phi = D\nabla^2\phi\,.$$

Note that this is valid only in an asymptotic sense. It is not possible to extend this to include higher-order spatial derivatives on the right-hand side. Generalized hydrodynamics can be obtained only by analysis of the asymptotic behavior of the response functions.

This comment may be relevant to the puzzling problem of long tails of correlation functions as noted for the velocity correlation of a Brownian particle (1.6.15). This sort of long-time tails was found first by *Alder* and *Wainright* [4.27] for velocity autocorrelation functions of a fluid using molecular dynamic calculations. It is essentially a hydrodynamic back-flow effect and the tail is proportional to $t^{-d/2}$, where d is the dimensionality. In three dimensions, the time integral of the diffusion constant converges and the contribution from the tail is very small. In lower dimensions, the time integrals diverge if such a long-time tail exists. This indicates that diffusion in low dimensional fluids may be characteristically different from ordinary diffusion.

The problem of long-time tails looks more serious if we want to generalize hydrodynamic equations. For example, the Burnett or super-Burnett equation is an extension of the ordinary diffusion equation [4.28]. It assumes the form

$$\frac{\partial}{\partial t} P = D_0 \nabla^2 P + D_2 (\nabla^2)^2 P + \dots . \tag{4.7.4}$$

By calculating the moments $\langle x^2(t)\rangle$ and $\langle x^4(t)\rangle$ from this equation, we find easily that

$$\begin{aligned} D_2 &= \lim_{t\to\infty} D_2(t) \\ &= \lim_{t\to\infty} \int_0^t \int_0^{t_1} \int_0^{t_2} \langle u_x(0)\, u_x(t_1)\, u_x(t_2)\, u_x(t_3)\rangle_c \, dt_1\, dt_2\, dt_3 , \end{aligned}$$

where the integrand is the fourth cumulant correlation of velocity. Because of the back-flow effect, $D_2(t) \sim t^{1/2}$ in three dimensions. Therefore the Burnett coefficients cannot be calculated by the velocity correlations.

This is not too surprising. This simply means that (4.7.4) has no *a priori* right to claim existence: it depends on the system considered. If the system is such that divergent velocity correlation arises at higher order, the Burnett equation does not exist. The admittance function (4.5.26) has no simple pole of the form

$$-\mathrm{i}\omega = D_0 (\mathrm{i}k)^2 + D_2 (\mathrm{i}k)^4 + \dots$$

because of its singularity at $\boldsymbol{k} = \omega = 0$.

We conjecture that the validity of hydrodynamic equations in macrodimensions is asymptotic in this sense. Generalizations must be made by careful analysis of admittance functions at $\boldsymbol{k} = \omega = 0$. The higher-order terms of an extended diffusion equation will contain both spatial and temporal derivatives. In other words, the equation will be nonlocal in space and time. The question is at present an outstanding problem [4.29].

We have repeatedly stressed the great separations in the orders of magnitude of characteristic time constants at various levels of coarse graining. The long-time tails indicate that clear separation may not always be possible. Then the stochastic nature of the problem can differ very much from standard models familiar to us.

5. Quantum Field Theoretical Methods in Statistical Mechanics

We have seen that macroscopic properties in a linear irreversible process are determined by the response function, the relaxation function, the complex admittance or the double-time correlation functions. This chapter briefly describes techniques for calculating these functions. Of course, there are many methods of calculation, each of which has its own merits and demerits and has particular key points to be considered. A simple example for the determination of the response function by using the kinetic theoretical method was given in Chap. 3.

In this chapter we shall not treat these methods devised and improved a long time ago, but will describe the new method originally found in quantum field theory and transferred into statistical mechanics. Hence we suppose that the method of second quantization is familiar to the reader. The new method is based on the use of various kinds of Green's functions which are related to both the response function and the equations of motion. Via the Green's functions, we have been able to discuss both the thermal equilibrium state and the irreversible process by the same theoretical system. We shall examine this situation by taking the plasma as an example, without entering into too many details like the distinction of electrons and holes due to the existence of a Fermi surface. The method of Green's functions is, of course, not always efficacious. Also we need not employ a steam-hammer to crack a nut: we must not forget that problems can also be solved by devising techniques within the customary methods.

5.1 Double-Time Green's Functions

Let us begin by introducing Green's functions. The family of Green's functions is classified into four groups, i.e., the retarded, the advanced, the causal and the temperature Green's functions, and in addition to this, each group is classified into groups of many-particle Green's functions, i.e., single-particle, two-particle Green's functions, and so on. For the description by double-time correlations referred to in Sect. 4.6, the double-time Green's functions, which are special cases of the many-particle Green's functions, are convenient. Up to Sect. 5.4 we shall deal with the family of these double-time Green's functions.

5.1.1 Retarded Green's Functions

In the linear relation (3.1.4), the response function $\Phi_{\mu\nu}(t)$ appeared only for positive times t in accordance with the causality. In Sect. 4.3 the response function was extended to negative times by using the statistical-mechanics result (4.2.14). However, such a generalized response function is inconvenient for extending the range of the time integral in the linear relation to $(-\infty, +\infty)$. Rather, introducing a function made by inserting the Heaviside unit step function (3.1.7)

$$G_{\mu\nu}^{\text{ret}}(t, t') = \theta(t-t')\left\langle \frac{1}{\text{i}\hbar}[\hat{B}_\mu(t), \hat{A}_\nu(t')]\right\rangle \tag{5.1.1}$$

is more convenient for that purpose, because comparison with (4.2.14) shows that this function agrees with $-\Phi_{\mu\nu}(t-t')$ for $t > t'$ and vanishes for $t < t'$. The linear relation is written in the form

$$B_\mu(t) - B_\mu^{\text{eq}} = -\int_{-\infty}^{\infty} dt' \sum_\nu G_{\mu\nu}^{\text{ret}}(t, t')\, X_\nu(t'). \tag{5.1.2}$$

The lower limit of time integration in the defining equation (3.2.5) of the complex admittance can also be extended:

$$\chi_{\mu\nu}(\omega) = -\int_{-\infty}^{\infty} dt\, \text{e}^{\text{i}\omega t}\, G_{\mu\nu}^{\text{ret}}(t, 0)\,. \tag{5.1.3}$$

This shows that the complex admittance is directly connected with the spectral function of (5.1.1)

$$G_{\mu\nu}^{\text{ret}}(t, t') = \int_{-\infty}^{\infty} \frac{d\omega}{2\pi} \exp[-\text{i}\omega(t-t')]\, K_{\mu\nu}^{\text{ret}}(\omega)\,; \tag{5.1.4}$$

$$K_{\mu\nu}^{\text{ret}}(\omega) = -\chi_{\mu\nu}(\omega)\,. \tag{5.1.5}$$

In this sense, the function (5.1.1) is expected to be a very convenient quantity. Further, it is connected with the spectral function (4.3.30) of the response function, continuing to negative times, Sect. 4.3,

$$\left\langle \frac{1}{\text{i}\hbar}[\hat{B}_\mu(t), \hat{A}_\nu(t')]\right\rangle = -\int_{-\infty}^{\infty} \frac{d\omega}{2\pi} \exp[-\text{i}\omega(t-t')]\, \Phi_{\mu\nu}[\omega]$$

by

$$K_{\mu\nu}^{\text{ret}}(\omega) = -\lim_{\varepsilon\to+0} \int_{-\infty}^{\infty} \frac{d\omega'}{2\pi\text{i}} \frac{\Phi_{\mu\nu}[\omega']}{\omega' - \omega - \text{i}\varepsilon} \tag{5.1.6}$$

as is obvious from (4.3.32).

Taking into account what we have seen and generalizing it, we define *the retarded Green's function of the anticommutator or of the commutator type*

$$G^{\mathrm{ret},\pm}_{A,B}(t,t') = \theta(t-t')\left\langle \frac{1}{\mathrm{i}\hbar}[\hat{A}(t),\hat{B}(t')]_{\pm}\right\rangle \tag{5.1.7}$$

for a pair of arbitrary operators $\hat{A}$ and $\hat{B}$, where $[\hat{A},\hat{B}]_{\pm} = \hat{A}\hat{B} \pm \hat{B}\hat{A}$. The functions (5.1.1) is the retarded Green's function of the commutator type for a pair formed of a dynamical variable $\hat{B}_{\mu}$ and a displacement $\hat{A}_{\nu}$. An anticommutator-type function can be employed not only as a single-particle Green's function for a system of fermions, Sect. 5.4, but also in discussing the symmetrized correlation introduced in Sect. 4.4.2. In accordance with the above example, by making the Fourier analysis

$$G^{\mathrm{ret},\pm}_{A,B}(t,t') = \int_{-\infty}^{\infty} \frac{d\omega}{2\pi} \exp[-\mathrm{i}\omega(t-t')]\, K^{\mathrm{ret},\pm}_{A,B}(\omega)\,, \tag{5.1.8}$$

$$\left\langle \frac{1}{\mathrm{i}\hbar}[\hat{A}(t),\hat{B}(t')]_{\pm}\right\rangle = \int_{-\infty}^{\infty} \frac{d\omega}{2\pi} \exp[-\mathrm{i}\omega(t-t')]\, \Lambda^{\pm}_{A,B}(\omega)\,, \tag{5.1.9}$$

we obtain the following relation corresponding to (5.1.6):

$$K^{\mathrm{ret},\pm}_{A,B}(\omega) = \lim_{\varepsilon\to+0} \int_{-\infty}^{\infty} \frac{d\omega'}{2\pi\mathrm{i}} \frac{\Lambda^{\pm}_{A,B}(\omega')}{\omega'-\omega-\mathrm{i}\varepsilon}\,. \tag{5.1.10}$$

Its proof is obvious from (4.3.31) or from

$$\int_{-\infty}^{\infty} dt\, \mathrm{e}^{\mathrm{i}\omega t}[\pm\theta(\pm t)] = \lim_{\varepsilon\to+0} \frac{\mathrm{i}}{\omega\pm\mathrm{i}\varepsilon}\,. \tag{5.1.11}$$

If the average ... stands for the canonical average, then

$$\Lambda^{+}_{A,B}(\omega) = \frac{2}{\hbar\omega} E_{\beta}(\hbar\omega)\Lambda^{-}_{A,B}(\omega) = \coth\left(\frac{\beta\hbar\omega}{2}\right)\Lambda^{-}_{A,B}(\omega) \tag{5.1.12}$$

holds, as is obvious from (4.4.4, 6). Therefore, when either the anticommutator type or the commutator type is obtained, the other must be calculated from it. In doing this, we need to solve (5.1.10): the method is given below.

In some cases we may calculate *the retarded Green's function of the product type*

$$\begin{aligned} G^{\mathrm{ret}}_{A,B}(t,t') &= \theta(t-t')\langle\hat{A}(t)\hat{B}(t')\rangle \\ &= \int_{-\infty}^{\infty} \frac{d\omega}{2\pi} \exp[-\mathrm{i}\omega(t-t')]\, K_{A,B}(\omega)\,, \end{aligned} \tag{5.1.13}$$

since, by using the spectral function $J_{A,B}(\omega)$ defined by (4.3.3) for the expectation value of the product $A(t)\,B(t')$, we obtain

$$K_{A,B}(\omega) = \lim_{\varepsilon\to+0}\int_{-\infty}^{\infty}\frac{d\omega'}{2\pi\mathrm{i}}\,\frac{J_{A,B}(\omega')}{\omega'-\omega-\mathrm{i}\varepsilon} \tag{5.1.14}$$

and hence

$$\mathrm{i}\hbar\,\Lambda^{\pm}_{A,B}(\omega) = (1 \pm \mathrm{e}^{-\beta\hbar\omega})\,J_{A,B}(\omega)\,, \tag{5.1.15}$$

as is obvious from (4.4.3, 5). However, as is seen in the example given at the beginning of this section and in the next three sections, the Green's function of the commutator or the anticommutator type is in many cases more convenient than that of the product type.

5.1.2 Advanced Green's Functions

The Heaviside unit step function inserted into the retarded Green's function is chosen to satisfy the causality in the linear relation (5.1.2). If the step function is inserted to satisfy not the causality stating that a present effect depends only on its past cause but instead the anticausality stating that a present effect depends only on its future cause, *the advanced Green's functions* arise

$$\begin{aligned} G^{\mathrm{adv},\pm}_{A,B}(t,t') &= -\,\theta(t'-t)\left\langle\frac{1}{\mathrm{i}\hbar}[\hat{A}(t),\hat{B}(t')]_{\pm}\right\rangle \\ &= \int_{-\infty}^{\infty}\frac{d\omega}{2\pi}\exp[-\mathrm{i}\omega(t-t')]\,K^{\mathrm{adv},\pm}_{A,B}(\omega)\,. \end{aligned} \tag{5.1.16}$$

According to (5.1.11), their spectral functions are given by

$$K^{\mathrm{adv},\pm}_{A,B}(\omega) = \lim_{\varepsilon\to+0}\int_{-\infty}^{\infty}\frac{d\omega'}{2\pi\mathrm{i}}\,\frac{\Lambda^{\pm}_{A,B}(\omega')}{\omega'-\omega+\mathrm{i}\varepsilon}\,. \tag{5.1.17}$$

Due to the difference in the method of insertion of the step function, the sign before ε in the denominator of (5.1.17) differs from that in the retarded Green's functions.

As shown in the next section, the retarded and advanced Green's functions satisfy the same differential equation, so that if one is determined, the other is obtained by much the same calculation. When both are obtained, we can solve the problem of determining the spectral function $\Lambda^{\pm}_{A,B}(\omega)$ by solving (5.1.10 or 17). Namely, we can prove that

$$K^{\mathrm{ret},\pm}_{A,B}(\omega) - K^{\mathrm{adv},\pm}_{A,B}(\omega) = \Lambda^{\pm}_{A,B}(\omega) \tag{5.1.18}$$

by using (3.6.7).

If $\Lambda^{\pm}_{A,B}(\omega')$ is not a pathological function and converges to zero sufficiently rapidly as $\omega' \to \pm\infty$, we can analytically continue the spectral functions (5.1.10, 17) into the complex plane by replacing the angular frequency ω with a complex number z. The resulting complex function is

$$K^{\pm}_{A,B}(z) = \int_{-\infty}^{\infty} \frac{d\omega'}{2\pi \mathrm{i}} \frac{\Lambda^{\pm}_{A,B}(\omega')}{\omega' - z} \tag{5.1.19}$$

and is analytic except on the real axis, because it is defined by the Cauchy integral. If the variable approaches from the upper or the lower side of the real axis, then

$$\begin{aligned} \lim_{\varepsilon\to+0} K^{+}_{A,B}(\omega \pm \mathrm{i}\varepsilon) &= K^{\left\{\substack{\mathrm{ret}\\ \mathrm{adv}}\right\},+}_{A,B}(\omega)\,, \\ \lim_{\varepsilon\to+0} K^{-}_{A,B}(\omega \pm \mathrm{i}\varepsilon) &= K^{\left\{\substack{\mathrm{ret}\\ \mathrm{adv}}\right\},-}_{A,B}(\omega)\,, \end{aligned} \tag{5.1.20}$$

whence, according to (5.1.18), a point ω at which these limiting values differ from each other is a singularity giving $\Lambda^{\pm}_{A,B}(\omega) \neq 0$:

$$\Lambda^{\pm}_{A,B}(\omega) = \lim_{\varepsilon\to+0} [K^{\pm}_{A,B}(\omega + \mathrm{i}\varepsilon) - K^{\pm}_{A,B}(\omega - \mathrm{i}\,\varepsilon)]\,. \tag{5.1.21}$$

The above discussion is similar to that on the complex admittance in Sect. 3.6; this analogy is anticipated from (5.1.5). According to the discussion given in Sect. 3.6, dispersion relations should hold between the real and imaginary parts of the spectral function of the Green's function. The reader is adivsed to try to discuss symmetry properties and to derive the dispersion relations. Here, we only remark that the retarded and advanced Green's functions are related differently to the causality.

Although we have stated that the retarded Green's function is physically very convenient, including the discontinuous factor in the step function is mathematically inconvenient. For instance, we cannot set $t = t'$ in $G^{\mathrm{ret}}_{A,B}(t, t')$. We must keep in the limits $t \to t' \pm 0$. Nevertheless, even with such a discontinuous function, we can calculate a quantity at the same time $t = t'$ like the expectation value of a product $\langle \hat{A}\hat{B} \rangle$. To do so, we need only to calculate using the spectral functions described above and in Sect. 4.4. The importance of spectral functions was pointed out by *Bogolyubov* and *Tyablikov* [5.1], who introduced the double-time Green's function [5.1–3].

5.2 Chain of Equations of Motion and the Decoupling Approximation

Let us determine the equations governing the temporal behavior of the Green's functions (5.1.7, 16) introduced in the preceding section. Since the operators $\hat{A}(t)$, etc., appearing in the defining equations are the Heisenberg operators (4.2.11), their motion is determined by the Heisenberg equation of motion

$$\frac{d}{dt}\hat{A}(t) = \frac{1}{i\hbar}[\hat{A}(t), \mathscr{H}] \tag{5.2.1}$$

where $\mathscr{H}$ stands for the Hamiltonian of the system (in this chapter without the operator symbol). Note that, as in (4.2.12), we have neglected interactions between the system and its environment such as the heat reservoir. By making use of (5.2.1) we obtain the equations determining the time variation of the Green's functions, which we shall simply call the *equations of motion.*

5.2.1 Chain of Equations of Motion

When we differentiate the defining equations (5.1.7, 16), the time derivatives of the step function give inhomogeneous terms proportional to the same Dirac δ function

$$\frac{d}{dt}[\pm\,\theta(\pm\,[t-t'])] = \delta(t-t')\,, \tag{5.2.2}$$

as intended when we inserted the minus sign into the advanced Green's function (5.1.16):

$$\begin{aligned}
\frac{d}{dt}G^{\mathrm{ret},\pm}_{A,B}(t,t') &= \delta(t-t')\left\langle \frac{1}{i\hbar}[\hat{A},\hat{B}]_{\pm}\right\rangle \\
&\quad + \theta(t-t')\left\langle \frac{1}{i\hbar}\left[\frac{1}{i\hbar}[\hat{A}(t),\mathscr{H}],\hat{B}(t')\right]_{\pm}\right\rangle, \\
\frac{d}{dt}G^{\mathrm{adv},\pm}_{A,B}(t,t') &= \delta(t-t')\left\langle \frac{1}{i\hbar}[\hat{A},\hat{B}]_{\pm}\right\rangle \\
&\quad + \left\{-\,\theta(t'-t)\left\langle \frac{1}{i\hbar}\left[\frac{1}{i\hbar}[\hat{A}(t),\mathscr{H}],\hat{B}(t')\right]_{\pm}\right\rangle\right\}.
\end{aligned} \tag{5.2.3}$$

Given explicit forms of the Hamiltonian $\mathscr{H}$ and the operator $\hat{A}$, the commutator $[\hat{A}(t),\mathscr{H}]/i\hbar$ can be calculated. If the resulting operator contains only the original operator $\hat{A}(t)$ as operators, the second terms on the right-hand side of the above equations are written in terms of $G^{\mathrm{ret},\pm}_{A,B}$ or $G^{\mathrm{adv},\pm}_{A,B}$, and the equations become closed. In addition, the distinction between the retarded and advanced functions enters only by the step function factor, so that the two equations (5.2.3) ought to have the same structure. Whether the retarded or the advanced Green's function is obtained by solving this equation depends on how the boundary condition is imposed with respect to time, i.e., on whether we set it as zero for $t < t'$ or for $t > t'$. The existence of the inhomogeneous term including the δ function $\delta(t-t')$ as a factor in (5.2.3) corresponds to a point source placed at the point $t = t'$ of the time axis and induces a discontinuity of the solution at $t \lessgtr t'$, hence the name, the Green's function.

The Hamiltonian $\mathscr{H}$ generally includes interaction terms among particles constituting the system, so that the commutator $[\hat{A}(t), \mathscr{H}]/i\hbar$ can seldom be written only with the operator $\hat{A}(t)$ but usually includes another more complex operator $\hat{A}_1(t)$. Then, the second terms of (5.2.3) contain in addition to $G^{\mathrm{ret},\pm}_{A,B}$ or $G^{\mathrm{adv},\pm}_{A,B}$ more complicated or rather higher-order Green's functions $G^{\mathrm{ret},\pm}_{A_1,B}$ or $G^{\mathrm{adv},\pm}_{A_1,B}$, respectively, and the equations do not close. Thus, if we construct the equation of motion for $G^{\mathrm{ret},\pm}_{A_1,B}$ or $G^{\mathrm{adv},\pm}_{A_1,B}$, then Green's functions of a still higher order will appear in the latter equation. Thus we obtain a *chain (or hierarchy) of equations of motion.* This situation originates in the interaction term in the Hamiltonian, and is always inevitable when discussing a many-body problem. A special case when the chain closes after a number of steps may occur when there is, as it were, a complete set of operators $\hat{A}(t)$, $\hat{A}_1(t)$, $\hat{A}_2(t), \ldots, \hat{A}_n(t)$ and the algebra of operators is closed within it. However, in general such a simple case occurs only as an approximation. In particular, with the thermodynamic limit and an infinite degree of freedom for our system, the chain of equations of motion continues indefinitely.

We have to determine the Green's function by solving this chain of equations of motion. This is precisely a many-body problem for which we have not yet found any general method of exact solution and can pick up at best a special form of solution, e.g., a solution varying synchronized with macroscopic variables such as described in Chap. 4. Therefore, we often adopt the *decoupling approximation.* Since the chain of equations of motion has such a structure that we have to know higher-order Green's functions in order to determine lower-order Green's functions, we introduce an assumption suitably expressing a Green's function of a certain order in terms of Green's functions of lower orders. Then the set of equations becomes closed with the first equation of motion that contains the Green's function of the same order. Neglecting the other equations of motion for the remaining Green's functions of higher order, the chain of equations is decoupled halfway. The accuracy of our approximation depends on which link of the chain is severed and also on what expression is used to express the higher-order Green's function in terms of the lower-order Green's functions for the decoupling. However, we have no general criterion, and have no choice but to be reconciled with our assumption by comparing the result obtained with experiments or with other existing theories. We shall give an example of the decoupling approximation below.

5.2.2 Complex Dielectric Function of a Plasma in a Decoupling Approximation

Let us calculate the complex dielectric function $\varepsilon(\boldsymbol{k}, \omega)$ of the plasma by making use of the Nozières-Pines formula obtained from (4.5.55) with (4.5.57 or 18).

For simplicity, we shall neglect electron spins and radiation field and make an approximation of positive ions smeared out uniformly in space. In this model, the electric charge density is given in terms of the second quantization by

$$\hat{\varrho}^{\text{ind}}(\boldsymbol{r}) = -e\left[\hat{\psi}^{\dagger}(\boldsymbol{r})\,\hat{\psi}(\boldsymbol{r}) - \frac{1}{V}\int_V d\boldsymbol{r}'\,\hat{\psi}^{\dagger}(\boldsymbol{r}')\,\hat{\psi}(\boldsymbol{r}')\right], \tag{5.2.4}$$

$\hat{\psi}(\boldsymbol{r})$ being the quantized wave function of electrons. By taking the Fourier transformation on the spatial coordinate $\boldsymbol{r}$ and changing to the momentum representation

$$\hat{\psi}(\boldsymbol{r}) = \sum_{\boldsymbol{k}} \frac{\exp(\mathrm{i}\,\boldsymbol{k}\cdot\boldsymbol{r})}{\sqrt{V}}\,\hat{a}_{\boldsymbol{k}}, \quad \hat{\varrho}^{\text{ind}}(\boldsymbol{r}) = \frac{1}{V}\sum_{\boldsymbol{k}} \exp(\mathrm{i}\,\boldsymbol{k}\cdot\boldsymbol{r})\,\hat{\varrho}_{\boldsymbol{k}}^{\text{ind}}, \tag{5.2.5}$$

then

$$\hat{\varrho}_{\boldsymbol{k}}^{\text{ind}} = \begin{cases} 0 & (\boldsymbol{k} = 0) \\ -e\sum_{\boldsymbol{q}} \hat{a}_{\boldsymbol{q}-\boldsymbol{k}/2}^{\dagger}\,\hat{a}_{\boldsymbol{q}+\boldsymbol{k}/2} & (\boldsymbol{k} \neq 0) \end{cases}. \tag{5.2.6}$$

These physical variables govern the dielectric phenomena which we are trying to analyze. Although it is sufficient for our purpose merely to construct the Green's function by regarding these variables as $\hat{A}$ and $\hat{B}$ in the preceding section, it is more convenient to consider the retarded Green's function for the summands of (5.2.6)

$$\begin{aligned} &G^{\text{ret},-}(\boldsymbol{k}_1, \boldsymbol{k}_2, t; \boldsymbol{k}_3, \boldsymbol{k}_4, t') \\ &= \theta(t-t')\left\langle \frac{1}{\mathrm{i}\hbar}[\hat{a}_{\boldsymbol{k}_1}^{\dagger}(t)\,\hat{a}_{\boldsymbol{k}_2}(t), \hat{a}_{\boldsymbol{k}_3}^{\dagger}(t')\,\hat{a}_{\boldsymbol{k}_4}(t')]\right\rangle \\ &= \int_{-\infty}^{\infty} \frac{d\omega}{2\pi} \exp[-\mathrm{i}\omega(t-t')]\,K^{\text{ret},-}(\boldsymbol{k}_1, \boldsymbol{k}_2; \boldsymbol{k}_3, \boldsymbol{k}_4; \omega). \end{aligned} \tag{5.2.7}$$

Then the Nozières-Pines formula is

$$\begin{aligned} \frac{1}{\varepsilon(\boldsymbol{k}, \omega)} - 1 &= \frac{4\pi e^2}{\boldsymbol{k}^2}\int_{-\infty}^{\infty} dt\, \mathrm{e}^{\mathrm{i}\omega t} \\ &\quad \times \lim_{V\to\infty} \frac{1}{V}\sum_{\boldsymbol{q},\boldsymbol{q}'} G^{\text{ret},-}\left(\boldsymbol{q}-\frac{\boldsymbol{k}}{2}, \boldsymbol{q}+\frac{\boldsymbol{k}}{2}, t; \boldsymbol{q}'+\frac{\boldsymbol{k}}{2}, \boldsymbol{q}'-\frac{\boldsymbol{k}}{2}, 0\right) \\ &= \frac{4\pi e^2}{\boldsymbol{k}^2}\lim_{V\to\infty}\frac{1}{V}\sum_{\boldsymbol{q},\boldsymbol{q}'} K^{\text{ret},-}\left(\boldsymbol{q}-\frac{\boldsymbol{k}}{2}, \boldsymbol{q}+\frac{\boldsymbol{k}}{2}; \boldsymbol{q}'+\frac{\boldsymbol{k}}{2}, \boldsymbol{q}'-\frac{\boldsymbol{k}}{2}; \omega\right). \end{aligned} \tag{5.2.8}$$

This expression corresponds to (5.1.5) discussed at the beginning of Sect. 5.1.

Now let us set up the equation of motion satisfied by the retarded Green's function (5.2.7). We shall take as the Hamiltonian of the system

$$\begin{aligned}\mathscr{H} &= \int_V d\boldsymbol{r} \frac{\hbar^2}{2m} \frac{\partial \hat{\psi}^\dagger(\boldsymbol{r})}{\partial \boldsymbol{r}} \cdot \frac{\partial \hat{\psi}(\boldsymbol{r})}{\partial \boldsymbol{r}} \\ &+ \frac{1}{2} \iint_V d\boldsymbol{r}\, d\boldsymbol{r}'\, \hat{\psi}^\dagger(\boldsymbol{r})\, \hat{\psi}^\dagger(\boldsymbol{r}')\, \Phi(\boldsymbol{r}-\boldsymbol{r}')\, \hat{\psi}(\boldsymbol{r}')\, \hat{\psi}(\boldsymbol{r}) \\ &= \sum_{\boldsymbol{k}} E(\boldsymbol{k})\, \hat{a}_{\boldsymbol{k}}^\dagger \hat{a}_{\boldsymbol{k}} + \frac{1}{2V} \sum_{\boldsymbol{k}_1, \boldsymbol{k}_2, \boldsymbol{k}_3, \boldsymbol{k}_4} \Phi_{\boldsymbol{k}_1-\boldsymbol{k}_4}\, \delta_{\boldsymbol{k}_1+\boldsymbol{k}_2, \boldsymbol{k}_3+\boldsymbol{k}_4}\, \hat{a}_{\boldsymbol{k}_1}^\dagger \hat{a}_{\boldsymbol{k}_2}^\dagger \hat{a}_{\boldsymbol{k}_3} \hat{a}_{\boldsymbol{k}_4}\,.\end{aligned} \tag{5.2.9}$$

The first term on the right-hand side is the kinetic energy of electrons and the second term, the Coulomb interactions between electrons:

$$E(\boldsymbol{k}) = \frac{\hbar^2 \boldsymbol{k}^2}{2m}, \quad \Phi(\boldsymbol{r}) = \frac{e^2}{r}, \quad \Phi_{\boldsymbol{k}} = \int d\boldsymbol{r} \exp(-\mathrm{i}\boldsymbol{k}\cdot\boldsymbol{r})\, \Phi(\boldsymbol{r}) = \frac{4\pi e^2}{\boldsymbol{k}^2}. \tag{5.2.10}$$

Since electrons obey Fermi statistics, their commutation relations are given by

$$[\hat{a}_{\boldsymbol{k}}, \hat{a}_{\boldsymbol{k}'}]_+ = 0, \quad [\hat{a}_{\boldsymbol{k}}, \hat{a}_{\boldsymbol{k}'}^\dagger]_+ = \delta_{\boldsymbol{k},\boldsymbol{k}'}, \quad [\hat{a}_{\boldsymbol{k}}^\dagger, \hat{a}_{\boldsymbol{k}'}^\dagger]_+ = 0. \tag{5.2.11}$$

By incorporating the effects of uniformly distributed positive charges into the Hamiltonian (5.2.9), we can prove that the Coulomb interactions between the positive charges and those between positive charges and electrons cancel the term with $\boldsymbol{k}_1 = \boldsymbol{k}_4$ in (5.2.9) for a sufficiently large system, i.e., in the thermodynamic limit [5.4]. In the following, let us take account of this effect by setting $\Phi_0 = 0$.

The proof proceeds as follows. Let us regard Φ_0 as nondivergent for the present. We evaluate the Coulomb interaction between electrons and the positive background as

$$\iint d\boldsymbol{r}\, d\boldsymbol{R} \frac{[-e\hat{\psi}^\dagger(\boldsymbol{r})\, \hat{\psi}(\boldsymbol{r})]\, \hat{N}e/V}{|\boldsymbol{r}-\boldsymbol{R}|} = -\frac{\hat{N}^2}{V} \Phi_0\,,$$

those between the positive charges as

$$\frac{1}{2} \iint d\boldsymbol{R}\, d\boldsymbol{R}' \frac{(\hat{N}e/V)^2}{|\boldsymbol{R}-\boldsymbol{R}'|} = \frac{\hat{N}^2}{2V} \Phi_0\,,$$

and the Coulomb interaction term with $\boldsymbol{k}_1 = \boldsymbol{k}_4$ in (5.2.9) as $(\hat{N}^2/2V)\,\Phi_0 - (\hat{N}/2V)\,\Phi_0$. On the other hand,

$$\begin{aligned}\frac{1}{2} \iint d\boldsymbol{r}\, d\boldsymbol{r}' \frac{\hat{\varrho}^{\mathrm{ind}}(\boldsymbol{r})\, \hat{\varrho}^{\mathrm{ind}}(\boldsymbol{r}')}{|\boldsymbol{r}-\boldsymbol{r}'|} &= \frac{1}{2V} \sum_{\boldsymbol{k}} \Phi_{\boldsymbol{k}}\, \hat{\varrho}_{\boldsymbol{k}}^{\mathrm{ind}}\, \hat{\varrho}_{-\boldsymbol{k}}^{\mathrm{ind}} \\ &= [\text{ Coulomb term in (5.2.9) with } \boldsymbol{k}_1 \neq \boldsymbol{k}_4] - \frac{\hat{N}}{2V} \Phi_0 - \frac{\hat{N}}{2} \sum_{\boldsymbol{k}} \Phi_{\boldsymbol{k}}\,.\end{aligned}$$

Thus

[Coulomb term in (5.2.9)] + (background effects)

$$= \frac{1}{2V} \sum_k \Phi_k \, \hat{\varrho}_k^{\text{ind}} \, \hat{\varrho}_{-k}^{\text{ind}} + \frac{\hat{N}}{2} \sum_k \Phi_k \, .$$

The last term $(\hat{N}/2) \sum \Phi_k$ is the self-energy of electrons. If we neglect this divergent self-energy as usual, no term with Φ_0 remain because $\hat{\varrho}_k^{\text{ind}} = 0$ for $\boldsymbol{k} = 0$.

By using (5.2.11), we evaluate the commutator in the equation of motion (5.2.3) for the retarded Green's function (5.2.7) to obtain

$$\begin{aligned} & \mathrm{i}\hbar \frac{d}{dt} G^{\text{ret},-}(\boldsymbol{k}_1, \boldsymbol{k}_2, t; \boldsymbol{k}_3, \boldsymbol{k}_4, t') \\ & = \delta(t-t') \left[\delta_{\boldsymbol{k}_2, \boldsymbol{k}_3} \langle \hat{a}_{\boldsymbol{k}_1}^\dagger \hat{a}_{\boldsymbol{k}_4} \rangle - \delta_{\boldsymbol{k}_1, \boldsymbol{k}_4} \langle \hat{a}_{\boldsymbol{k}_3}^\dagger \hat{a}_{\boldsymbol{k}_2} \rangle \right] \\ & \quad + \left[E(\boldsymbol{k}_2) - E(\boldsymbol{k}_1)\right] G^{\text{ret},-}(\boldsymbol{k}_1, \boldsymbol{k}_2, t; \boldsymbol{k}_3, \boldsymbol{k}_4, t') \\ & \quad + \frac{1}{V} \sum_{\boldsymbol{k}_1', \boldsymbol{k}_2', \boldsymbol{k}_3', \boldsymbol{k}_4'} \Phi_{\boldsymbol{k}_2' - \boldsymbol{k}_3'} (\delta_{\boldsymbol{k}_1, \boldsymbol{k}_1'} \, \delta_{\boldsymbol{k}_2 + \boldsymbol{k}_2', \boldsymbol{k}_3' + \boldsymbol{k}_4'} - \delta_{\boldsymbol{k}_2, \boldsymbol{k}_4'} \, \delta_{\boldsymbol{k}_1' + \boldsymbol{k}_2', \boldsymbol{k}_3', + \boldsymbol{k}_1}) \\ & \quad \times G^{\text{ret},-}(\boldsymbol{k}_1', \boldsymbol{k}_2, \boldsymbol{k}_3', \boldsymbol{k}_4', t; \boldsymbol{k}_3, \boldsymbol{k}_4, t') \, . \end{aligned} \tag{5.2.12}$$

As was stated earlier, the Green's function of the next order higher by one

$$\begin{aligned} & G^{\text{ret},-}(\boldsymbol{k}_1', \boldsymbol{k}_2', \boldsymbol{k}_3', \boldsymbol{k}_4', t; \boldsymbol{k}_3, \boldsymbol{k}_4, t') \\ & = \theta(t-t') \left\langle \frac{1}{\mathrm{i}\hbar} [\hat{a}_{\boldsymbol{k}_1'}^\dagger(t) \, \hat{a}_{\boldsymbol{k}_2'}^\dagger(t) \, \hat{a}_{\boldsymbol{k}_3'}(t) \, \hat{a}_{\boldsymbol{k}_4'}(t), \hat{a}_{\boldsymbol{k}_3}^\dagger(t') \, \hat{a}_{\boldsymbol{k}_4}(t')] \right\rangle \end{aligned} \tag{5.2.13}$$

has appeared due to the interaction. If we formulate the equation of motion for the latter Green's function (5.2.13), we shall have an equation containing another Green's function of the next order higher by one, so that the chain grows by one link.

Even if we extend the length of the chain, more complicated equations still arise; therefore, let us sever the chain at this point as an example of the crudest decoupling approximation. Namely, let us consider expressing the Green's function (5.2.13) approximately in terms of the lower-order Green's function (5.2.7). A usual way is to assume the replacement of extra pairs of creation and annihilation operators contained in the higher-order Green's function by their averages, and this method corresponds to the molecular field approximation in the theory of magnetism [5.5]. For (5.2.13), we choose a pair $\hat{a}^\dagger(t) \, \hat{a}(t)$ from the two creation operators $\hat{a}_{\boldsymbol{k}_1'}^\dagger(t)$ and $\hat{a}_{\boldsymbol{k}_2'}^\dagger(t)$ and the two annihilation operators $\hat{a}_{\boldsymbol{k}_3'}(t)$ and $\hat{a}_{\boldsymbol{k}_4'}(t)$, and replace it by its expectation value $\langle \hat{a}^\dagger(t) \, \hat{a}(t) \rangle$. Since there are four ways of choosing a pair, by summing over them, we make the following approximation

$$\begin{aligned} &G^{\mathrm{ret},-}(\boldsymbol{k}_1', \boldsymbol{k}_2', \boldsymbol{k}_3', \boldsymbol{k}_4', t; \boldsymbol{k}_3, \boldsymbol{k}_4, t') \\ &= \langle \hat{a}^\dagger_{\boldsymbol{k}_2'} \hat{a}_{\boldsymbol{k}_3'} \rangle G^{\mathrm{ret},-}(\boldsymbol{k}_1', \boldsymbol{k}_4', t; \boldsymbol{k}_3, \boldsymbol{k}_4, t') + \langle \hat{a}^\dagger_{\boldsymbol{k}_1'} \hat{a}_{\boldsymbol{k}_4'} \rangle G^{\mathrm{ret},-}(\boldsymbol{k}_2', \boldsymbol{k}_3', t; \boldsymbol{k}_3, \boldsymbol{k}_4, t') \\ &- \langle \hat{a}^\dagger_{\boldsymbol{k}_1'} \hat{a}_{\boldsymbol{k}_3'} \rangle G^{\mathrm{ret},-}(\boldsymbol{k}_2', \boldsymbol{k}_4', t; \boldsymbol{k}_3, \boldsymbol{k}_4, t') - \langle \hat{a}^\dagger_{\boldsymbol{k}_2'} \hat{a}_{\boldsymbol{k}_4'} \rangle G^{\mathrm{ret},-}(\boldsymbol{k}_1', \boldsymbol{k}_3', t; \boldsymbol{k}_3, \boldsymbol{k}_4, t') . \end{aligned} \tag{5.2.14}$$

The negative sign inserted into the third and fourth terms on the right-hand side comes from interchanging the operators before replacing the pair with the expectation value, thereby retaining the Fermi statistics. If the system is invariant with respect to the spatial translation in the thermodynamic limit, it can be proved that

$$\langle \hat{a}^\dagger_{\boldsymbol{k}} \hat{a}_{\boldsymbol{k}'} \rangle = \delta_{\boldsymbol{k}\boldsymbol{k}'} \langle \hat{a}^\dagger_{\boldsymbol{k}} \hat{a}_{\boldsymbol{k}} \rangle = \delta_{\boldsymbol{k}\boldsymbol{k}'} n_{\boldsymbol{k}} \tag{5.2.15}$$

by using (5.2.11). Here $n_{\boldsymbol{k}}$ yields the average number density of electrons occupying a state $\boldsymbol{k}$ in thermal equilibrium. Substituting (5.2.14, 15) into (5.2.12) gives as the equation of motion in the crudest decoupling approximation

$$\begin{aligned} &\mathrm{i}\hbar \frac{d}{dt} G^{\mathrm{ret},-}(\boldsymbol{k}_1, \boldsymbol{k}_2, t; \boldsymbol{k}_3, \boldsymbol{k}_4, t') \\ &= \delta(t-t')\, \delta_{\boldsymbol{k}_1\boldsymbol{k}_4} \delta_{\boldsymbol{k}_2\boldsymbol{k}_3} (n_{\boldsymbol{k}_1} - n_{\boldsymbol{k}_2}) + [E(\boldsymbol{k}_2) - E(\boldsymbol{k}_1)]\, G^{\mathrm{ret},-}(\boldsymbol{k}_1, \boldsymbol{k}_2, t; \boldsymbol{k}_2, \boldsymbol{k}_4, t') \\ &\quad + (n_{\boldsymbol{k}_1} - n_{\boldsymbol{k}_2})\, \Phi_{\boldsymbol{k}_1-\boldsymbol{k}_2} \frac{1}{V} \sum_{\boldsymbol{k}_1', \boldsymbol{k}_2'} \delta_{\boldsymbol{k}_1-\boldsymbol{k}_2, \boldsymbol{k}_1'-\boldsymbol{k}_2'}\, G^{\mathrm{ret},-}(\boldsymbol{k}_1', \boldsymbol{k}_2', t; \boldsymbol{k}_3, \boldsymbol{k}_4, t') \\ &\quad - \left[\frac{1}{V} \sum_{\boldsymbol{k}'} n_{\boldsymbol{k}'} (\Phi_{\boldsymbol{k}'-\boldsymbol{k}_2} - \Phi_{\boldsymbol{k}'-\boldsymbol{k}_1}) \right] G^{\mathrm{ret},-}(\boldsymbol{k}_1, \boldsymbol{k}_2, t; \boldsymbol{k}_3, \boldsymbol{k}_4, t') \\ &\quad - (n_{\boldsymbol{k}_1} - n_{\boldsymbol{k}_2}) \frac{1}{V} \sum_{\boldsymbol{k}_1', \boldsymbol{k}_2'} \Phi_{\boldsymbol{k}_1-\boldsymbol{k}_1'}\, \delta_{\boldsymbol{k}_1-\boldsymbol{k}_2, \boldsymbol{k}_1'-\boldsymbol{k}_2'}\, G^{\mathrm{ret},-}(\boldsymbol{k}_1', \boldsymbol{k}_2', t; \boldsymbol{k}_3, \boldsymbol{k}_4, t') . \end{aligned} \tag{5.2.16}$$

This is the closed equation for the required lowest-order Green's function (5.2.7).

We need to solve (5.2.16) to obtain the Green's function, but (5.2.16) is still very complicated. The last two terms that originate from the two terms with negative sign in (5.2.14) yield only correction terms to $E(\boldsymbol{k})$, etc., so that we neglect them here for simplicity. As a consequence, effects of the Fermi statistics remain only in the form of the electron distribution $n_{\boldsymbol{k}}$. This procedure corresponds to the well-known *random phase approximation* developed by *Bohm* and *Pines* [5.6] in the many-body problem at absolute zero. At a finite temperature we may call it a kind of random phase approximation in the sense that it is a kind of Hartree approximation linearized in the external force field [5.7], i.e., it leads to the Vlasov equation linearized in the external field (Sect. 5.3) [5.8].

To calculate the complex dielectric function (5.2.8), the spectral function $K^{\mathrm{ret},-}(\boldsymbol{k}_1, \boldsymbol{k}_2; \boldsymbol{k}_3, \boldsymbol{k}_4; \omega)$ is more convenient than the Green's function; the same is true for solving the equation of motion (5.2.16). This is because an algebraic equation arises by Fourier transforming (5.2.16) with respect to time, since the time differentiation $i\hbar\, d/dt$ is replaced by $\hbar\omega$. So far we have given equations for the retarded Green's function but, as noted in the beginning, equations for the advanced Green's function can also be obtained in the same way: the equation of motion, in particular, assumes the same form. The distinction between the retarded and advanced functions arises in the difference in boundary conditions for the time variable; or in the terminology of the spectral function, when the angular frequency ω is replace by a complex number z, the distinction lies in the different analytic behavior in the upper half or the lower half complex plane, as is obvious from the discussion in Sect. 3.6.

It is more convenient to consider the complex function defined in (5.1.19) and satisfying (5.1.20). The equation satisfied by the latter function ought to coincide with that obtained by substituting ω by z in the Fourier transform of the equation of motion for the Green's function. In the above-mentioned approximation, it is written in the following form:

$$\left[\hbar z + E\left(\boldsymbol{q}-\frac{\boldsymbol{k}}{2}\right) - E\left(\boldsymbol{q}+\frac{\boldsymbol{k}}{2}\right)\right] K^{-}\left(\boldsymbol{q}-\frac{\boldsymbol{k}}{2}, \boldsymbol{q}+\frac{\boldsymbol{k}}{2}; \boldsymbol{q}'+\frac{\boldsymbol{k}'}{2}, \boldsymbol{q}'-\frac{\boldsymbol{k}'}{2}; z\right)$$
$$= (n_{\boldsymbol{q}-\boldsymbol{k}/2} - n_{\boldsymbol{q}+\boldsymbol{k}/2}) \left[\delta_{\boldsymbol{q}\boldsymbol{q}'}\,\delta_{\boldsymbol{k}\boldsymbol{k}'} + \frac{\Phi_{\boldsymbol{k}}}{V}\sum_{\boldsymbol{q}''} K^{-}\right.$$
$$\left.\times\left(\boldsymbol{q}''-\frac{\boldsymbol{k}}{2}, \boldsymbol{q}''+\frac{\boldsymbol{k}}{2}; \boldsymbol{q}'+\frac{\boldsymbol{k}'}{2}, \boldsymbol{q}'-\frac{\boldsymbol{k}'}{2}; z\right)\right]. \tag{5.2.17}$$

We have renamed $\boldsymbol{k}_1 = \boldsymbol{q} - \boldsymbol{k}/2$, $\boldsymbol{k}_2 = \boldsymbol{q} + \boldsymbol{k}/2$, $\boldsymbol{k}_3 = \boldsymbol{q}' + \boldsymbol{k}'/2$, and $\boldsymbol{k}_4 = \boldsymbol{q}' - \boldsymbol{k}'/2$ following (5.2.8). If we move z out of the real axis as in (5.1.20), we may divide both sides of the above equation by $hz + E(\boldsymbol{q} - \boldsymbol{k}/2) - E(\boldsymbol{q} + \boldsymbol{k}/2)$:

$$K^{-}\left(\boldsymbol{q}-\frac{\boldsymbol{k}}{2}, \boldsymbol{q}+\frac{\boldsymbol{k}}{2}; \boldsymbol{q}'+\frac{\boldsymbol{k}'}{2}, \boldsymbol{q}'-\frac{\boldsymbol{k}'}{2}; z\right)$$
$$= K_0^{-}\left(\boldsymbol{q}-\frac{\boldsymbol{k}}{2}, \boldsymbol{q}+\frac{\boldsymbol{k}}{2}; \boldsymbol{q}+\frac{\boldsymbol{k}}{2}, \boldsymbol{q}-\frac{\boldsymbol{k}}{2}; z\right)$$
$$\times\left[\delta_{\boldsymbol{q}\boldsymbol{q}'}\,\delta_{\boldsymbol{k}\boldsymbol{k}'} + \frac{\Phi_{\boldsymbol{k}}}{V}\sum_{\boldsymbol{q}''} K^{-}\left(\boldsymbol{q}''-\frac{\boldsymbol{k}}{2}, \boldsymbol{q}''+\frac{\boldsymbol{k}}{2}; \boldsymbol{q}'+\frac{\boldsymbol{k}'}{2}, \boldsymbol{q}'-\frac{\boldsymbol{k}'}{2}; z\right)\right]. \tag{5.2.18}$$

Here, K_0^{-} is the spectral function for noninteracting electrons:

$$K_0^{-}\left(\boldsymbol{q}-\frac{\boldsymbol{k}}{2}, \boldsymbol{q}+\frac{\boldsymbol{k}}{2}; \boldsymbol{q}'+\frac{\boldsymbol{k}'}{2}, \boldsymbol{q}'-\frac{\boldsymbol{k}'}{2}; z\right)$$
$$= \delta_{\boldsymbol{q}\boldsymbol{q}'}\,\delta_{\boldsymbol{k}\boldsymbol{k}'}\frac{n_{\boldsymbol{q}-\boldsymbol{k}/2} - n_{\boldsymbol{q}+\boldsymbol{k}/2}}{\hbar z + E(\boldsymbol{q}-\boldsymbol{k}/2) - E(\boldsymbol{q}+\boldsymbol{k}/2)}, \tag{5.2.19}$$

where the electron distribution $n_{\boldsymbol{k}}$ generally includes interaction effects. Summing both sides of (5.2.18) over $\boldsymbol{q}$ gives a simple algebraic equation for $\sum K^-$ which can be solved easily. Substituting this solution into the right-hand side of (5.2.18) gives K^- itself, but $\sum K^-$ is sufficient for calculating the complex dielectric function (5.2.8). After some rearrangement

$$\varepsilon[\boldsymbol{k},\omega] = 1 - \frac{4\pi e^2}{\boldsymbol{k}^2} \lim_{\varepsilon\to+0} \lim_{V\to\infty} \frac{1}{V} \times \sum_{\boldsymbol{q},\boldsymbol{q}'} K_0^- \left(\boldsymbol{q}-\frac{\boldsymbol{k}}{2}, \boldsymbol{q}+\frac{\boldsymbol{k}}{2};\ \boldsymbol{q}'+\frac{\boldsymbol{k}}{2}, \boldsymbol{q}'-\frac{\boldsymbol{k}}{2}; \omega+\mathrm{i}\,\varepsilon\right). \tag{5.2.20}$$

This approximate expression was first obtained by Lindhard [5.9] by solving the kinetic equation. Although it was not explicitly mentioned before in (5.2.8), the limit $\varepsilon\to+0$ in (5.1.20) must be taken after the thermodynamic limit $V\to+\infty$. This holds also for the above equation, because the analyticity providing a basis for (5.1.20) is, as discussed in Sect. 3.6, based on the fact that effects of a force disappear after a sufficiently long time, i.e., irreversibility can be expected only after taking the thermodynamic limit.

According to (5.2.20, 19), we can calculate the complex dielectric function if we know the electron distribution $n_{\boldsymbol{k}}$. To obtain the expectation value of a product of operators such as $n_{\boldsymbol{k}} = \langle \hat{a}_{\boldsymbol{k}}^\dagger \hat{a}_{\boldsymbol{k}} \rangle$, we need to know the spectral function of the single-particle Green's function, Sect. 5.4. Here we shall not go through the trouble of setting up the equation of motion for the single-particle Green's function and making a particular decoupling of its linked higher-order equations, but simply substitute for the distribution $n_{\boldsymbol{k}}$ the thermal equilibrium value for noninteracting electrons. This approximation is made quite often. As shown in Sect. 5.4, the Fermi distribution within the free electron approximation for the distribution $n_{\boldsymbol{k}}$ takes the form

$$n_{\boldsymbol{k}} = \frac{1}{\exp\{[E(\boldsymbol{k})-\mu]/kT\}+1}\,.$$

When Fermi statistics is not effective we may take for $n_{\boldsymbol{k}}$ the Maxwell-Boltzmann distribution. The result of the calculation for the wave vector $\boldsymbol{k}\approx 0$ becomes

$$\varepsilon'[\boldsymbol{k},\omega] = 1 - \left(\frac{\omega_\mathrm{p}}{\omega}\right)^2 \left[1 + 2\frac{\boldsymbol{k}^2}{m\omega^2}\bar{E} + O(\boldsymbol{k}^4)\right],$$
$$\varepsilon''[\boldsymbol{k},\omega] = 2\frac{ne^2}{\hbar|\boldsymbol{k}|^3}\sqrt{\frac{m(2\pi)^3}{kT}}\exp\left(-\frac{(\hbar\omega)^2/E(\boldsymbol{k})+E(\boldsymbol{k})}{4kT}\right)\sinh\left(\frac{\hbar\omega}{2kT}\right), \tag{5.2.21}$$

where $\omega_\mathrm{p} = \sqrt{4\pi n e^2/m}$ stands for the Langmuir plasma frequency and $\bar{E} = (3/2)\,kT$, the average energy per particle.

5.3 Relation to the Kinetic Equation

In the preceding section, we did not directly construct the Green's function corresponding to the electric charge density (5.2.6), but considered the Green's function for $\hat{q}^{\dagger}_{q-k/2}\,\hat{a}_{q+k/2}$ without summation over $\boldsymbol{q}$. This procedure means, in reality, that we have switched from hydrodynamic quantities to kinetic-theoretical quantities by demanding a more detailed description. This can be seen by examining the relationship between the equation-of-motion method for the Green's function discussed in the previous section and the kinetic-theoretical method using the Boltzmann transport equation and the like.

5.3.1 Klimontovich Operator

The quantum-theoretical quantity corresponding to the one-particle distribution function in the classical molecular kinetic theory of gases is the *Wigner* phase space distribution [5.10], and is the statistical ensemble average of an operator introduced by *Klimontovich* [5.11]

$$\begin{aligned}\hat{F}_1(\boldsymbol{r},\boldsymbol{p},t) &= \int_V d\boldsymbol{R}\,\exp(-\mathrm{i}\boldsymbol{q}\cdot\boldsymbol{R})\,\hat{\psi}^{\dagger}\left(\boldsymbol{r}-\frac{\boldsymbol{R}}{2},t\right)\hat{\psi}\left(\boldsymbol{r}+\frac{\boldsymbol{R}}{2},t\right)\\ &= \sum_k \exp(\mathrm{i}\boldsymbol{k}\cdot\boldsymbol{r})\,\hat{a}^{\dagger}_{q-k/2}(t)\,\hat{a}_{q+k/2}(t)\,.\end{aligned} \tag{5.3.1}$$

This is written as a Heisenberg operator and we have set $\boldsymbol{p}=\hbar\boldsymbol{q}$. At a glance it is clear that it satisfies the normalization condition for the one-particle distribution

$$\int_V d\boldsymbol{r}\,\frac{1}{V}\sum_q \hat{F}_1(\boldsymbol{r},\boldsymbol{p},t)=\hat{N} \tag{5.3.2}$$

and that the electric charge distribution, for instance, can be written in the form

$$\hat{\varrho}^{\mathrm{ind}}(\boldsymbol{r},t)=\int_V d\boldsymbol{r}'\,\frac{1}{V}\sum_q\left[-e\,\delta(\boldsymbol{r}-\boldsymbol{r}')+\frac{e}{V}\right]\hat{F}_1(\boldsymbol{r}',\boldsymbol{p},t)\,. \tag{5.3.3}$$

The form of the latter equation is obtained by expressing (5.2.4) in terms of the distribution function. Remembering that in the thermodynamical limit $V\to\infty$,

$$\frac{1}{V}\sum_q\cdots\to\int\frac{d\boldsymbol{p}}{(2\pi\hbar)^3}\cdots,$$

an analogy to classical kinetic theory is clearer. To pass from a quantity depending on one wave vector like the electric charge density (5.2.6) to a quantity depending on two wave vector $\hat{a}^{\dagger}_{q-k/2}\,\hat{a}_{q+k/2}$ is thus changing to a quantity depending on the momentum $\boldsymbol{p}$ in addition to the space coordinate $\boldsymbol{r}$, as is shown in (5.3.1). This implies more precision in the description.

Assuming that an external time-dependent field of force is imposed upon our system with the Hamiltonian $\mathscr{H}$:

$$\mathscr{H}^{\text{ext}}(t) = \int_V d\boldsymbol{r}\,\hat{\psi}^{\dagger}(\boldsymbol{r})\,U^{\text{ext}}(\boldsymbol{r},t)\,\hat{\psi}(\boldsymbol{r}) = \int_V d\boldsymbol{r}\,\frac{1}{V}\sum_{\boldsymbol{q}} U^{\text{ext}}(\boldsymbol{r},t)\,\hat{F}_1(\boldsymbol{r},\boldsymbol{p})\,, \quad (5.3.4)$$

regarding the above as corresponding to (4.1.3) and applying (5.1.2), then the Wigner distribution function deviates from its equilibrium value

$$F_1^{\text{eq}}(\boldsymbol{p}) = \langle \hat{a}^{\dagger}_{\boldsymbol{q}}\,\hat{a}_{\boldsymbol{q}}\rangle = n_{\boldsymbol{q}} \quad (5.3.5)$$

in the following way

$$\delta F_1(\boldsymbol{r},\boldsymbol{p},t) = \int_{-\infty}^{\infty} dt \int d\boldsymbol{r}'\,\frac{1}{V}\sum_{\boldsymbol{q}'} G^{\text{ret},-}_{F,F}(\boldsymbol{r},\boldsymbol{p},t;\boldsymbol{r}',\boldsymbol{p}',t')\,U^{\text{ext}}(\boldsymbol{r}',t')\,. \quad (5.3.6)$$

Where $\boldsymbol{p} = \hbar\boldsymbol{q}$ and $\boldsymbol{p}' = \hbar\boldsymbol{q}'$, and we have introduced the retarded Green's function

$$G^{\text{ret},-}_{F,F}(\boldsymbol{r},\boldsymbol{p},t;\boldsymbol{r}',\boldsymbol{p}',t') = \theta(t-t')\left\langle \frac{1}{\mathrm{i}\hbar}[\hat{F}_1(\boldsymbol{r},\boldsymbol{p},t),\,\hat{F}_1(\boldsymbol{r}',\boldsymbol{p}',t')]\right\rangle. \quad (5.3.7)$$

5.3.2 Self-Consistent Field Approximation

By virtue of (5.3.6), the construction of the equation of motion for the Green's function (5.3.7) introduced above is nothing but the construction of the equation of motion for the Wigner distribution function for a system affected by an external field. Namely, (5.3.6) should be a solution of the latter equation obtained by retaining terms up to the first order in the external field.

Let us take the example of a plasma from the previous section. If we express the Green's function (5.3.7) in terms of (5.2.7), we can use the calculation given there. According to the definition of the Klimontovich operator (5.3.1),

$$G^{\text{ret},-}_{F,F}(\boldsymbol{r},\boldsymbol{p},t;\boldsymbol{r}',\boldsymbol{p}',t')$$
$$= \sum_{\boldsymbol{k},\boldsymbol{k}'} \mathrm{e}^{\mathrm{i}(\boldsymbol{k}\cdot\boldsymbol{r}-\boldsymbol{k}'\cdot\boldsymbol{r}')}\,G^{\text{ret},-}\left(\boldsymbol{q}-\frac{\boldsymbol{k}}{2},\boldsymbol{q}+\frac{\boldsymbol{k}}{2},t;\boldsymbol{q}'+\frac{\boldsymbol{k}'}{2},\boldsymbol{q}'-\frac{\boldsymbol{k}'}{2},t'\right), \quad (5.3.8)$$

whence (5.3.6) can be written as

$$\delta F_1(\boldsymbol{r},\boldsymbol{p},t) = \int_{-\infty}^{\infty}\frac{d\omega}{2\pi}\,\frac{1}{V}\sum_{\boldsymbol{k}} \mathrm{e}^{\mathrm{i}(\boldsymbol{k}\cdot\boldsymbol{r}-\omega t)}$$
$$\times \sum_{\boldsymbol{k}',\boldsymbol{q}'} K^{\text{ret},-}\left(\boldsymbol{q}-\frac{\boldsymbol{k}}{2},\boldsymbol{q}+\frac{\boldsymbol{k}}{2};\boldsymbol{q}'+\frac{\boldsymbol{k}'}{2},\boldsymbol{q}'-\frac{\boldsymbol{k}'}{2};\omega\right) U^{\text{ext}}_{\boldsymbol{k}',\omega} \quad (5.3.9)$$

where we have introduced

$$U_{k,\omega}^{\text{ext}} = \int_{-\infty}^{\infty} dt \int_V d\boldsymbol{r} \exp\left[-\mathrm{i}(\boldsymbol{k}\cdot\boldsymbol{r}-\omega t)\right] U^{\text{ext}}(\boldsymbol{r},t)\,. \tag{5.3.10}$$

If we differentiate (5.3.9) with respect to t, $\hbar\omega/\mathrm{i}\hbar$ appears in front of $K^{\text{ret},-}$ on the right-hand side. We eliminate $\hbar\omega K^{\text{ret},-}$ by means of the equation obtained by setting $z=\omega+\mathrm{i}0$ in (5.2.17), i.e., we make the transformation (5.3.9) of the equation of motion (5.2.17). Then, by using a technique such as demonstrated below, where $\boldsymbol{p}''=\hbar\boldsymbol{q}''$,

$$\frac{1}{\mathrm{i}\hbar}\int_{-\infty}^{\infty}\frac{d\omega}{2\pi}\frac{1}{V}\sum_{\boldsymbol{k}}\exp\left[\mathrm{i}(\boldsymbol{k}\cdot\boldsymbol{r}-\omega t)\right]\sum_{\boldsymbol{q}',\boldsymbol{k}'}(n_{\boldsymbol{q}-\boldsymbol{k}/2}-n_{\boldsymbol{q}+\boldsymbol{k}/2})\,\Phi_{\boldsymbol{k}}$$

$$\times\frac{1}{V}\sum_{\boldsymbol{q}''}K^{\text{ret},-}\left(\boldsymbol{q}''-\frac{\boldsymbol{k}}{2},\boldsymbol{q}''+\frac{\boldsymbol{k}}{2};\boldsymbol{q}'+\frac{\boldsymbol{k}'}{2},\boldsymbol{q}'-\frac{\boldsymbol{k}'}{2};\omega\right)U_{\boldsymbol{k}',\omega}^{\text{ext}}$$

$$=\frac{1}{\mathrm{i}\hbar}\int_{-\infty}^{\infty}\frac{d\omega}{2\pi}\frac{1}{V}\sum_{\boldsymbol{k}}\exp\left[\mathrm{i}(\boldsymbol{k}\cdot\boldsymbol{r}-\omega t)\right]\Phi_{\boldsymbol{k}}(n_{\boldsymbol{q}-\boldsymbol{k}/2}-n_{\boldsymbol{q}+\boldsymbol{k}/2})$$

$$\times\frac{1}{V}\sum_{\boldsymbol{q}''}\int_{-\infty}^{\infty}dt'\int_V d\boldsymbol{r}'\exp\left[-\mathrm{i}(\boldsymbol{k}\cdot\boldsymbol{r}'-\omega t')\right]\delta F_1(\boldsymbol{r}',\boldsymbol{p}'',t')$$

$$=\frac{1}{\mathrm{i}\hbar}\int_V d\boldsymbol{r}'\frac{1}{V}\sum_{\boldsymbol{k}}\exp\left[\mathrm{i}\,\boldsymbol{k}\cdot(\boldsymbol{r}-\boldsymbol{r}')\right]\int d\boldsymbol{R}\exp(-\mathrm{i}\boldsymbol{k}\cdot\boldsymbol{R})\,\Phi(\boldsymbol{R})$$

$$\times\left[F_1^{\text{eq}}\left(\boldsymbol{p}-\frac{\hbar\boldsymbol{k}}{2}\right)-F_1^{\text{eq}}\left(\boldsymbol{p}+\frac{\hbar\boldsymbol{k}}{2}\right)\right]\frac{1}{V}\sum_{\boldsymbol{q}''}\delta F_1(\boldsymbol{r}',\boldsymbol{p}'',t)$$

$$=\frac{1}{\mathrm{i}\hbar}\int_V d\boldsymbol{r}'\int d\boldsymbol{R}\,\Phi(\boldsymbol{R})\frac{1}{V}\sum_{\boldsymbol{k}}\exp\left[\mathrm{i}\boldsymbol{k}\cdot(\boldsymbol{r}-\boldsymbol{r}'-\boldsymbol{R})\right]$$

$$\times\left[\exp\left(-\frac{\boldsymbol{k}}{2}\cdot\frac{\partial}{\partial\boldsymbol{q}}\right)-\exp\left(\frac{\boldsymbol{k}}{2}\cdot\frac{\partial}{\partial\boldsymbol{q}}\right)\right]F_1^{\text{eq}}(\boldsymbol{p})\frac{1}{V}\sum_{\boldsymbol{q}''}\delta F_1(\boldsymbol{r}',\boldsymbol{p}'',t)$$

$$=\frac{1}{\mathrm{i}\hbar}\int_V d\boldsymbol{r}'\int d\boldsymbol{R}\,\Phi(\boldsymbol{R})\left[\delta\left(\boldsymbol{r}-\boldsymbol{r}'-\boldsymbol{R}+\frac{\mathrm{i}}{2}\frac{\partial}{\partial\boldsymbol{q}}\right)-\delta\left(\boldsymbol{r}-\boldsymbol{r}'-\boldsymbol{R}-\frac{\mathrm{i}}{2}\frac{\partial}{\partial\boldsymbol{q}}\right)\right]$$

$$\times F_1^{\text{eq}}(\boldsymbol{p})\frac{1}{V}\sum_{\boldsymbol{q}''}\delta F_1(\boldsymbol{r}',\boldsymbol{p}'',t)$$

$$=\frac{1}{\mathrm{i}\hbar}\int_V d\boldsymbol{r}'\left[\Phi\left(\boldsymbol{r}-\boldsymbol{r}'+\frac{\mathrm{i}}{2}\frac{\partial}{\partial\boldsymbol{q}}\right)-\Phi\left(\boldsymbol{r}-\boldsymbol{r}'-\frac{\mathrm{i}}{2}\frac{\partial}{\partial\boldsymbol{q}}\right)\right]$$

$$\times F_1^{\text{eq}}(\boldsymbol{p})\frac{1}{V}\sum_{\boldsymbol{q}''}\delta F_1(\boldsymbol{r}',\boldsymbol{p}'',t)$$

$$= \frac{1}{i\hbar} \int_V dr' \left[\exp\left(\frac{i}{2}\frac{\partial}{\partial \boldsymbol{q}} \cdot \frac{\partial}{\partial \boldsymbol{r}}\right) - \exp\left(-\frac{i}{2}\frac{\partial}{\partial \boldsymbol{q}} \cdot \frac{\partial}{\partial \boldsymbol{r}}\right)\right]$$

$$\times \Phi(\boldsymbol{r} - \boldsymbol{r}') \frac{1}{V} \sum_{\boldsymbol{q}'} \delta F_1(\boldsymbol{r}', \boldsymbol{p}', t)\, F_1^{\mathrm{eq}}(\boldsymbol{p})$$

$$= \frac{2}{\hbar} \sin\left(\frac{\hbar}{2}\frac{\partial}{\partial \boldsymbol{r}} \cdot \frac{\partial}{\partial \boldsymbol{p}}\right) \left[\int_V dr' \frac{1}{V} \sum_{\boldsymbol{q}'} \Phi(\boldsymbol{r} - \boldsymbol{r}')\, \delta F_1(\boldsymbol{r}', \boldsymbol{p}', t)\right] F_1^{\mathrm{eq}}(\boldsymbol{p}), \tag{5.3.11}$$

we obtain

$$\left\{\frac{\partial}{\partial t} + \frac{2}{\hbar} \sin\left[\frac{\hbar}{2}\left(\frac{\partial}{\partial \boldsymbol{p}}\right)_E \cdot \frac{\partial}{\partial \boldsymbol{r}}\right] E\left(\frac{\boldsymbol{p}}{\hbar}\right)\right\} \delta F_1(\boldsymbol{r}, \boldsymbol{p}, t)$$

$$- \frac{2}{\hbar} \sin\left(\frac{\hbar}{2}\frac{\partial}{\partial \boldsymbol{r}} \cdot \frac{\partial}{\partial \boldsymbol{p}}\right) U^{\mathrm{eff}}(\boldsymbol{r}, t)\, F_1^{\mathrm{eq}}(\boldsymbol{p}) = 0\,. \tag{5.3.12}$$

Here $(\partial/\partial \boldsymbol{p})_E$ is understood to act upon $\boldsymbol{p}$ contained in $E(\boldsymbol{p}/\hbar)$ only. The potential of the mean field has been introduced:

$$U^{\mathrm{eff}}(\boldsymbol{r}, t) = U^{\mathrm{ext}}(\boldsymbol{r}, t) + \int_V dr' \frac{1}{V} \sum_{\boldsymbol{q}'} \Phi(\boldsymbol{r} - \boldsymbol{r}')\, \delta F_1(\boldsymbol{r}', \boldsymbol{p}', t)\,. \tag{5.3.13}$$

Equation (5.3.12) is the quantal equation of motion equivalent to the equation of motion for the Green's function (5.2.17) in the decoupling approximation discussed in the previous section. It is a linearized equation with respect to deviation from equilibrium $\delta F_1(\boldsymbol{r}, \boldsymbol{p}, t) = F_1(\boldsymbol{r}, \boldsymbol{p}, t) - F_1^{\mathrm{eq}}(\boldsymbol{p})$, corresponding to the linear approximation (5.3.6 or 9) with respect to the external field. In the kinetic stage a description by the double-time correlation becomes generally a description by means of a linearized kinetic equation, e.g., the linear equation (5.3.12). However, as we have made the crudest decoupling approximation, the resultant equation lacks the collision term usually included on the right-hand side of the kinetic equation. To obtain the collision term, we need an approximation in which the chain of equations of motion is decoupled at a link higher by at least one step, because the cross section of a collision of particles begins with second-order terms in the two-particle interaction potential.

The Wigner distribution function reduces in the classical limit $\hbar \to 0$ to the one-particle distribution function. Therefore it is a quantity convenient for determining a deviation from the classical value in the form of a power series in $\hbar$. In the limit $\hbar \to 0$, we may remove the sine symbol including the differential operators in the kinetic equation (5.3.12):

$$\left(\frac{\partial}{\partial t} + \frac{\partial E(\boldsymbol{p}/\hbar)}{\partial \boldsymbol{p}} \cdot \frac{\partial}{\partial \boldsymbol{r}}\right) \delta F_1(\boldsymbol{r}, \boldsymbol{p}, t) - \frac{\partial U^{\mathrm{eff}}(\boldsymbol{r}, t)}{\partial \boldsymbol{r}} \cdot \frac{\partial}{\partial \boldsymbol{p}} F_1^{\mathrm{eq}}(\boldsymbol{p}) = 0\,.$$

This equation is a linearized form of the well-known *self-consistent field equation*

$$\left(\frac{\partial}{\partial t}+\frac{\partial E(\boldsymbol{p}/\hbar)}{\partial \boldsymbol{p}}\cdot\frac{\partial}{\partial \boldsymbol{r}}-\frac{\partial U^{\mathrm{eff}}(\boldsymbol{r},t)}{\partial \boldsymbol{r}}\cdot\frac{\partial}{\partial \boldsymbol{p}}\right)F_1(\boldsymbol{r},\boldsymbol{p},t)=0\,, \tag{5.3.14}$$

which is called the *Vlasov equation* in plasma theory [5.12]. Equation (5.3.12) is the quantal form of the linearized self-consistent field equation and is convenient in discussing quantum corrections when the sine function is expanded in powers of $\hbar$.

5.3.3 Plasma Oscillation

Relation (5.3.9) is exact within the linear-response theory. Nevertheless, in the decoupling approximation described in the previous section it can be rewritten in a more physically intuitive form. Although we have not attempted it so far, the solution of (5.2.18) is, with (5.2.20), given by

$$\begin{aligned}
&K^{\mathrm{ret},-}\left(\boldsymbol{q}-\frac{\boldsymbol{k}}{2},\boldsymbol{q}+\frac{\boldsymbol{k}}{2};\boldsymbol{q}'+\frac{\boldsymbol{k}'}{2},\boldsymbol{q}'-\frac{\boldsymbol{k}'}{2};\omega\right)\\
&=K_0^{\mathrm{ret},-}\left(\boldsymbol{q}-\frac{\boldsymbol{k}}{2},\boldsymbol{q}+\frac{\boldsymbol{k}}{2};\boldsymbol{q}'+\frac{\boldsymbol{k}'}{2},\boldsymbol{q}'-\frac{\boldsymbol{k}'}{2};\omega\right)\\
&\quad+K_0^{\mathrm{ret},-}\left(\boldsymbol{q}-\frac{\boldsymbol{k}}{2},\boldsymbol{q}+\frac{\boldsymbol{k}}{2};\boldsymbol{q}+\frac{\boldsymbol{k}}{2},\boldsymbol{q}-\frac{\boldsymbol{k}}{2};\omega\right)\frac{\delta_{\boldsymbol{k}\boldsymbol{k}'}}{V}\frac{\Phi_{\boldsymbol{k}}}{\varepsilon[\boldsymbol{k},\omega]}\\
&\quad\times K_0^{\mathrm{ret},-}\left(\boldsymbol{q}'-\frac{\boldsymbol{k}'}{2},\boldsymbol{q}'+\frac{\boldsymbol{k}'}{2};\boldsymbol{q}'+\frac{\boldsymbol{k}'}{2},\boldsymbol{q}'-\frac{\boldsymbol{k}'}{2};\omega\right).
\end{aligned} \tag{5.3.15}$$

Substituting this into (5.3.9) and using (5.2.20) once more in the process yields

$$\begin{aligned}
\delta F_1(\boldsymbol{r},\boldsymbol{p},t)=&\int_{-\infty}^{\infty}\frac{d\omega}{2\pi}\frac{1}{V}\sum_{\boldsymbol{k}}\exp\left[\mathrm{i}(\boldsymbol{k}\cdot\boldsymbol{r}-\omega t)\right]\\
&\times\sum_{\boldsymbol{q}',\boldsymbol{k}'}K_0^{\mathrm{ret},-}\left(\boldsymbol{q}-\frac{\boldsymbol{k}}{2},\boldsymbol{q}+\frac{\boldsymbol{k}}{2};\boldsymbol{q}'+\frac{\boldsymbol{k}'}{2},\boldsymbol{q}'-\frac{\boldsymbol{k}'}{2};\omega\right)\frac{U^{\mathrm{ext}}_{\boldsymbol{k}',\omega}}{\varepsilon[\boldsymbol{k}',\omega]}\,.
\end{aligned} \tag{5.3.16}$$

By comparing this approximate equation with the exact one (5.3.9), it emerges that in the decoupling approximation discussed in the previous section, effects of the interaction between electrons can be included in the screening of the external field via the complex dielectric function $\varepsilon[\boldsymbol{k},\omega]$. Note here that the electronic distribution $n_{\boldsymbol{k}}$ included in $K_0^{\mathrm{ret},-}$ is affected in general by the interaction, as mentioned previously.

Expression (5.3.16) suggests wave phenomena. The integral over angular frequencies can be performed by adding a very large semicircle in the lower half of the complex plane if $t > 0$, to change it into a contour integral: ω is regarded as a complex variable, while keeping the functional form of the integrand. If singularities of the integrand in the lower half-plane are only poles, we need only to evaluate residues at these poles and obtain a superposition of plane waves with angular frequencies and damping factors corresponding to the positions of poles. Of interest are waves with long lifetimes such as those which can be observed macroscopically. It is obvious from (5.1.10 or 5) and the discussion given in Sect. 3.6 that the spectral function $K^{\mathrm{ret},-}$ of the retarded Green's function is analytic in the upper half-plane. But if it is analytically continued into the lower half-plane, it ought to have singularities unless it is a constant. Indeed, the first factor $K_0^{\mathrm{ret},-}$ in (5.3.16) has a first-order pole at $\omega = [E(\boldsymbol{q}+\boldsymbol{k}/2) - E(\boldsymbol{q}-\boldsymbol{k}/2)]/\hbar - \mathrm{i}\,0$, as is made clear by considering ω as a complex variable after setting $z = \omega + \mathrm{i}\,0$ in (5.2.19). This pole is not interesting at the present since it refers to waves of very high frequency corresponding to the excitation of an electron-hole pair. The possibility of wave of low frequency, long wavelength and low damping depends on the existence and position of poles of the second factor ε^{-1} or zeros of the dielectric function $\varepsilon[\boldsymbol{k},\omega]$. Especially when the external field U^{ext} is very small and, moreover, consists of low-frequency parts only, then contributions from these zeros alone will give finite waves. When we write $\omega = \bar{\omega} - \mathrm{i}\,\gamma$ by separating the real and imaginary parts of the complex number ω, we are interested in $0 < \gamma \ll |\bar{\omega}|$. Assuming the existence of such a zero,

$$\begin{aligned} 0 &= \varepsilon[\boldsymbol{k},\bar{\omega}-\mathrm{i}\,\gamma] = \varepsilon'[\boldsymbol{k},\bar{\omega}-\mathrm{i}\,\gamma] + \mathrm{i}\,\varepsilon''[\boldsymbol{k},\bar{\omega}-\mathrm{i}\,\gamma] \\ &= \varepsilon'[\boldsymbol{k},\bar{\omega}] - \mathrm{i}\,\gamma\frac{\partial\varepsilon'[\boldsymbol{k},\bar{\omega}]}{\partial\bar{\omega}} + \mathrm{i}\,\varepsilon''[\boldsymbol{k},\bar{\omega}]\,, \\ &\therefore\ \varepsilon'[\boldsymbol{k},\bar{\omega}] = 0\,, \quad \gamma = \frac{\varepsilon''[\boldsymbol{k},\bar{\omega}]}{\partial\varepsilon'[\boldsymbol{k},\bar{\omega}]/\partial\bar{\omega}}\,. \end{aligned} \tag{5.3.17}$$

Here ε' and ε'' stand for the functional forms of the real and imaginary parts of the complex dielectric function ε, respectively. By using the result (5.2.21) evaluated at the long-wavelength region $\boldsymbol{k} \approx 0$ in the previous section, we obtain

$$\begin{aligned} \bar{\omega}^2 &= \omega_{\mathrm{p}}^2 + \frac{4\hat{E}}{\hbar^2}E(\boldsymbol{k}) + O(\boldsymbol{k}^4)\,, \\ \gamma &= \sqrt{\frac{\pi}{2}}\,\frac{\sinh(\beta\hbar\omega_{\mathrm{p}}/2)}{\beta\hbar(r_{\mathrm{D}}|\boldsymbol{k}|)^3}\exp\left(-\frac{1}{2(r_{\mathrm{D}}|\boldsymbol{k}|)^2}\right). \end{aligned} \tag{5.3.18}$$

Here $\bar{\omega}$ denotes the well-known *plasma frequency*, and γ the *Landau damping constant* [5.13].

In the above equation, $r_{\mathrm{D}} = \sqrt{kT/(4\pi n e^2)}$ stands for the *Debye screening radius* [5.14]. The static dielectric constant can easily be calculated for $\boldsymbol{k} \approx 0$

from (5.2.20). It takes the form

$$\varepsilon[\boldsymbol{k},0]=1+\frac{4\pi e^2}{\boldsymbol{k}^2}\left(\frac{\partial n}{\partial\mu}\right)_T . \tag{5.3.19}$$

If we use the Maxwell-Boltzmann distribution, the second term on the right-hand side becomes $(r_D|\boldsymbol{k}|)^{-2}$, so that $\Phi_k/\varepsilon[\boldsymbol{k},0]=4\pi e^2/(\boldsymbol{k}^2+r_D^{-2})$ gives the screened Coulomb potential.

5.4 Single-Particle Green's Function and the Causal Green's Function

The Green's functions described hitherto are such that, as in (5.1.1) for instance, the operators $\hat{A}$ and $\hat{B}$ are physical quantities with classical homologues. For the Green's function equivalent to the complex admittance discussed in Chap. 3 by virtue of (5.1.5) the operators $\hat{A}$ and $\hat{B}$ correspond to the displacement or the current discussed in irreversible thermodynamics, i.e., to variables in the hydrodynamic description. In Sects. 5.2, 3 we introduced somewhat more microscopic Green's functions by using operators corresponding to variables in the kinetic theoretical description. However, any Green's function could be transferred into classical statistical mechanics in the limit $\hbar\to 0$ or in the high-temperature limit. On the other hand, the quantum mechanical quantity of the probability amplitude has no homologue in classical mechanics. If we use the operator representing the probability amplitude, i.e., the quantized wave function, we can construct a new kind of Green's function which cannot be transferred into classical statistical mechanics. We ought to be able to describe phenomena on a more microscopic basis with these new Green's functions, and obtain, for example, information about properties of elementary excitations in a material.

5.4.1 Single-Particle Green's Functions

Among the Green's functions constructed by using the quantized wave functions or the creation and annihilation operators themselves as the operators $\hat{A}$ and $\hat{B}$, those made of a pair of creation and annihilation operators, e.g., those corresponding to (5.1.7, 16)

$$\begin{aligned}G^{\mathrm{ret},\pm}(\boldsymbol{k},t;\boldsymbol{k}',t')&=\theta(t-t')\left\langle\frac{1}{\mathrm{i}\hbar}[\hat{a}_{\boldsymbol{k}}(t),\hat{a}^{\dagger}_{\boldsymbol{k}'}(t')]_{\pm}\right\rangle,\\ G^{\mathrm{adv},\pm}(\boldsymbol{k},t;\boldsymbol{k}',t')&=-\theta(t'-t)\left\langle\frac{1}{\mathrm{i}\hbar}[\hat{a}_{\boldsymbol{k}}(t),\hat{a}^{\dagger}_{\boldsymbol{k}'}(t')]_{\pm}\right\rangle\end{aligned} \tag{5.4.1}$$

are called the *single-particle Green's functions.* Single-particle Green's functions of the type (5.1.13) could be derived, but as expected from the commutation relations for the creation and annihilation operators, the Green's function of the commutator type (5.4.1) are more convenient for particles obeying Bose statistics, while those of the anticommutator type (5.4.1) are suitable for particles obeying Fermi statistics. The reason is that the inhomogeneous term on the right-hand side of the equation of motion (5.2.3) is simplified to $\delta(t-t')\,\delta_{k,k'}$. For the Green's functions like (5.1.13), the inhomogeneous term includes the factor $a_k\,a_{k'}^\dagger = \delta_{k,k'}(1 \mp n_k)$, so that we are compelled to follow a complicated procedure: first we determine the Green's function by assuming n_k as known, then derive an equation for n_k by using the spectral function of this Green's function and solve it to determine n_k, then finally determine the Green's function itself.

For the single-particle Green's function, calculations are often simplified by using the grand canonical ensemble rather than the canonical ensemble (4.2.7) to obtain the average

$$\langle\ldots\rangle = \frac{\mathrm{Tr}\,\{\mathrm{e}^{-\beta \mathsf{H}}\ldots\}}{\mathrm{Tr}\,\{\mathrm{e}^{-\beta \mathsf{H}}\}}\,, \qquad \mathsf{H} = \mathscr{H} - \mu\hat{N}\,, \quad \hat{N} = \sum_k \hat{a}_k^\dagger\,\hat{a}_k\,. \tag{5.4.2}$$

Still, the motion is governed by the Hamiltonian $\mathscr{H}$ in this case, too:

$$\hat{a}_k(t) = \exp(\mathrm{i}t\mathscr{H}/\hbar)\,\hat{a}_k\exp(-\mathrm{i}t\mathscr{H}/\hbar)\,, \qquad \hat{a}_{k'}^\dagger(t') = \exp(\mathrm{i}t'\mathscr{H}/\hbar)\,\hat{a}_{k'}^\dagger\exp(-\mathrm{i}t'\mathscr{H}/\hbar)\,. \tag{5.4.3}$$

(The reader should carefully distinguish between $\mathscr{H}$ and H in the following formulas.) If the total number of particles $\hat{N}$ is conserved by this motion, i.e., if $[\hat{N},\mathscr{H}] = 0$, then the single-particle Green's function (5.4.1) depends on time in the form of the difference $t-t'$ as hitherto. It is convenient to consider auxiliary Green's functions $\bar{G}^{\mathrm{ret},\pm}$ and $\bar{G}^{\mathrm{adv},\pm}$ which are constructed by using in place of the Heisenberg operators (5.4.3) auxiliary operators

$$\check{a}_k(t) = \exp(\mathrm{i}t\mathsf{H}/\hbar)\,\hat{a}_k\exp(-\mathrm{i}t\mathsf{H}/\hbar)\,, \qquad \check{a}_{k'}^\dagger(t') = \exp(\mathrm{i}t'\mathsf{H}/\hbar)\,\hat{a}_{k'}^\dagger\exp(-\mathrm{i}t'\mathsf{H}/\hbar)\,, \tag{5.4.4}$$

i.e., the functions obtained by substituting all $\mathscr{H}$ in $G^{\mathrm{ret},\pm}$ and $G^{\mathrm{adv},\pm}$ by H. Since, as is easily proved by using the commutation relations,

$$\begin{aligned}&\hat{N}\,\hat{a}_k = \hat{a}_k(\hat{N}-1)\,, \quad \hat{N}\,\hat{a}_{k'}^\dagger = \hat{a}_{k'}^\dagger(\hat{N}+1)\,,\\ \therefore\ &\exp(-\alpha\hat{N})\,\hat{a}_k\exp(\alpha\hat{N}) = \exp(\alpha)\,\hat{a}_k\,,\\ &\exp(-\alpha\hat{N})\,\hat{a}_{k'}^\dagger\exp(\alpha\hat{N}) = \exp(-\alpha)\,\hat{a}_{k'}^\dagger\end{aligned} \tag{5.4.5}$$

are valid, for $[\mathscr{H}, \hat{N}] = 0$ then

$$\breve{a}_{\boldsymbol{k}}(t) = \mathrm{e}^{\mathrm{i}\mu t/\hbar}\,\hat{a}_{\boldsymbol{k}}(t)\,, \quad \breve{a}^{\dagger}_{\boldsymbol{k}'}(t') = \mathrm{e}^{-\mathrm{i}\mu t'/\hbar}\,\hat{a}^{\dagger}_{\boldsymbol{k}'}(t')\,. \tag{5.4.6}$$

Thus the relationships between the true Green's functions $G^{\mathrm{ret},\pm}$ and $G^{\mathrm{adv},\pm}$, obtained by using the grand canonical ensemble, and the auxiliary Green's functions obtained by using H for the motion, are simplified to

$$\begin{aligned} \bar{G}^{\mathrm{ret},\pm}(\boldsymbol{k}, t; \boldsymbol{k}', t') &= \exp\left[\mathrm{i}\mu(t-t')/\hbar\right] G^{\mathrm{ret},\pm}(\boldsymbol{k}, t; \boldsymbol{k}', t')\,, \\ \bar{G}^{\mathrm{adv},\pm}(\boldsymbol{k}, t; \boldsymbol{k}', t') &= \exp\left[\mathrm{i}\mu(t-t')/\hbar\right] G^{\mathrm{adv},\pm}(\boldsymbol{k}, t; \boldsymbol{k}', t')\,, \end{aligned} \tag{5.4.7}$$

respectively. The relationships between the corresponding spectral functions are given by

$$\begin{aligned} \bar{K}^{\mathrm{ret},\pm}\left(\boldsymbol{k}, \boldsymbol{k}'; \omega - \frac{\mu}{\hbar}\right) &= K^{\mathrm{ret},\pm}(\boldsymbol{k}, \boldsymbol{k}'; \omega)\,, \\ \bar{K}^{\mathrm{adv},\pm}\left(\boldsymbol{k}, \boldsymbol{k}'; \omega - \frac{\mu}{\hbar}\right) &= K^{\mathrm{adv},\pm}(\boldsymbol{k}, \boldsymbol{k}'; \omega)\,, \end{aligned} \tag{5.4.8}$$

respectively: only the origin of the angular frequency is shifted.

5.4.2 Single-Particle Green's Functions for Free Particles

As the simplest example of calculating the Green's function, let us determine the form for noninteracting particles. We assume as the Hamiltonian the following form, which is obtained, for instance, by removing the interaction term from (5.2.9) for the plasma:

$$\mathscr{H}_0 = \sum_{\boldsymbol{k}} E(\boldsymbol{k})\,\hat{a}^{\dagger}_{\boldsymbol{k}}\,\hat{a}_{\boldsymbol{k}}\,, \quad \therefore\ \mathsf{H}_0 = \sum_{\boldsymbol{k}} \left[E(\boldsymbol{k}) - \mu\right]\hat{a}^{\dagger}_{\boldsymbol{k}}\,\hat{a}_{\boldsymbol{k}}\,. \tag{5.4.9}$$

As for (5.4.5) we can easily prove

$$\begin{aligned} \exp(\alpha\,\mathsf{H}_0)\,\hat{a}_{\boldsymbol{k}}\exp(-\alpha\,\mathsf{H}_0) &= \exp(-\alpha\left[E(\boldsymbol{k}) - \mu\right])\,\hat{a}_{\boldsymbol{k}}\,, \\ \exp(\alpha\,\mathsf{H}_0)\,\hat{a}^{\dagger}_{\boldsymbol{k}'}\exp(-\alpha\,\mathsf{H}_0) &= \exp(\alpha\left[E(\boldsymbol{k}) - \mu\right])\,\hat{a}^{\dagger}_{\boldsymbol{k}'}\,. \end{aligned} \tag{5.4.10}$$

Using these equations and cyclically changing operators in the trace operation, then

$$\begin{aligned} \langle \hat{a}^{\dagger}_{\boldsymbol{k}'}\,\hat{a}_{\boldsymbol{k}}\rangle_0 &= \frac{\mathrm{Tr}\{\exp(-\beta\,\mathsf{H}_0)\,\hat{a}^{\dagger}_{\boldsymbol{k}'}\,\hat{a}_{\boldsymbol{k}}\}}{\mathrm{Tr}\{\exp(-\beta\,\mathsf{H}_0)\}} = \frac{\mathrm{Tr}\{\hat{a}_{\boldsymbol{k}}\exp(-\beta\,\mathsf{H}_0)\,\hat{a}^{\dagger}_{\boldsymbol{k}'}}{\mathrm{Tr}\{\exp(-\beta\,\mathsf{H}_0)\}} \\ &= \exp(-\beta\left[E(\boldsymbol{k}) - \mu\right])\,\langle \hat{a}_{\boldsymbol{k}}\,\hat{a}^{\dagger}_{\boldsymbol{k}'}\rangle_0\,. \end{aligned} \tag{5.4.11}$$

By using the commutation relation, we get the expectation value of the product

$$\langle \hat{a}_{k'}^{\dagger}\, \hat{a}_{k} \rangle_0 = \exp\left(-\beta\left[E(\boldsymbol{k})-\mu\right]\right)\left(\delta_{\boldsymbol{k}\boldsymbol{k}'} \mp \langle \hat{a}_{k'}^{\dagger}\, \hat{a}_{k} \rangle_0\right),$$

$$\therefore \langle \hat{a}_{k'}^{\dagger}\, \hat{a}_{k} \rangle_0 = \delta_{\boldsymbol{k}\boldsymbol{k}'}\, f(E(\boldsymbol{k})), \quad f(E) = \frac{1}{\exp\left[\beta(E-\mu)\right] \pm 1}. \tag{5.4.12}$$

Here the upper sign corresponds to fermions, the lower sign to bosons. According to (5.4.10),

$$\bar{G}_0^{\mathrm{ret},\pm}(\boldsymbol{k}, t; \boldsymbol{k}', t') = \theta(t-t')\frac{\delta_{\boldsymbol{k}\boldsymbol{k}'}}{\mathrm{i}\hbar}\exp\{-\mathrm{i}\left[E(\boldsymbol{k})-\mu\right](t-t')/\hbar\} \tag{5.4.13}$$

using the same double sign. Therefore,

$$\bar{K}_0^{\pm}(\boldsymbol{k}, \boldsymbol{k}'; z) = \frac{\delta_{\boldsymbol{k}\boldsymbol{k}'}}{\hbar z + \mu - E(\boldsymbol{k})} \tag{5.4.14}$$

corresponds by virtue of (5.1.11) to the function (5.1.19) obtained by the analytic continuation of the spectral function into the complex plane. Shifting the origin by $\mu/\hbar$ as in (5.4.8) gives

$$K_0^{\pm}(\boldsymbol{k}, \boldsymbol{k}'; z) = \frac{\delta_{\boldsymbol{k}\boldsymbol{k}'}}{\hbar z - E(\boldsymbol{k})}. \tag{5.4.15}$$

According to (5.1.21) the spectral function of the expectation value of the anticommutator or of the commutator (5.1.9) becomes

$$\Lambda_0^{\pm}(\boldsymbol{k}, \boldsymbol{k}'; \omega) = -2\pi\mathrm{i}\,\delta_{\boldsymbol{k}\boldsymbol{k}'}\,\delta(\hbar\omega - E(\boldsymbol{k})). \tag{5.4.16}$$

The results of (5.4.15, 16) suggests that the *single-particle energy* $E(\boldsymbol{k})$ is determined from poles of the spectral function $K^{\pm}$ or from maximum points of the spectral function $\Lambda^{\pm}$, respectively. Indeed, elementary excitations or quasiparticles appearing in many-particle system are so defined [5.15, 16].

5.4.3 Causal Green's Functions

In addition to the retarded and advanced Green's functions discussed so far, the *causal Green's function* also exists. That corresponding to (5.1.7, 13, 16) is defined by

$$G_{A,B}^{\mathrm{caus}}(t, t') = \left\langle \frac{1}{\mathrm{i}\hbar} T_t\{\hat{A}(t)\,\hat{B}(t')\} \right\rangle. \tag{5.4.17}$$

Each of the operators $\hat{A}$ and $\hat{B}$ is constructed from the products of a certain number of the creation and annihilation operators. Further, T_t denotes the *Wick time-ordering operator* [5.17]

$$T_t\{\hat{A}(t)\,\hat{B}(t')\} = \theta(t-t')\,\hat{A}(t)\,\hat{B}(t') \mp \theta(t'-t)\,\hat{B}(t')\,\hat{A}(t), \tag{5.4.18}$$

where the double sign corresponds to the sign obtained by regarding the creation and annihilation operators as anticommutable for fermions or as commutable for bosons when the product $\hat{A}\hat{B}$ is rearranged into $\hat{B}\hat{A}$. The spectral function of the causal Green's function is determined in the following form:

$$K_{A,B}^{\text{caus}}(\omega) = \int_{-\infty}^{\infty} \frac{d\omega'}{2\pi\mathrm{i}} \left[\frac{\mathscr{P}}{\omega' - \omega} + \mathrm{i}\pi\delta(\omega' - \omega) \begin{Bmatrix} \tanh \\ \coth \end{Bmatrix} \left(\frac{\beta\hbar\omega'}{2} \right) \right] \Lambda_{A,B}(\omega'), \tag{5.4.19}$$

where $\Lambda_{A,B}(\omega)$ is given by using the spectral function $J_{A,B}(\omega)$ in (5.1.15) as

$$\mathrm{i}\hbar\Lambda_{A,B}(\omega) = (1 \pm \mathrm{e}^{-\beta\hbar\omega}) J_{A,B}(\omega). \tag{5.4.20}$$

Both the hyperbolic functions in (5.4.19) and the double sign in the above equation correspond to the double sign in (5.4.18).

The causal Green's function was introduced for the diagram technique at absolute zero temperature devised by *Feynman* [5.18] in quantum field theory, but was found to be unusable for that technique at finite temperature [5.2]. Moreover, as is obvious from the form (5.4.19), its spectral function cannot be continued analytically into the complex plane, so it is mathematically inconvenient compared with the retarded or advanced Green's functions. Nevertheless, as shown in Sect. 5.6, it can be reconsidered as the temperature Green's function by taking the reciprocal temperature instead of time as a variable.

The equation of motion for the causal Green's function has the same form as (5.2.3). The causal Green's function can be identified by the boundary condition derived from the relation obtained via (4.4.2)

$$\langle \hat{B}(t')\hat{A}(t)\rangle = \langle \hat{A}(t - \mathrm{i}\hbar\beta)\hat{B}(t')\rangle \tag{5.4.21}$$

as follows:

$$G_{A,B}^{\text{caus}}(t, t') = \mp G_{A,B}^{\text{caus}}(t - \mathrm{i}\hbar\beta, t') \quad (t < t'). \tag{5.4.22}$$

5.5 Basic Formula for Perturbational Expansion

The computational techniques in quantum field theory were greatly advanced by the invention of Feynman diagram techniques. These techniques, which correspond to a treatment at absolute zero, made the perturbational expansion easy to grasp, as was referred to in the previous section. The techniques were generalized by *Matsubara* [5.19] to finite temperatures. In contrast to the failure with the double-time Green's function, the generalization was made possible by introducing the idea of the temperature Green's function. We shall describe the latter technique in this and the following sections.

5.5.1 Perturbational Expansion of the Equilibrium Density Matrix

Let us assume that the Hamiltonian of our system is of the form of a sum of an unperturbed term $\mathscr{H}_0$ and a perturbing term $g\,\mathscr{H}^{\text{int}}$, g denoting a coupling constant characteristic of the intensity of perturbation. We want to expand the density matrix for the grand canonical ensemble in powers of the coupling constant g. We assume the total number of particles $\hat{N}$ to commute with $\mathscr{H}_0$ and $\mathscr{H}^{\text{int}}$. As in (5.4.2)

$$\mathsf{H} = \mathsf{H}_0 + g\,\mathscr{H}^{\text{int}}, \quad \mathsf{H}_0 = \mathscr{H}_0 - \mu\hat{N}, \tag{5.5.1}$$

introducing the average in the unperturbed system

$$\langle\ldots\rangle_0 = \frac{\operatorname{Tr}\{\exp(-\beta\,\mathsf{H}_0\ldots)\}}{\operatorname{Tr}\{\exp(-\beta\,\mathsf{H}_0)\}}. \tag{5.5.2}$$

For the perturbational expansion of the unitary transformation $\exp(-\mathrm{i}t\mathscr{H}/\hbar)$, which induces the motion of the system, it is convenient to use the interaction framework. Let us copy this procedure regarding $\exp(-\tau\mathsf{H})$. We set

$$\mathrm{e}^{-\tau\mathsf{H}} = \mathrm{e}^{-\tau\mathsf{H}_0}\hat{\mathscr{S}}(\tau) \tag{5.5.3}$$

and expand the operator $\hat{\mathscr{S}}(\tau)$. Differentiating with respect to τ yields

$$\frac{d}{d\tau}\hat{\mathscr{S}}(\tau) = -g\,\mathscr{H}^{\text{int}}(\tau)\,\hat{\mathscr{S}}(\tau), \quad \hat{\mathscr{S}}(0) = 1. \tag{5.5.4}$$

Here we have defined an operator

$$\hat{A}(\tau) = \mathrm{e}^{\tau\mathsf{H}_0}\hat{A}\,\mathrm{e}^{-\tau\mathsf{H}_0} \tag{5.5.5}$$

corresponding to the operator in the interaction picture. Transforming the differential equation (5.5.4) into the integral equation,

$$\hat{\mathscr{S}}(\tau) = 1 - g\int_0^\tau d\tau'\,\mathscr{H}^{\text{int}}(\tau')\,\hat{\mathscr{S}}(\tau').$$

Solving this by successive substitution gives

$$\begin{aligned}\hat{\mathscr{S}}(\tau) = 1 + \sum_{n=1}^{\infty}(-g)^n\int_0^\tau d\tau_n\int_0^\tau d\tau_{n-1}\,\theta(\tau_n-\tau_{n-1})\ldots\\ \times\int_0^\tau d\tau_1\,\theta(\tau_2-\tau_1)\,\mathscr{H}^{\text{int}}(\tau_n)\,\mathscr{H}^{\text{int}}(\tau_{n-1})\ldots\mathscr{H}^{\text{int}}(\tau_1),\end{aligned} \tag{5.5.6}$$

where θ is the Heaviside unit step function (3.1.7) and has been inserted to make the upper limits of integrals of uniform value. To write this in a clearer form, let us introduce the *Dyson time-ordering operator* [5.20] P_τ modified by changing the variable from time to τ:

$$
\begin{aligned}
&P_\tau \{\hat{A}_1(\tau_1)\hat{A}_2(\tau_2)\dots\hat{A}_n(\tau_n)\} \\
&\quad = \sum_P \theta(\tau_{j_n}-\tau_{j_{n-1}})\,\theta(\tau_{j_{n-1}}-\tau_{j_{n-2}})\dots\theta(\tau_{j_2}-\tau_{j_1}) \\
&\quad \times \hat{A}_{j_n}(\tau_{j_n})\hat{A}_{j_{n-1}}(\tau_{j_{n-1}})\dots\hat{A}_{j_1}(\tau_{j_1}),
\end{aligned}
\tag{5.5.7}
$$

where P stands for the permutation operator changing the number series $(1,2,\dots,n)$ into $(j_1,j_2,\dots,j_n)$ and $\sum_P$ denotes the summation over all $n!$ permutations. Since the unit step functions are inserted, only a summand such that $\tau_{j_n}>\tau_{j_{n-1}}>\dots>\tau_{j_1}$ remains among the summands. By virtue of the operator P_τ the product of operators $A_1(\tau_1)\dots A_n(\tau_n)$ is arranged in increasing order of τ from right to left. Equation (5.5.6) may be written as

$$
\begin{aligned}
\hat{\mathscr{S}}(\tau) &= 1+\sum_{n=1}^{\infty}\frac{(-g)^n}{n!}\int_0^\tau d\tau_n\int_0^\tau d\tau_{n-1}\dots\int_0^\tau d\tau_1\,P_\tau[\mathscr{H}^{\text{int}}(\tau_n)\dots\mathscr{H}^{\text{int}}(\tau_1)] \\
&= P_\tau \exp\left[-g\int_0^\tau d\tau'\,\mathscr{H}^{\text{int}}(\tau')\right].
\end{aligned}
\tag{5.5.8}
$$

Since the order of operators is arranged correctly by virtue of P_τ, we may neglect the noncommutability of operators on the right of P_τ and treat operators as c numbers. We shall see below that this makes our description of formulas extremely simple and is a great help in disentangling complicated considerations. The *Wick time-ordering operator* T_τ, obtained by replacing time with the variable τ in (5.4.18) for T_t, is generally defined so that in (5.5.7) for P_τ each term is multiplied by an extra factor $+1$ if the corresponding permutation P is even or ∓ 1 if odd as in (5.4.18). The interaction $\mathscr{H}^{\text{int}}$ contains the creation and annihilation operators for fermions: they must appear in the form of pairs, i.e., the total number of particles is conserved and P_τ may be replaced by T_τ for the products appearing in (5.5.8)

$$
\hat{\mathscr{S}}(\tau) = T_\tau \exp\left[-g\int_0^\tau d\tau'\,\mathscr{H}^{\text{int}}(\tau')\right].
\tag{5.5.9}
$$

This yields the basic formula for perturbational expansion.

5.5.2 Perturbational Expansion of the Thermodynamic Potential

According to (5.5.3) the grand canonical partition function can be written as

$$
\mathrm{e}^{-\beta J} = \mathrm{Tr}\{\mathrm{e}^{-\beta \mathsf{H}}\} = \mathrm{Tr}\{\mathrm{e}^{-\beta \mathsf{H}_0}\}\langle\hat{\mathscr{S}}(\beta)\rangle_0
\tag{5.5.10}
$$

and hence the corresponding thermodynamic potential J is given by

$$J = J_0 - kT \ln \langle \hat{\mathscr{S}}(\beta) \rangle_0, \quad J_0 = -kT \ln \mathrm{Tr}\{\mathrm{e}^{-\beta \mathsf{H}_0}\}. \tag{5.5.11}$$

Here J_0 is the potential for the unperturbed system and ought to be evaluated easily. Note that only the average over the unperturbed system appears in (5.5.11).

5.6 Temperature Green's Function

In the previous section we have taken advantage of the analogy between the unitary operator $\exp(-\mathrm{i}t\mathscr{H}/\hbar)$ and the density matrix $\exp(-\beta\mathscr{H})$ and copied the perturbational expansion formula for the former in constructing the latter. For the former, a diagram technique for the perturbational expansion was devised by making use of the causal Green's function. *Matsubara* translated the causal Green's function from the time variable into the reciprocal temperature τ [5.19].

5.6.1 Temperature Green's Functions (Matsubara-Green's Functions)

The temperature Green's function corresponding to the causal Green's function (5.4.17) is defined by

$$\mathscr{G}_{A,B}(\tau, \tau') = -\langle T_\tau \{\check{A}(\tau)\, \check{B}(\tau')\}\rangle, \tag{5.6.1}$$

where there is an operator

$$\check{A}(\tau) = \mathrm{e}^{\tau \mathsf{H}} \hat{A}\, \mathrm{e}^{-\tau \mathsf{H}} = \hat{\mathscr{S}}^{-1}(\tau)\, \hat{A}(\tau)\, \hat{\mathscr{S}}(\tau) \tag{5.6.2}$$

corresponding to the auxiliary operator (5.4.4). Let us rewrite the temperature Green's function (5.6.1) into a form convenient for perturbational expansion. Generalizing (5.5.9), we set for $\tau_1 > \tau_2$

$$\hat{\mathscr{S}}(\tau_1, \tau_2) = T_\tau \exp\left[-g \int_{\tau_2}^{\tau_1} d\tau' \mathscr{H}^{\mathrm{int}}(\tau')\right]. \tag{5.6.3}$$

If we choose an arbitrary τ' such that $\tau_1 > \tau' > \tau_2$, then

$$\hat{\mathscr{S}}(\tau_1, \tau_2) = \hat{\mathscr{S}}(\tau_1, \tau')\, \hat{\mathscr{S}}(\tau', \tau_2) \tag{5.6.4}$$

thanks to the existence of the operator T_τ. Since $\hat{\mathscr{S}}(\tau)$ of (5.5.9) is equal to $\hat{\mathscr{S}}(\tau, 0)$, we set $\tau_2 = 0$ and rewrite τ' into the new τ_2 and thus obtain

$$\hat{\mathscr{S}}(\tau_1, \tau_2) = \hat{\mathscr{S}}(\tau_1)\, \hat{\mathscr{S}}^{-1}(\tau_2) \quad (\tau_1 > \tau_2). \tag{5.6.5}$$

By using this expression and (5.4.2, 5.3, 6.2, 5.10), we can write (5.6.1) for $\beta > \tau > \tau'$ as

$$\mathcal{G}_{A,B}(\tau, \tau') = -\langle \mathcal{S}^{-1}(\tau)\hat{A}(\tau)\hat{\mathcal{S}}(\tau)\hat{\mathcal{S}}^{-1}(\tau')\hat{B}(\tau')\hat{\mathcal{S}}(\tau')\rangle$$
$$= -\frac{\langle \hat{\mathcal{S}}(\beta, \tau)\hat{A}(\tau)\hat{\mathcal{S}}(\tau, \tau')\hat{B}(\tau')\hat{\mathcal{S}}(\tau', 0)\rangle_0}{\langle \hat{\mathcal{S}}(\beta)\rangle_0}. \quad (5.6.6)$$

Similarly, if $\beta > \tau' > \tau$,

$$\mathcal{G}_{A,B}(\tau, \tau') = \pm\frac{\langle \hat{\mathcal{S}}(\beta, \tau')\hat{B}(\tau')\hat{\mathcal{S}}(\tau', \tau)\hat{A}(\tau)\hat{\mathcal{S}}(\tau, 0)\rangle_0}{\langle \hat{\mathcal{S}}(\beta)\rangle_0} \quad (5.6.7)$$

because the sign $\mp$ appears when we rearrange the product $\hat{A}\hat{B}$ into $\hat{B}\hat{A}$ as for (5.4.18). By using the operator T_τ, these results can be written in the single form

$$\mathcal{G}_{A,B}(\tau, \tau') = -\frac{\langle T_\tau\{\hat{A}(\tau)\hat{B}(\tau')\hat{\mathcal{S}}(\beta)\}\rangle_0}{\langle \hat{\mathcal{S}}(\beta)\rangle_0}. \quad (5.6.8)$$

That is to say, factorizing $\hat{\mathcal{S}}(\beta, 0)$ by virtue of (5.6.4) into the product of three $\hat{\mathcal{S}}$'s such as in (5.6.6 or 7) and inserting $\hat{A}$ and $\hat{B}$ into them, we obtain (5.6.6 or 7), respectively. In (5.6.8), the expectation values for the unperturbed system are employed in the same way as in the thermodynamic potential (5.5.11).

5.6.2 Fourier Analysis of the Temperature Green's Function

In the theory of double-time Green's functions, their Fourier transform, i.e., the spectral functions, played an important role. It was demonstrated by *Abrikosov* et al. [5.21] how a Fourier analysis can be devised from the temperature Green's function, which plays again an important role.

First, let us ascertain that (5.6.1) is a function of the difference $\tau - \tau'$. By definition, for $\beta > \tau - \tau'$ we can write it as

$$\mathcal{G}_{A,B}(\tau, \tau') = -\frac{\mathrm{Tr}\{\exp[(\tau - \tau' - \beta)\mathsf{H}]\hat{A}\exp[-(\tau - \tau')\mathsf{H}]\hat{B}\}}{\mathrm{Tr}\{\exp(-\beta\mathsf{H}\}}. \quad (5.6.9)$$

Since the Hamiltonian $\mathcal{H}$ has a ground state and hence eigenvalues of H are bounded below, the condition $\beta > \tau - \tau' > 0$ is necessary and sufficient for the convergence of the trace. For $\beta > \tau' - \tau > 0$,

$$\mathcal{G}_{A,B}(\tau, \tau') = \pm\frac{\mathrm{Tr}\{\exp[(\tau - \tau')\mathsf{H}]\hat{A}\exp[-(\tau - \tau' + \beta)\mathsf{H}]\hat{B}\}}{\mathrm{Tr}\{\exp(-\beta\mathsf{H})\}} \quad (5.6.10)$$

and $\beta > \tau' - \tau > 0$ is necessary and sufficient for convergence. According to (5.6.9, 10) it is obvious that the expression (5.6.1) depends on τ and τ' only through the difference $\tau - \tau'$.

Next, setting $\tau' = 0$ means considering $\beta > \tau > 0$ for (5.6.9) and $0 > \tau > -\beta$ for (5.6.10). Further, if we replace τ in (5.6.9) with $\tau + \beta$, then the new τ belongs to the range $0 > \tau > -\beta$, whence, comparing the result with (5.6.10), we find that

$$\mathscr{G}_{A,B}(\tau, 0) = \mp \mathscr{G}_{A,B}(\tau + \beta, 0) \quad (0 > \tau > -\beta) \tag{5.6.11}$$

holds. This relation, which corresponds to the boundary condition (5.4.22) for the causal Green's function, imposes an important condition on the temperature Green's function defined in a finite interval $\beta > \tau > -\beta$.

A function of τ defined in this finite interval can be expended into a Fourier series

$$\begin{aligned} \mathscr{G}_{A,B}(\tau, \tau') &= kT \sum_{n=-\infty}^{\infty} \exp[-\mathrm{i}\,\xi_n(\tau - \tau')]\, \mathscr{K}_{A,B}(\mathrm{i}\,\xi_n), \\ \mathscr{K}_{A,B}(\mathrm{i}\,\xi_n) &= \frac{1}{2} \int_{-\beta}^{\beta} d\tau\, \mathrm{e}^{\mathrm{i}\,\xi_n \tau} \mathscr{G}_{A,B}(\tau, 0), \end{aligned} \tag{5.6.12}$$

where ξ_n must be determined so as to satisfy condition (5.6.11). If in (5.6.12) $\tau' = 0$ and $\tau + \beta$ replaces τ, provided that $\beta > \tau + \beta > 0$, i.e., $0 > \tau > -\beta$, then

$$\mathscr{G}_{A,B}(\tau + \beta, 0) = kT \sum_{n=-\infty}^{\infty} \mathrm{e}^{-\mathrm{i}\,\xi_n \tau}\, \mathrm{e}^{-\mathrm{i}\,\xi_n \beta}\, \mathscr{K}_{A,B}(\mathrm{i}\,\xi_n).$$

Since this expression must be equal to (5.6.12) multiplied by ∓ 1,

$$\sum_{n=-\infty}^{\infty} (1 \pm \mathrm{e}^{-\mathrm{i}\,\xi_n \beta})\, \mathrm{e}^{-\mathrm{i}\,\xi_n \tau}\, \mathscr{K}_{A,B}(\mathrm{i}\,\xi_n) = 0$$

must hold for an arbitrary τ, and hence ξ_n has to be determined by

$$\mathrm{e}^{-\mathrm{i}\xi_n \beta} = \mp 1. \tag{5.6.13}$$

That is to say, corresponding to the double sign in (5.4.18),

$$\xi_n = \begin{cases} (2n+1)\,\pi kT \\ 2n\pi kT \end{cases} \quad (n = 0, \pm 1, \pm 2, \ldots). \tag{5.6.14}$$

Which of these values should be taken depends on the structure of the operators $\hat{A}$ and $\hat{B}$. When the operators $\hat{A}$ and $\hat{B}$ are, for instance, the creation and annihilation operators, respectively, as in the case of the single-particle Green's function, then $\xi_n = (2n+1)\,\pi\, kT$ for fermions and $\xi_n = 2n\pi\, kT$ for bosons.

Why the factor $kT = 1/\beta$ has been specially separated on the right-hand side of the first equation of (5.6.12) is as follows. According to (5.6.14) $\Delta\xi_n = \xi_{n+1} - \xi_n = 2\pi\, kT$, and therefore, if we take the limit $T \to 0$ in order to reduce this expression so quantum field theory techniques can be used, then

$$kT\sum_n \ldots = \sum_n \frac{\Delta\xi_n}{2\pi} \ldots \to \int_{-\infty}^{\infty} \frac{d\xi}{2\pi} \ldots$$

and accordingly, introducing new variables t and ω such that $\tau = \mathrm{i}t/\hbar$ and $\xi = -\mathrm{i}\hbar\omega$,

$$kT\sum_{n=-\infty}^{\infty} \exp[-\mathrm{i}\xi_n(\tau-\tau')] \ldots \to \int_{-\mathrm{i}\infty}^{\mathrm{i}\infty} \frac{d\omega}{2\pi} \exp[-\mathrm{i}\omega(\tau-\tau')]\frac{\hbar}{\mathrm{i}} \ldots . \tag{5.6.15}$$

Provided that the contour along the imaginary axis in (5.6.15) can be rotated clockwise by 90° and turned into the real axis, (5.6.12) reduces to the Fourier transformation of the double-time Green's function. One should note the difference by a factor $\mathrm{i}/\hbar$ between the definition of the causal Green's function (5.4.17) and that of the temperature Green's function (5.6.1).

Finally, note that if we separate the integral in the second equation of (5.6.12) into two integrals over $\beta > \tau > 0$ and $0 > \tau > -\beta$ and use (5.6.11, 13) in the second of the two integrals, we can rewrite it as

$$\mathcal{K}_{A,B}(\mathrm{i}\xi_n) = \int_0^\beta d\tau\, \mathrm{e}^{\mathrm{i}\xi_n\tau} \mathcal{G}_{A,B}(\tau, 0) . \tag{5.6.16}$$

5.6.3 Single-Particle Temperature Green's Function for Noninteracting Particles

The Hamiltonian for noninteracting particles is given by (5.4.9). Thus, remembering (5.4.10) and making use of

$$\langle \hat{a}_k \hat{a}_{k'}^\dagger \rangle_0 = \delta_{kk'}[1 \mp f(E(\boldsymbol{k}))] \tag{5.6.17}$$

derived from (5.4.11, 12), we immediately obtain

$$\begin{aligned} \mathcal{G}_0(\boldsymbol{k}, \tau; \boldsymbol{k}', \tau') &= -\langle T_\tau\{\hat{a}_{\boldsymbol{k}}(\tau)\hat{a}_{\boldsymbol{k}'}^\dagger(\tau')\}\rangle_0 \\ &= -\delta_{\boldsymbol{kk}'} \exp(-[E(\boldsymbol{k})-\mu](\tau-\tau') \\ &\quad \times \{\theta(\tau-\tau')[1 \mp f(E(\boldsymbol{k}))] \mp \theta(\tau'-\tau) f(E(\boldsymbol{k}))\} . \end{aligned} \tag{5.6.18}$$

Taking (5.6.13) into account, we can determine the Fourier coefficient (5.6.16):

$$\mathcal{K}_0(\boldsymbol{k}, \boldsymbol{k}'; \mathrm{i}\xi_n) = \int_0^\beta d\tau\, \mathrm{e}^{\mathrm{i}\xi_n\tau} \mathcal{G}_0(\boldsymbol{k}, \tau; \boldsymbol{k}', 0) = \frac{\delta_{\boldsymbol{kk}'}}{\mathrm{i}\xi_n + \mu - E(\boldsymbol{k})} . \tag{5.6.19}$$

The upper signs in (5.6.17, 18) correspond to fermions and the lower ones to bosons. Notice the analogy between (5.6.19) and (5.4.14). According to the transformation (5.6.15), at absolute zero (5.6.19) reduces to (5.4.14), while at a finite temperature, this analogy furnishes the key to relating the temperature Green's function to the double-time Green's function.

5.6.4 Abrikosov-Gor'kov-Dzyaloshinskii-Fradkin Theorem

We want to demonstrate first that one can easily show the correspondence between the double-time Green's function and the temperature Green's function for single-particle Green's functions. The Fourier transform of the single-particle Green's function

$$\begin{aligned}\mathscr{G}(\boldsymbol{k},\tau;\boldsymbol{k}',\tau') &= -\langle T_\tau\{\check{a}_{\boldsymbol{k}}(\tau)\,\check{a}^\dagger_{\boldsymbol{k}'}(\tau')\}\rangle \\ &= kT\sum_{n=-\infty}^{\infty}\exp[-\mathrm{i}\,\xi_n(\tau-\tau')]\,\mathscr{K}(\boldsymbol{k},\boldsymbol{k}';\mathrm{i}\,\xi_n)\end{aligned} \tag{5.6.20}$$

can be written by the technique given in Sect. 4.3 and 5.1 in the form

$$\begin{aligned}\mathscr{K}(\boldsymbol{k},\boldsymbol{k}';\mathrm{i}\,\xi_n) &= \int_0^\beta d\tau\, \mathrm{e}^{\mathrm{i}\,\xi_n\tau}\,\mathscr{G}(\boldsymbol{k},\tau;\boldsymbol{k}',0) \\ &= \int_{-\infty}^{\infty}\frac{d\omega}{2\pi\mathrm{i}}\,\frac{\bar{\Lambda}^\pm(\boldsymbol{k},\boldsymbol{k}';\omega)}{\omega-\mathrm{i}\,\xi_n/\hbar},\end{aligned} \tag{5.6.21}$$

where the double sign $\pm$ attached to $\bar{\Lambda}$ corresponds to $\eta = \mp 1$, and

$$\begin{aligned}&\mathrm{i}\hbar\bar{\Lambda}^\pm(\boldsymbol{k},\boldsymbol{k}';\omega) = (1-\eta\,\mathrm{e}^{-\beta\hbar\omega})\,\bar{J}(\boldsymbol{k},\boldsymbol{k}';\omega)\,, \\ &\bar{J}(\boldsymbol{k},\boldsymbol{k}';\omega) = \int\int_{-\infty}^{\infty} d\bar{E}\,d\bar{E}'\exp[\beta(J-\bar{E})]\,2\pi\hbar\,\delta(\bar{E}+\hbar\omega-\bar{E}') \\ &\qquad\times \mathrm{Tr}\,[\delta(\bar{E}-\mathsf{H})\,\hat{a}_{\boldsymbol{k}}\,\delta(\bar{E}'-\mathsf{H})\,\hat{a}^\dagger_{\boldsymbol{k}'}]\,.\end{aligned} \tag{5.6.22}$$

Here $\eta = -1$ corresponds to the case for fermions, while $\eta = +1$ to that for bosons.

The double-time Green's function corresponding to this temperature Green's function is the single-particle Green's function given by (5.4.1), i.e., $\bar{G}^{\mathrm{ret},\pm}$ or $\bar{G}^{\mathrm{adv},\pm}$, in which $\mathscr{H}$ is replaced by H of (5.4.2). The spectral function of this double-time Green's function is given by (5.1.10, 17) or their analytical continuation (5.1.19) in which $\mathscr{H}$ is replaced by H

$$K(\boldsymbol{k},\boldsymbol{k}';z) = \int_{-\infty}^{\infty}\frac{d\omega}{2\pi\mathrm{i}}\,\frac{\bar{\Lambda}^\pm(\boldsymbol{k},\boldsymbol{k}';\omega)}{\omega-z}\,. \tag{5.6.23}$$

Comparing this equation with (5.6.21), we find at once the validity of

$$\mathscr{K}(\boldsymbol{k},\boldsymbol{k}';\mathrm{i}\,\xi_n) = K\left(\boldsymbol{k},\boldsymbol{k}';\frac{\mathrm{i}}{\hbar}\,\xi_n\right). \tag{5.6.24}$$

If the double-time Green's function is obtained, the Fourier coefficient of the corresponding temperature Green's function is determined by analytically continuing the spectral function of the former Green's function and evaluating it at $z = \mathrm{i}\,\xi_n/\hbar$. The problem of determining the temperature Green's function from the double-time Green's function is easily solved in this way, but we are now treating the converse problem of determining the double-time Green's function from the temperature Green's function. From (5.6.24), we cannot determine the value of a complex function K except at points of a series $z = \mathrm{i}\,\xi_n/\hbar$ $(n = 0, \pm 1, \pm 2, \ldots)$. Only with a special case can a complex function be determined from its value at points of a series. According to the theory of complex functions, a function is uniquely determined by values at points of a series with accumulation points in a region within which the function is analytic. We stated in Sect. 5.1 that the function K is analytic except on the real axis. If we consider the half-plane $\mathrm{Im}\; z > 0$, for instance, K is analytic, and the point series $\mathrm{i}\,\xi_n/\hbar$ $(n = 1, 2, 3, \ldots)$ has an accumulation point at infinity. Therefore, the function K on the upper half complex plane is uniquely determined by (5.6.24). The same holds for the lower half complex plane. Thus, the double-time Green's function can be determined from the temperature Green's function. We call this the Abrikosov-Gor'kov-Dzyaloshinskii-Fradkin theorem [5.21, 22]. A nonuniqueness such that we may replace z by $z\,[\exp(\beta h z)]^a$ with an arbitrary number a is proved to be removed if we require that K vanishes as $|z| \to \infty$ [5.23].

This theorem is also valid for the double-temperature Green's function (5.6.1) corresponding to the double-time Green's function (5.1.7 or 16), i.e., when the operators $\hat{A}$ and $\hat{B}$ are not necessarily an annihilation and a creation operator, but are their products in which annihilation and creation operators are paired. In this case, we need not distinguish G from $\bar{G}$, because the chemical potential correction like (5.4.7) due to (5.4.6) is canceled out. When G is not distinguished from $\bar{G}$ as in this case, we can prove the following equation in the same way as for (5.6.21):

$$\mathscr{K}_{A,B}(\mathrm{i}\,\xi_n) = \int_{-\infty}^{\infty} \frac{d\omega}{2\pi\mathrm{i}} \frac{\Lambda^{\pm}_{A,B}(\omega)}{\omega - \mathrm{i}\,\xi_n/\hbar}. \tag{5.6.25}$$

In this equation $\Lambda^{\pm}_{A,B}(\omega)$ agrees with the same term in (5.1.19). Thus, it is obvious that we may use the Abrikosov-Gor'kov-Dzyaloshinskii-Fradkin theorem as it stands.

We can easily derive relations like (5.6.24) for the two-particle Green's function or general multiparticle Green's functions, but so far have not been able to solve the problem of determining a multivariable complex function from its value at points of a series in general. For a *two-particle Green's function*

$$\mathscr{G}(\boldsymbol{k}_1, \tau_1, \boldsymbol{k}_2, \tau_2; \boldsymbol{k}_3, \tau_3, \boldsymbol{k}_4, \tau_4) = +\langle T_\tau\,\{\check{a}_{\boldsymbol{k}_1}(\tau_1)\,\check{a}_{\boldsymbol{k}_2}(\tau_2)\,\check{a}^{\dagger}_{\boldsymbol{k}_3}(\tau_3)\,\check{a}^{\dagger}_{\boldsymbol{k}_4}(\tau_4)\}\rangle, \tag{5.6.26}$$

analytical continuation to the corresponding quadruple-time Green's function is known but is very complicated. However, as seen above, we can easily deal with the double-temperature Green's function corresponding to the double-time Greens's function (5.2.7)

$$\mathscr{G}(\boldsymbol{k}_1, \boldsymbol{k}_2, \tau; \boldsymbol{k}_3, \boldsymbol{k}_4, \tau') = -\langle T_\tau \{\check{a}^\dagger_{\boldsymbol{k}_1}(\tau)\, \check{a}_{\boldsymbol{k}_2}(\tau)\, \check{a}^\dagger_{\boldsymbol{k}_3}(\tau')\, \check{a}_{\boldsymbol{k}_4}(\tau')\}\rangle$$

$$= kT \sum_{n=-\infty}^{\infty} \exp[-\mathrm{i}\,\xi_n(\tau-\tau')]\, \mathscr{K}(\boldsymbol{k}_1, \boldsymbol{k}_2; \boldsymbol{k}_3, \boldsymbol{k}_4; \mathrm{i}\,\xi_n)\,, \tag{5.6.27}$$

which is a special case of the quadruple-temperature Green's function (5.6.26):

$$\mathscr{G}(\boldsymbol{k}_1, \boldsymbol{k}_2, \tau; \boldsymbol{k}_3, \boldsymbol{k}_4, \tau') = \lim_{\substack{\Delta \to +0 \\ \Delta' \to +0}} \mathscr{G}(\boldsymbol{k}_2, \tau, \boldsymbol{k}_4, \tau'; \boldsymbol{k}_1, \tau+\Delta, \boldsymbol{k}_3, \tau'+\Delta')\,. \tag{5.6.28}$$

5.7 Diagram Technique

We have derived the basic formula (5.5.11) for the perturbational expansion of the thermodynamic potential in Sect. 5.5 and the basic formula (5.6.8) for the temperature Green's function in Sect. 5.6. Now let us consider a method of producing a Feynman diagram, originally devised for quantum field theory, for each term of a series obtained by substituting the expanded form of (5.5.9) into one of the basic formulas mentioned above. Let us take as an example the Hamiltonian (5.2.9) employed in discussing the plasma. We assume that in (5.5.1) H_0 has the form (5.4.9) and the interaction is

$$g\,\mathscr{H}^{\mathrm{int}}(\tau) = \frac{1}{2V} \sum_{\boldsymbol{k}_1, \boldsymbol{k}_2, \boldsymbol{k}_1', \boldsymbol{k}_2'} \Phi_{\boldsymbol{k}_2-\boldsymbol{k}_2'}\, \delta_{\boldsymbol{k}_1+\boldsymbol{k}_2, \boldsymbol{k}_1'+\boldsymbol{k}_2'}\, \hat{a}^\dagger_{\boldsymbol{k}_2}(\tau)\, \hat{a}^\dagger_{\boldsymbol{k}_1}(\tau)\, \hat{a}_{\boldsymbol{k}_1'}(\tau)\, \hat{a}_{\boldsymbol{k}_2'}(\tau)\,. \tag{5.7.1}$$

5.7.1 Bloch-De Dominicis Theorem

First let us prove a theorem which gives us a method of calculating the average of a product of the creation and annihilation operators appearing in each term of the perturbational expansion series. All the operators $\hat{A}$, $\hat{B}$ and the like are assumed to be either the creation operator $\hat{a}^\dagger$ or the annihilation operator $\hat{a}$. If we define $\eta = -1$ for fermions and $\eta = +1$ for bosons, then $\eta^2 = 1$ and the commutation relations can be written in the form

$$\hat{A}\hat{B} = \eta\,\hat{B}\hat{A} + \overline{AB}\,, \tag{5.7.2}$$

where we have defined the c numbers $\overline{AB}$ as follows:

$$\overline{a_k\, a_{k'}} = 0\,, \quad \overline{a^\dagger_k\, a^\dagger_{k'}} = 0\,, \quad \overline{a_k\, a^\dagger_{k'}} = \delta_{kk'}\,, \quad \overline{a^\dagger_{k'}\, a_k} = -\,\eta\,\delta_{kk'}\,. \tag{5.7.3}$$

If $\delta(A)=1$ when $\hat{A}$ is a creation operator, $\delta(A)=-1$ when $\hat{A}$ is an annihilation operator, and $E(A)=E(\boldsymbol{k})$ when $\hat{A}$ is either $\hat{a}_{\boldsymbol{k}}^{\dagger}$ or $\hat{a}_{\boldsymbol{k}}$, we obtain from the formula (5.4.10) the relation

$$\hat{A}\exp(-\tau\mathsf{H}_0)=\exp\{\delta(A)\ \tau[E(A)-\mu]\}\exp(-\tau\mathsf{H}_0)\hat{A}. \tag{5.7.4}$$

Let us calculate the expectation value $\langle\hat{A}_n\hat{A}_{n-1}\ldots\hat{A}_2\hat{A}_1\rangle_0$ with the notations defined above. Our course of calculation is a generalization of the rearranging method (5.4.11) used to obtain (5.4.12). First, moving $\hat{A}_n$ to the right of $\hat{A}_1$ by repeatedly using the commutation relation (5.7.2), then

$$\begin{aligned}\langle\hat{A}_n\hat{A}_{n-1}\ldots\hat{A}_1\rangle_0&=\overline{A_nA_{n-1}}\langle\hat{A}_{n-2}\ldots\hat{A}_1\rangle_0+\eta\langle\hat{A}_{n-1}\hat{A}_nA_{n-2}\ldots\hat{A}_1\rangle_0\\&=\sum_{m=1}^{n-1}\eta^{n-m-1}\overline{A_nA_m}\langle\hat{A}_{n-1}\ldots\hat{A}_{m+1}\hat{A}_{m-1}\ldots\hat{A}_1\rangle_0\\&\quad+\eta^{n-1}\langle\hat{A}_{n-1}\ldots\hat{A}_1\hat{A}_n\rangle_0.\end{aligned}$$

The last term, which alone contains a product of n operators, can be rewritten by virtue of (5.7.4) in the form

$$\begin{aligned}\langle\hat{A}_{n-1}\ldots\hat{A}_1\hat{A}_n\rangle_0&=\frac{\mathrm{Tr}\{\hat{A}_n\exp(-\beta\mathsf{H}_0)\hat{A}_{n-1}\ldots\hat{A}_1\}}{\mathrm{Tr}\{\exp(-\beta\mathsf{H}_0)\}}\\&=\exp\{\delta(A_n)\beta[E(A_n)-\mu]\}\langle\hat{A}_n\hat{A}_{n-1}\ldots\hat{A}_1\rangle_0.\end{aligned}$$

Transferring this term to the left-hand side finally yields

$$\begin{aligned}&\langle\hat{A}_n\hat{A}_{n-1}\ldots\hat{A}_1\rangle_0\\&\quad=\sum_{m=1}^{n-1}\frac{\eta^{n-m-1}\overline{A_nA_m}}{1-\eta^{n-1}\exp(\delta(A_n)\beta[E(A_n)-\mu])}\langle\hat{A}_{n-1}\ldots\hat{A}_{m+1}\hat{A}_{m-1}\ldots\hat{A}_1\rangle_0.\end{aligned} \tag{5.7.5}$$

The right-hand side includes only the expectation values, in each of which the number of operators has decreased by two. Therefore, if we repeatedly use the above equation, we arrive at either the expectation values of one operator

$$\langle\hat{a}_{\boldsymbol{k}}\rangle_0=0,\quad\langle\hat{a}_{\boldsymbol{k}}^{\dagger}\rangle_0=0 \tag{5.7.6}$$

when n is odd, or the expectation values of the products of two operators when n is even. For $n=2$, (5.7.5) is written as

$$\langle\hat{A}_2\hat{A}_1\rangle_0=\frac{\overline{A_2A_1}}{1-\eta\exp\{\delta(A_2)\beta[E(A_2)-\mu]\}}, \tag{5.7.7}$$

which gives (5.4.12, 6.17), and also

$$\langle \hat{a}_k \hat{a}_{k'} \rangle_0 = 0\,, \quad \langle \hat{a}_k^\dagger \hat{a}_{k'}^\dagger \rangle_0 = 0 \tag{5.7.8}$$

by virtue of (5.7.3). According to (5.7.7), (5.7.5) can be written in the form

$$\langle \hat{A}_n \hat{A}_{n-1} \dots \hat{A}_1 \rangle_0 = \begin{cases} 0 \quad (n = \text{odd}) \\ \sum\limits_{m=1}^{n-1} \eta^{n-m-1} \langle \hat{A}_n \hat{A}_m \rangle_0 \langle \hat{A}_{n-1} \dots A_{m+1} \hat{A}_{m-1} \dots \hat{A}_1 \rangle_0 \\ \qquad (n = \text{even})\,. \end{cases} \tag{5.7.9}$$

In this form, the difference arising from different particle statistics manifests itself only through the factors η^{n-m-1}. These factors may appear by moving $\hat{A}_m$ in the product of operators $\hat{A}_n (\hat{A}_{n-1} \dots \hat{A}_{m+1}) \hat{A}_m (\hat{A}_{m-1} \dots \hat{A}_1)$ to the left by interchanging it with $n-m-1$ operators $(\hat{A}_{n-1} \dots \hat{A}_{m+1})$. In this way we produce the product $\hat{A}_n \hat{A}_m (\hat{A}_{n-1} \dots \hat{A}_{m+1}) (\hat{A}_{m-1} \dots \hat{A}_1)$, provided that the second term on the right-hand side of the commutation relation (5.7.2) is neglected, i.e., that we take into account the anticommutativity or the commutativity only. Formula (5.7.6) shows that we may replace the pair $\hat{A}_n \hat{A}_m$ thus produced on the left end of the product by its expectation value $\langle \hat{A}_n \hat{A}_m \rangle_0$. The sum is taken over all possible choices of $\hat{A}_m$. For $n = 4$, then immediately

$$\langle \hat{A}_4 \hat{A}_3 \hat{A}_2 \hat{A}_1 \rangle_0 = \langle \hat{A}_4 \hat{A}_3 \rangle_0 \langle \hat{A}_2 \hat{A}_1 \rangle_0 + \eta \langle \hat{A}_4 \hat{A}_2 \rangle_0 \langle \hat{A}_3 \hat{A}_1 \rangle_0 + \eta^2 \langle \hat{A}_4 \hat{A}_1 \rangle_0 \langle \hat{A}_3 \hat{A}_2 \rangle_0\,. \tag{5.7.10}$$

If n is an even number greater than or equal to 6, we need to apply (5.7.9) again to find finally the following general formula

$$\langle \hat{A}_n \hat{A}_{n-1} \dots \hat{A}_1 \rangle_0 = \sum_Q \eta^{\nu(Q)} \langle \hat{A}_{m_n} \hat{A}_{m_{n-1}} \rangle_0 \dots \langle \hat{A}_{m_4} \hat{A}_{m_3} \rangle_0 \langle \hat{A}_{m_2} \hat{A}_{m_1} \rangle_0\,, \tag{5.7.11}$$

where Q denotes a permutation which transform the progression $(n, n-1, \dots, 1)$ into another $(m_n, m_{n-1}, \dots, m_2, m_1)$ satisfying the conditions

$$m_n = n > m_{n-2} > \dots > m_4 > m_2\,, \tag{5.7.12}$$

$$m_n > m_{n-1}\,, m_{n-2} > m_{n-3}\,, \dots, m_4 > m_3\,, m_2 > m_1\,, \tag{5.7.13}$$

and the sum $\sum_Q$ is taken over all possible such permutations. Further, $\nu(Q)$ is found to be the number of interchanges into which the permutation Q is resolved. For instance, the permutation Q transforming the progression (6, 5, 4, 3, 2, 1) into (6, 3, 5, 1, 4, 2) is resolved as $Q = (6)\,(53124) = (54)\,(52)\,(51)\,(53)$, so that $\nu(Q) = 4$. The notation (54) (52) (51) (53) means that we first make the interchange (53) in the progression (6, 5, 4, 3, 2, 1) to obtain (6, 3, 4, 5, 2, 1), next make the interchange (51) in this result, etc., and finally make the interchange (54). It is clear that one factor η is obtained per one interchange. The relation (5.7.11) is called the *Bloch-De Dominicis therorem* [5.24].

According to (5.7.4), (5.5.5), corresponding to an operator in the interaction representation, can be written as

$$\hat{A}(\tau) = \exp\{\delta(A)\,\tau[E(a)-\mu]\}\,\hat{A}\,. \tag{5.7.14}$$

Since this is equal to $\hat{A}$ multiplied by a c number, we may replace the operator $\hat{A}_m$ with $\hat{A}_m(\tau_m)$ in the Bloch-De Dominics theorem. Assuming that $\tau_n \geqq \tau_{n-1} \geqq \ldots \geqq \tau_2 \geqq \tau_1$, then

$$\begin{aligned}&\langle \hat{A}_n(\tau_n)\,\hat{A}_{n-1}(\tau_{n-1})\ldots\hat{A}_1(\tau_1)\rangle_0\\&\quad=\sum_Q \eta^{\nu(Q)}\langle \hat{A}_{m_n}(\tau_{m_n})\,\hat{A}_{m_{n-1}}(\tau_{m_{n-1}})\rangle_0\ldots\langle \hat{A}_{m_2}(\tau_{m_2})\,\hat{A}_{m_1}(\tau_{m_1})\rangle_0\,.\end{aligned} \tag{5.7.15}$$

Since the magnitudes of τ's are arranged in order of the progression $(n, n-1, \ldots, 1)$, the conditions (5.7.12, 13) may be stated in terms of τ's as

$$\tau_{m_n} = \tau_n \geqq \tau_{m_{n-2}} \geqq \ldots \geqq \tau_{m_4} \geqq \tau_{m_2}\,, \tag{5.7.16}$$

$$\tau_{m_n} \geqq \tau_{m_{n-1}}, \ldots, \tau_{m_4} \geqq \tau_{m_3}, \tau_{m_2} \geqq \tau_{m_1}\,. \tag{5.7.17}$$

On the right-hand side of (5.7.15), we need only to take into account the expectation values of pairs of the creation and annihilation operators by virtue of the results of (5.7.8). According to this, the left-hand side of (5.7.15) remains nonzero only when it contains $n/2$ creation operators and $n/2$ annihilation operators.

5.7.2 Perturbational Expansion of $\langle \hat{\mathscr{S}}(\beta)\rangle_0$

Now we apply the above result to the expanded form of the expectation value of (5.5.9)

$$\langle \hat{\mathscr{S}}(\beta)\rangle_0 = 1 + \sum_{n=1}^{\infty} \frac{(-g)^n}{n!} \int_0^\beta d\tau_n \int_0^\beta d\tau_{n-1} \ldots \int_0^\beta d\tau_1 \langle T_\tau\{\mathscr{H}^{\text{int}}(\tau_n)\ldots\mathscr{H}^{\text{int}}(\tau_1)\}\rangle_0 \tag{5.7.18}$$

by assuming the concrete form of interaction (5.7.1). We have to treat expectation values of the following type:

$$\begin{aligned}&\langle a^\dagger(2n)\,\hat{a}^\dagger(2n-1)\,\hat{a}(2n-1)\,\hat{a}(2n)\\&\quad\cdot\hat{a}^\dagger(2n-2)\ldots\hat{a}(4)\cdot\hat{a}^\dagger(2)\,\hat{a}^\dagger(1)\,\hat{a}(1)\,\hat{a}(2)\rangle_0\,,\end{aligned} \tag{5.7.19}$$

where $\hat{a}^\dagger(j)\,\hat{a}^\dagger(j-1)\,\hat{a}(j-1)\,\hat{a}(j)$ is an abbreviation of the product of operators $\hat{a}^\dagger_{k_j}(\tau_j)\,\hat{a}^\dagger_{k_{j-1}}(\tau_{j-1})\,\hat{a}_{k'_{j-1}}(\tau_{j-1})\,a_{k'_j}(\tau_j)$ appearing in (5.7.1) provided that $\tau_j = \tau_{j-1}$. We rewrite (5.7.19) by making use of (5.7.15). The point is to calculate the factor $\eta^{\nu(Q)}$.

By repeating interchanges an even number of times, the average (5.7.19) can be rewritten as

$$\langle \hat{a}(2n)\,\hat{a}^\dagger(2n)\cdot\hat{a}(2n-1)\,\hat{a}^\dagger(2n-1) \cdot\hat{a}(2n-2)\ldots\hat{a}^\dagger(3)\cdot\hat{a}(2)\,\hat{a}^\dagger(2)\cdot\hat{a}(1)\,\hat{a}^\dagger(1)\rangle_0 . \tag{5.7.20}$$

In this procedure no factor η appears. Let us rewrite this expression in the form of a term in the sum on the right-hand side of (5.7.15). The operator corresponding to $\hat{A}_{m_n}(\tau_{m_n})$ in (5.7.15) is $\hat{a}^\dagger(2n)$ in (5.7.20). If its partner operator corresponding to $\hat{A}_{m_{n-1}}(\tau_{m_{n-1}})$ in (5.7.15) is supposed to be $\hat{a}(j)$ in (5.7.20), we move the product $\hat{a}(j)\,\hat{a}^\dagger(j)$ as a whole leftwards to the right side of $\hat{a}^\dagger(2n)$. A product of two operators can thus be moved by interchanges repeated an even number of times, without producing the factor η. Next, if an operator having $\hat{a}^\dagger(j)$ as its partner is supposed to be $\hat{a}(k)$, we move $\hat{a}(k)\,\hat{a}^\dagger(k)$ leftwards to the right side of $\hat{a}^\dagger(j)$. We repeat this procedure. Since in (5.7.19) the condition $\tau_{2n}\geqq\tau_{2n-1}\geqq\ldots\geqq\tau_2\geqq\tau_1$ is satisfied by virtue of the operator T_τ in the expansion (5.7.18), the condition (5.7.16) always ought to be satisfied by this procedure. When we have proceeded as far as a certain pair $\hat{a}(l)\,\hat{a}^\dagger(l)$ and moved it, it is possible that the partner of $\hat{a}^\dagger(l)$ is $\hat{a}(2n)$. Then we have to move $\hat{a}(2n)$ on the left end in (5.7.20) to the right-hand side of $\hat{a}^\dagger(l)$. The latter procedure requires an odd number of interchanges because of the existence of $\hat{a}^\dagger(2n)$, so that one factor η is produced:

$$\eta\,\langle \hat{a}^\dagger(2n)\,\hat{a}(j)\cdot\hat{a}^\dagger(j)\,\hat{a}(k)\cdot\hat{a}^\dagger(k)\ldots\hat{a}(l)\cdot\hat{a}^\dagger(l)\,\hat{a}(2n) \cdot\hat{a}(m)\ldots\hat{a}^\dagger(1)\rangle_0 . \tag{5.7.21}$$

We may replace each paired set of operators with its expectation value. If our rearrangement is not complete at this point, we repeat this procedure with respect to the remaining product $\hat{a}(m)\ldots\hat{a}^\dagger(1)$. When our rearrangement is finished, a number of rings like $2n\to j\to k\to\ldots\to l\to 2n$, $m\to\ldots\to m,\ldots$ should have been made, and since one factor η is produced per ring, a factor $\eta^{\lambda(Q)}$ results where $\lambda(Q)$ denotes the number of rings. This factor is a part of the factor $\eta^{\nu(Q)}$ in (5.7.15). Thus, we can write (5.7.21) in the form

$$\eta\,\langle \hat{a}^\dagger(2n)\,\hat{a}(j)\rangle_0\,\langle \hat{a}^\dagger(j)\,\hat{a}(k)\rangle_0\,\langle \hat{a}^\dagger(k)\ldots\hat{a}(l)\rangle_0\,\langle \hat{a}^\dagger(l)\,\hat{a}(2n)\rangle_0 \times\eta\,\langle \hat{a}^\dagger(m)\ldots\hat{a}(m)\rangle_0\,\eta\ldots\eta\ldots . \tag{5.7.22}$$

This expression is not yet in the form of a summand in (5.7.15), because the conditions (5.7.17) are not necessarily satisfied. Since in (5.7.15) an operator with a larger value of τ in a pair always lies on the left, while in (5.7.22) a creation operator always lies on the left, either case $\tau_p>\tau_q$ or $\tau_p<\tau_q$ may occur in a factor, say, $\langle \hat{a}^\dagger(p)\,\hat{a}(q)\rangle_0$. If $\tau_p>\tau_q$, we need no further rearrangement because the conditions are met, while if $\tau_p<\tau_q$, it must be rewritten into $\langle \hat{a}(q)\, \hat{a}^{\dagger}(p)\rangle_0$ to obtain the form of (5.7.15). In the latter case, one factor η is produced and contributes to $\eta^{\nu(Q)}$. However, it is difficult to evaluate the number of factors arising this way. Fortunately, by using the unperturbed single-particle Green's function we can write (5.6.18) as

$$\theta(\tau' - \tau)\, \langle \hat{a}^{\dagger}_{k'}(\tau')\, \hat{a}_{k}(\tau)\rangle_0 + \theta(\tau - \tau')\, \eta\, \langle a_{k}(\tau)\, \hat{a}^{\dagger}_{k'}(\tau')\rangle_0$$
$$= -\, \eta\, \mathscr{G}_0(\boldsymbol{k}, \tau; \boldsymbol{k}', \tau') \qquad (5.7.23)$$

as for (5.4.18). Therefore, in each term of the sum in (5.7.15) the expectation value of a pair multiplied by η contained not in the factor $\eta^{\lambda(Q)}$ but in the factor $\eta^{\nu(Q)}$ is $-\eta\mathscr{G}_0$. Here note that a pair with $\tau = \tau'$ appears in the perturbational expansion series in the same $\mathscr{H}^{\text{int}}$. For such a pair, the creation operator should always be placed on the left side of the annihilation operator. Thus we must use $-\eta\, \mathscr{G}_0(\boldsymbol{k}, \tau; \boldsymbol{k}', \tau + \Delta)$ and take the limit $\Delta \to +0$ at the end of the calculations. For example, in the first-order perturbation term, using (5.7.10),

$$\langle \hat{a}^{\dagger}_{k_2}(\tau)\, \hat{a}^{\dagger}_{k_1}(\tau)\, \hat{a}_{k'_1}(\tau)\, \hat{a}_{k'_2}(\tau)\rangle_0$$
$$= \eta\, \langle \hat{a}^{\dagger}_{k_2}(\tau)\, \hat{a}_{k'_1}(\tau)\rangle_0\, \langle \hat{a}^{\dagger}_{k_1}(\tau)\, \hat{a}_{k'_2}(\tau)\rangle_0 + \eta^2\, \langle \hat{a}^{\dagger}_{k_2}(\tau)\, \hat{a}_{k'_2}(\tau)\rangle_0\, \langle \hat{a}^{\dagger}_{k_1}(\tau)\, \hat{a}_{k'_1}(\tau)\rangle_0$$
$$= \eta\, [-\, \eta\, \mathscr{G}_0(\boldsymbol{k}'_1, \tau; \boldsymbol{k}_2, \tau + \Delta)]\, [-\, \eta\, \mathscr{G}_0(\boldsymbol{k}'_2, \tau; \boldsymbol{k}_1, \tau + \Delta')]$$
$$+\, \eta^2\, [-\, \eta\, \mathscr{G}_0(\boldsymbol{k}'_2, \tau; \boldsymbol{k}_2, \tau + \Delta)]\, [-\, \eta\, \mathscr{G}_0(\boldsymbol{k}'_1, \tau; \boldsymbol{k}_1, \tau + \Delta')] \qquad (5.7.24)$$

and the limits $\Delta \to +0$ and $\Delta' \to +0$ need to be taken only at the end. This example illustrates that the factor $-\eta$ of $-\eta\, \mathscr{G}_0$ may be omitted in evaluating the expectation value of the form (5.7.19), because the factor appears $2n$ times and $(-\eta)^{2n} = 1$.

5.7.3 Correspondence with Feynman Diagrams

We are now able to make each term of the perturbational expansion series (5.7.18) correspond to a Feynman diagram. First, to correspond to a single-particle Green's function $\mathscr{G}_0(\boldsymbol{k}, \tau; \boldsymbol{k}', \tau')$, let us draw a solid line called the *particle line.* It proceeds from an initial point $(\boldsymbol{k}', \tau')$ corresponding to a creation operator $\hat{a}^{\dagger}_{k'}(\tau')$ to a terminal point $(\boldsymbol{k}, \tau)$ corresponding to an annihilation operator $\hat{a}_{k}(\tau)$ (Fig. 5.1 a). Next, corresponding to a matrix element $V^{-1}\, \Phi_{k_2 - k'_2}\, \delta_{k_1 + k_2,\, k'_1 + k'_2}$ of the interaction (5.7.1), let us draw a broken

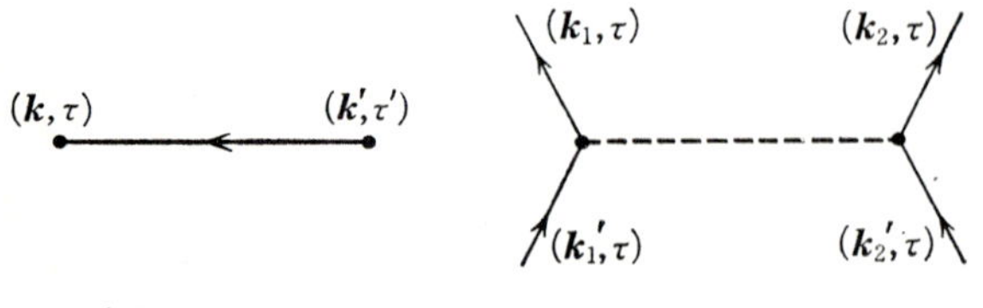

(a) (b)

Fig. 5.1. (**a**) Particle line, and (**b**) interaction line

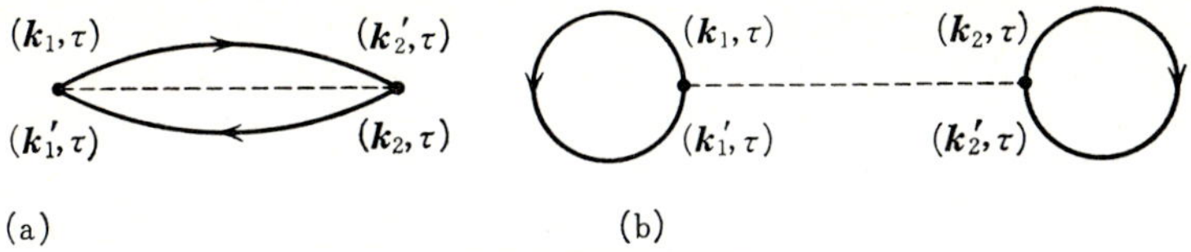

Fig. 5.2 a, b. Diagram of first order, with one ring (**a**) and with two rings (**b**)

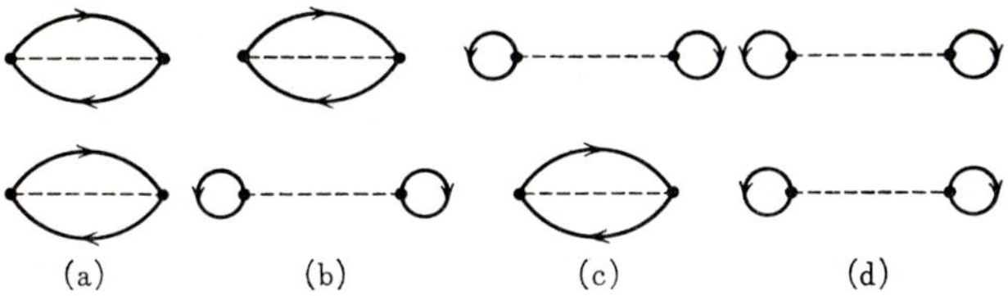

Fig. 5.3 a – d. Disconnected second-order diagrams

line called the *interaction line.* It starts from a vertex, which is the terminal point $(\boldsymbol{k}_1', \tau)$ of a particle line and the initial point $(\boldsymbol{k}_1, \tau)$ of another particle line, and ends at another vertex, which is the terminal point $(\boldsymbol{k}_2', \tau)$ of a particle line and the initial point $(\boldsymbol{k}_2, \tau)$ of another particle line (Fig. 5.1 b). Each term of the perturbational expansion series (5.7.18) has a corresponding diagram such that particle lines are connected to each other at vertices to form rings and interaction lines are inserted between the vertices.

For example, for the first-order perturbation term

$$-g\int_0^\beta d\tau\,\langle\mathscr{H}^{\text{int}}(\tau)\rangle_0$$
$$=-\frac{1}{2}\int_0^\beta d\tau\sum_{\boldsymbol{k}_1,\boldsymbol{k}_2,\boldsymbol{k}_1',\boldsymbol{k}_2'}\frac{\Phi_{\boldsymbol{k}_2-\boldsymbol{k}_2'}}{V}\,\delta_{\boldsymbol{k}_1+\boldsymbol{k}_2,\,\boldsymbol{k}_1'+\boldsymbol{k}_2'}\,\langle\hat{a}^\dagger_{\boldsymbol{k}_2}(\tau)\,\hat{a}^\dagger_{\boldsymbol{k}_1}(\tau)\,\hat{a}_{\boldsymbol{k}_1'}(\tau)\,\hat{a}_{\boldsymbol{k}_2'}(\tau)\rangle_0\,,$$

two kinds of diagrams appear corresponding to two terms in (5.7.24). Since the first term has one η, η of $-\eta\,\mathscr{G}_0$ excepted, it corresponds to a figure with one ring composed of two particle lines (Fig. 5.2 a). The second term has the factor η^2 and corresponds to a diagram having two rings of particle lines (Fig. 5.2 b). For the interaction (5.7.1), the latter term vanishes because of both Kronecker's delta $\delta_{\boldsymbol{k},\boldsymbol{k}'}$ in (5.6.18) and the property $\Phi_0=0$ referred to in Sect. 5.2.2.

There are 17 diagrams associated with second-order perturbational terms, and 13 of them are *connected diagrams,* in which all rings made of particle lines are connected by interaction lines in one cluster, while the remaining 4 are *disconnected diagrams* split into two parts. When the τ axis is vertical, 4 disconnected second-order diagrams consist of two kinds of first-order diagrams placed in a column, and correspond to 4 ways of association (Fig. 5.3). When a diagram splits into more than 2 parts, its contribution to the perturbational term becomes a product of values associated with disconnected parts. Therefore, the contributions from the terms corresponding to Fig. 5.3b–d vanish by virtue of $\Phi_0 = 0$. The situation for perturbational terms of higher orders is similar.

Conversely, an expansion term can be obtained from a diagram. To obtain terms of the n'th order, we first draw all possible types of diagrams which are made by putting together $2n$ particle lines and n interaction lines. For each diagram, we make a product of factors, i.e., of $\mathscr{G}_0$ for each particle line and of a matrix element $V^{-1}\Phi\,\delta$ for each interaction line, multiplying this product with a factor $\eta^{\lambda(Q)}$ corresponding to a number $\lambda(Q)$ of rings made of particle lines, and summing up over all wave vectors $\boldsymbol{k}$ appearing in the diagram. Finally, we carry out the operation in the expansion formula (5.7.18)

$$\frac{(-1)^n}{n!}\int_0^\beta d\tau_n \int_0^\beta d\tau_{n-1} \ldots \int_0^\beta d\tau_1 \,.$$

Then the expressions thus obtained are summed up over all nth-order diagrams. When we consider this sum, we need to consider only one of the topologically equivalent diagrams, because diagrams which are made equivalent to each other by mere permutation of wave vectors contribute the same value to the sum. Thus the factor 1/2 in $\mathscr{H}^{\text{int}}$ can be omitted and $n!$ replaced by some factor in the denominator of the last operation. Details are discussed below.

5.7.4 Matsubara Formula

To evaluate the thermodynamic potential (5.5.11), terms corresponding to all diagrams need not be determined for each order in the expansion series. Let us show that the logarithmic function in (5.5.11) is taken by choosing only special diagrams.

If a Feynman diagram associated with a perturbational term of the nth order is one in which p_1 first-order connected diagrams, p_2 second-order connected diagrams and so forth, l parts in all, are placed in a column, then

$$p_1 + 2p_2 + 3p_3 + \ldots = n; \quad p_1 + p_2 + p_3 + \ldots = l. \tag{5.7.25}$$

Then the contribution of this diagram to the nth order perturbational term is

$$\frac{(-g)^n}{n!}\left[\int_0^\beta d\tau_1 \langle \mathscr{H}^{\text{int}}(\tau_1)\rangle_0^c\right]^{p_1}\left[\int_0^\beta d\tau_2 \int_0^\beta d\tau_1 \langle T_\tau\{\mathscr{H}^{\text{int}}(\tau_2)\,\mathscr{H}^{\text{int}}(\tau_1)\}\rangle_0^c\right]^{p_2}$$
$$\times\left[\int_0^\beta d\tau_3 \ldots\right]^{p_3} \ldots, \tag{5.7.26}$$

where $\langle\ldots\rangle_0^c$ stands for the expectation value belonging to a special type of connected diagram within the given Feynman diagram. Since connected diagrams of the same kind give the same value, provided that summation over wave numbers and the τ integration is carried out, (5.7.26) has the simple form as a product of factors with exponents p_1, p_2, etc. Feynman diagrams giving the same value as (5.7.26) are obtained by a rearrangement of $n\mathscr{H}^{\text{int}}$. The number of possible rearrangements is

$$n!\bigg/\left[\prod_{m=1}^{\infty} p_m!\,(m!)^{p_m}\right]. \tag{5.7.27}$$

This is also the number of Feynman diagrams topologically equivalent to each other. The nth-order perturbational term is obtained by multiplying (5.7.26) with the factor (5.7.27) and summing over all terms associated with topologically different diagrams. In this way, by introducing the sum of contributions from terms associated with connected diagrams of the m'th order

$$C_m = \frac{(-g)^m}{m!}\int_0^\beta d\tau_m \ldots \int_0^\beta d\tau_1 \,\langle T_\tau\{\mathscr{H}^{\text{int}}(\tau_m)\ldots\mathscr{H}^{\text{int}}(\tau_1)\}\rangle_0^{\text{con}}, \tag{5.7.28}$$

resulting in

$$\sum_{\substack{p_1=0\\(p_1+2p_2+3p_3+\ldots=n)}}^{\infty}\;\sum_{p_2=0}^{\infty}\;\sum_{p_3=0}^{\infty}\ldots\prod_{m=1}^{\infty}\frac{1}{p_m!}\,C_m^{p_m} \tag{5.7.29}$$

in which we interpret the first term with $n = 0$ as 1, because $p_1 = p_2 = \ldots = 0$. If $n = 1$, then $p_1 = 1, p_2 = p_3 = \ldots = 0$, and the above expression gives C_1, i.e., the contribution associated with the connected diagram in Fig. 5.2a. For $n = 2$, $p_1 = 2$, $p_2 = p_3 = \ldots = 0$ and $p_1 = 0$, $p_2 = 1$, $p_3 = \ldots = 0$, and the above expression becomes $(1/2!)\,C_1^2 + C_2$. Here C_2 denotes the sum of contributions associated with 13 connected second-order diagrams referred to earlier. Further $(1/2!)\,C_1^2$ is the sum of contributions associated with the disconnected diagram which is made by placing the diagram appearing in C_1 in a column, as in Fig. 5.3a.

The expansion series (5.7.18) of $\langle\hat{\mathscr{S}}(\beta)\rangle_0$ is the sum of the expressions (5.7.29) over all n, and is given by

$$\langle\hat{\mathscr{S}}(\beta)\rangle_0 = \exp\left(\sum_{m=1}^{\infty} C_m\right) \tag{5.7.30}$$

because the summing over n is equivalent to omitting the condition $p_1 + 2p_2 + 3p_3 + \ldots = n$ in (5.7.29). Taking the logarithms of the above equation yields the thermodynamic potential (5.5.11). To rewrite this into a more tractable form, we need only the value obtained by picking out only those terms in the expansion series (5.7.18) which are associated with connected diagrams

$$\langle\hat{\mathscr{S}}(\beta)\rangle_0^{\text{con}} = 1 + \sum_{m=1}^{\infty} C_m\,. \tag{5.7.31}$$

The result is

$$J = J_0 - kT[\langle\hat{\mathscr{S}}(\beta)\rangle_0^{\text{con}} - 1]. \tag{5.7.32}$$

This is the beautiful formula derived by *Matsubara* [5.19].

5.8 Dyson Equation

This section deals with the correspondence between the perturbational expansion formula for the temperature Green's function and the Feynman diagrams. It is more convenient to determine the temperature Green's function first and then calculate various averages, especially the thermodynamic potential, than to evaluate the thermodynamic potential directly from the grand partition function as done in the previous section.

5.8.1 Single-Particle Temperature Green's Function

The single-particle temperature Green's function for an unperturbed system was determined in (5.6.18). For a perturbed system

$$\begin{aligned}\mathscr{G}(\boldsymbol{k}, \tau; \boldsymbol{k}', \tau') &= -\langle T_\tau\{\check{a}_{\boldsymbol{k}}(\tau)\,\check{a}^\dagger_{\boldsymbol{k}'}(\tau')\}\rangle \\ &= -\frac{\langle T_\tau\{\hat{a}_{\boldsymbol{k}}(\tau)\,\hat{a}^\dagger_{\boldsymbol{k}'}(\tau')\,\hat{\mathscr{S}}(\beta)\}\rangle_0}{\langle\hat{\mathscr{S}}(\beta)\rangle_0}\end{aligned} \tag{5.8.1}$$

according to (5.6.1, 8). The denominator was discussed in the preceding section. We can make a similar argument on the perturbational expansion of the numerator

$$\begin{aligned}-\langle T_\tau\{\hat{a}_{\boldsymbol{k}}(\tau)\,\hat{a}^\dagger_{\boldsymbol{k}'}(\tau')\,\hat{\mathscr{S}}(\beta)\}\rangle_0 &= \mathscr{G}_0(\boldsymbol{k}, \tau; \boldsymbol{k}', \tau') - \sum_{n=1}^{\infty}\frac{(-g)^n}{n!}\int_0^\beta d\tau_n \ldots \\ &\ldots \int_0^\beta d\tau_1\,\langle T_\tau\{\hat{a}_{\boldsymbol{k}}(\tau)\,\hat{a}^\dagger_{\boldsymbol{k}'}(\tau')\,\mathscr{H}^{\text{int}}(\tau_n)\ldots\mathscr{H}^{\text{int}}(\tau_1)\}\rangle_0\end{aligned} \tag{5.8.2}$$

by using the Bloch-De Dominicis theorem.

A comparison with the expansion formula (5.7.18) shows that two *external lines* have been added to the Feynman diagrams, i.e., a particle line with the terminal point $(\boldsymbol{k}, \tau)$ and another particle line with the initial point $(\boldsymbol{k}', \tau')$, because the extra factors of the annihilation operator $\hat{a}_{\boldsymbol{k}}(\tau)$ and the creation operator $\hat{a}^\dagger_{\boldsymbol{k}'}(\tau')$ are inserted. In the first term of the expansion (5.8.2) these two external lines are considered to be connected to each other. Let us call particle lines, other than the two external lines in a diagram, *interior lines.* The Feynman diagrams associated with first-order terms in the expansion series (5.8.2) consist of two kinds of disconnected and

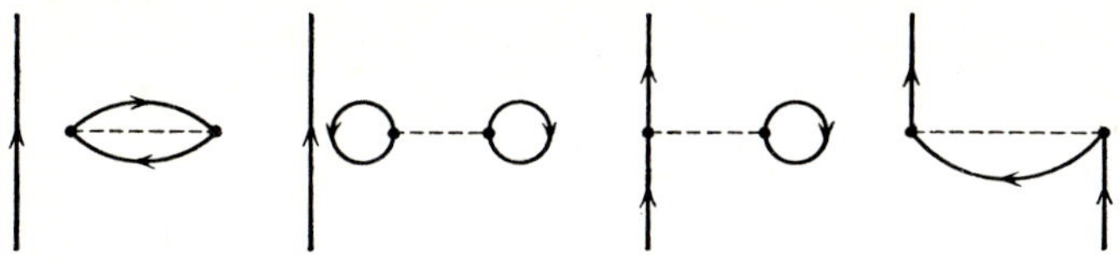

Fig. 5.4. (**a**) Disconnected diagram with one ring, (**b**) disconnected diagram with two rings, (**c**) connected diagram with one ring, and (**d**) diagram with no ring

connected diagrams. The disconnected diagrams are obtained by placing one of the first-order diagrams shown in Fig. 5.2 next to the joined external lines. The connected diagrams are obtained by cutting one of the interior lines in the first-order diagrams shown in Fig. 5.2 and connecting these lines with the external lines (Fig. 5.4). The diagrams associated with second-order terms in expansion (5.8.2) consist of many connected diagrams (examples of which are given below), in addition to the two kinds of disconnected diagrams: those consisting of the second-order diagrams in Fig. 5.3 placed next to an external line and those consisting of the first-order diagrams in Fig. 5.2 added to the connected diagrams in Fig. 5.4.

Let us consider here the contribution from disconnected diagrams. Among the terms of the $(m+n)$'th order in the expansion (5.8.2), the contribution from a disconnected diagram including a special type of connected diagram of the mth order becomes

$$-\frac{(-g)^{m+n}}{(m+n)!}\left[\int_0^\beta d\tau_m \ldots \int_0^\beta d\tau_1 \langle T_\tau \{\hat{a}_{\boldsymbol{k}}(\tau)\, \hat{a}^\dagger_{\boldsymbol{k}'}(\tau')\, \mathscr{H}^{\text{int}}(\tau_m) \ldots \mathscr{H}^{\text{int}}(\tau_1)\}\rangle_0^{\text{c}}\right]$$
$$\times\left[\int_0^\beta d\tau_n \ldots \int_0^\beta d\tau_1 \langle T_\tau \{\mathscr{H}^{\text{int}}(\tau_n) \ldots \mathscr{H}^{\text{int}}(\tau_1)\}\rangle_0\right], \tag{5.8.3}$$

where the first factor within the square brackets denotes the contribution from the part connected with the external lines, and the second factor within the square brackets, the contribution from the disconnected parts. Although the latter ought to have a structure such as given by (5.7.26), we have not written it down explicitly. As in the discussion above, rearranging $m+n$ operators $\mathscr{H}^{\text{int}}$ yields Feynman diagrams similar to that associated with (5.8.3), the number of which is the value (5.7.27) multiplied by $(m+n)!/(m!\,n!)$. Thus, by fixing the connected part of the mth order and summing it over all kinds of disconnected parts and over all $n = 0, 1, 2, \ldots$, the sum of contributions from disconnected parts ought to yield the same value as the expansion (5.7.18), so that such a sum of (5.8.3) becomes

$$-\frac{(-g)^m}{m!}\int_0^\beta d\tau_m \ldots \int_0^\beta d\tau_1 \langle T_\tau \{\hat{a}_{\boldsymbol{k}}(\tau)\, \hat{a}^\dagger_{\boldsymbol{k}'}(\tau')\, \mathscr{H}^{\text{int}}(\tau_m) \ldots \mathscr{H}^{\text{int}}(\tau_1)\}\rangle_0^{\text{c}} \langle \hat{\mathscr{S}}(\beta)\rangle_0 \,. \tag{5.8.4}$$

Since the expansion (5.8.2) is the sum of the above expression over all kinds of connected diagrams and over all $m = 0, 1, 2, \ldots$, then

$$\mathscr{G}(\boldsymbol{k}, \tau; \boldsymbol{k}', \tau') = \mathscr{G}_0(\boldsymbol{k}, \tau; \boldsymbol{k}', \tau') - \sum_{m=1}^{\infty} \frac{(-g)^m}{m!} \int_0^{\beta} d\tau_m \ldots$$
$$\ldots \int_0^{\beta} d\tau_1 \langle T_\tau \{\hat{a}_{\boldsymbol{k}}(\tau)\, \hat{a}^\dagger_{\boldsymbol{k}'}(\tau')\, \mathscr{H}^{\mathrm{int}}(\tau_m) \ldots \mathscr{H}^{\mathrm{int}}(\tau_1)\}\rangle_0^{\mathrm{con}}$$
$$= - \langle T_\tau \{\hat{a}_{\boldsymbol{k}}(\tau)\, \hat{a}^\dagger_{\boldsymbol{k}'}(\tau')\, \hat{\mathscr{S}}(\beta)\}\rangle_0^{\mathrm{con}}\,, \tag{5.8.5}$$

provided that connected diagrams connected with the external lines are picked out, denoted by $\langle\,\rangle^{\mathrm{con}}$, as we did in the previous section, picking out those unconnected with the external lines. Namely, the calculation may proceed neglecting all disconnected diagrams. Contributions from disconnected parts of each diagram cancel the denominator of (5.8.1).

Moreover, since in the second member of (5.8.5) expressions obtained by rearranging m operators $\mathscr{H}^{\mathrm{int}}$ give the same value, we may omit $m!$ in the denominator by taking into account only those contributions associated with topologically distinct diagrams. This is a simplification which did not occur in the case of the thermodynamic potential (5.7.32). For (5.7.28), there are no external lines which have to be fixed and the interior lines form closed rings, so that by fixing one among $m\,\mathscr{H}^{\mathrm{int}}$ operators and rearranging the remaining $m-1$ operators $\mathscr{H}^{\mathrm{int}}$, $(m-1)!$ topologically equivalent diagrams result; so $m!$ in the denominator is merely replaced by m even if we consider only topologically distinct diagrams. This situation makes the graphical summation described below possible.

5.8.2 Graphical Summation

In the expansion formula (5.8.5) for a single-particle temperature Green's function, we need to consider only connected diagrams of various forms. By partially summing up over all terms associated with diagrams including parts of a special type, i.e., by the *method of graphical summation*, we can derive valid relationship among single-particle Green's functions and the like.

To practise this diagram technique it is simplest to relate Fourier coefficients of the temperature Green's function to diagrams, because the single-particle Green's functions (5.8.1) and other Green's functions are diagonalized in the matrix representation with $(\boldsymbol{k}, \tau)$ as subscripts. Let us consider the Fourier coefficient $\mathscr{K}(\boldsymbol{k}, \boldsymbol{k}'; \mathrm{i}\,\xi_n)$ for the single-particle temperature Green's function (5.6.20). For a uniform system, we may set $\boldsymbol{k}' = \boldsymbol{k}$ by virtue of the spatial translational symmetry:

$$\mathscr{K}(\boldsymbol{k}, \boldsymbol{k}'; \mathrm{i}\,\xi_n) = \delta_{\boldsymbol{k},\boldsymbol{k}'}\, \mathscr{K}(\boldsymbol{k}, \mathrm{i}\,\xi_n)\,. \tag{5.8.6}$$

This permits us to assign to a single-particle line only one wave vector $\boldsymbol{k}$ instead of two vectors, i.e., $\boldsymbol{k}'$ at its initial point and $\boldsymbol{k}$ at its terminal point.

Furthermore, it is convenient to represent the interparticle interaction by a boson field. We assume as ususal an interaction of the form (5.7.1). Since the interaction (5.7.1) contains a δ function representing the wave vector conservation, it can be written in the form

$$g\,\mathscr{H}^{\mathrm{int}}(\tau) = \int_0^\beta d\tau' \frac{1}{2} \sum_{\boldsymbol{k}_1, \boldsymbol{k}_2} \sum_{\boldsymbol{q}} \frac{\Phi_{\boldsymbol{q}}}{V} \delta(\tau - \tau') \times \hat{a}^\dagger_{\boldsymbol{k}_2}(\tau + \Delta)\, \hat{a}^\dagger_{\boldsymbol{k}_1}(\tau' + \Delta')\, \hat{a}_{\boldsymbol{k}_1+\boldsymbol{q}}(\tau')\, \hat{a}_{\boldsymbol{k}_2-\boldsymbol{q}}(\tau)\,. \tag{5.8.7}$$

Here Δ and Δ' must be set as $+0$ at the end of the calculations. They have been introduced to place the creation operators to the left of the annihilation operators. The δ-function $\delta(\tau-\tau')$ is inserted so the interaction can be written in such a form that it appears to be produced by the exchange of a quantum of a quantized boson field. The Coulomb interaction (5.2.10) originates essentially from the exchange of longitudinal photons, and for an arbitrary finite $\Phi_{\boldsymbol{k}}$, this representation is always possible. It is known that

$$\mathscr{D}_0(\boldsymbol{r}, \tau; \boldsymbol{r}', \tau') = \Phi(\boldsymbol{r} - \boldsymbol{r}')\, \delta(\tau - \tau') \tag{5.8.8}$$

is proportional to the nonrelativistic approximation of the single-particle Green's function when the quantum of the field is free. By this technique, we can specify the direction of an interaction line in the same way as of a particle line.

Now, by transforming $\mathscr{G}_0(\boldsymbol{k}, \tau; \boldsymbol{k}', \tau')$ associated with a particle line in Sect. 5.7.3 into its Fourier coefficient (5.6.19) by means of (5.6.20), the exponential function $\exp(\mathrm{i}\,\xi_n \tau')$ can be assigned to the initial point $(\boldsymbol{k}', \tau')$ of the particle line and $\exp(-\mathrm{i}\,\xi_n \tau)$ to the terminal point $(\boldsymbol{k}, \tau)$. Similarly, based on the Fourier expansion of the δ-function

$$\delta(\tau - \tau') = kT \sum_{n=-\infty}^{\infty} \exp[-\mathrm{i}\,\zeta_n(\tau - \tau')], \quad \zeta_n = 2n\pi kT, \tag{5.8.9}$$

we assign $\exp(\mathrm{i}\,\zeta_n \tau')$ to the initial point and $\exp(-\mathrm{i}\,\zeta_n \tau)$ to the terminal point of the interaction line. A point at which an interaction line meets particle lines is called the vertex in a diagram. The τ integral of $\int d\tau\, \mathscr{H}^{\mathrm{int}}(\tau)$ in $\hat{\mathscr{S}}(\beta)$ of (5.8.5) and the τ' integral inserted into $\mathscr{H}^{\mathrm{int}}(\tau)$ in (5.8.7) lead to a kind of conservation law

$$kT \int_0^\beta d\tau \exp[\pm\,\mathrm{i}\,(\xi_{n''} - \xi_{n'} - \zeta_n)] = \delta_{n'', n'+n} \tag{5.8.10}$$

at the two vertices of both ends of an interaction line, taking into account the exponential functions from particle lines. The reason why ζ_n is chosen for the field quantum of the Bose type is, as is obvious from (5.8.10), to induce $\xi_{n'}$ and $\xi_{n''}$ of the particle to be either Fermi or Bose type. Since these variables ξ_n and ζ_n have a dimension of kT, i.e., of energy, the conservation law mentioned above states that at each vertex, i.e., when a particle either emits or absorbs a field quantum, energy is, so to speak, conserved. This corresponds to the fact that the conservation law of wave vectors represents the conservation of quasimomenta (Fig. 5.5 below). For simplicity, hereafter let us call ξ_n, ζ_n, etc., the quasienergy.

5.8.3 Feynman Rules

We can prescribe the following rules for determining perturbational terms of the mth order for the Fourier coefficient $\mathscr{K}(\boldsymbol{k}, \mathrm{i}\,\xi_n)$ of the single-particle Green's function.

1) Draw all topologically distinct connected diagrams composed of m interaction lines and $2m+1$ particle lines including the two (one if $m=0$) external lines and having vertices of the type shown in Fig. 5.5. Determine a mathematical expression associated with each diagram by the following items, and sum up those expressions associated with all diagrams.

2) For each diagram, affix the sign factor $(-1)^m$ originating from $(-g)^m$ in the perturbational expansion (5.8.5).

3) Assign a direction to each particle line and each interaction line, and assign a wave vector and a quasienergy to each line. In doing so, make wave vectors and quasienergies entering into or departing from each vertex obey the conservation laws, and write dependent ones in terms of sum or difference (Fig. 5.5).

4) Affix the following factor to a particle line with a wave vector $\boldsymbol{k}$ and a quasienergy ξ_n

$$\mathscr{K}_0(\boldsymbol{k}, \mathrm{i}\,\xi_n) = \frac{1}{\mathrm{i}\,\xi_n + \mu - E(\boldsymbol{k})}\,. \tag{5.8.11}$$

5) Affix the factor Φ_q/V to an interaction line with the assigned wave vector $\boldsymbol{q}$ and quasienergy ζ_n.

6) If a diagram includes λ closed rings of particle lines, multiply by the factor η^λ described in Sect. 5.7.2 ($\eta=-1$ for fermions and $\eta=+1$ for bosons).

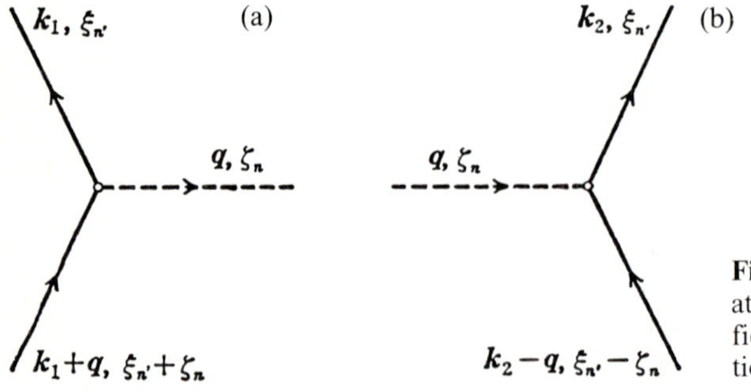

Fig. 5.5 a, b. Conservation laws at a vertex. (**a**) Emission of a field quantum, and (**b**) absorption of a field quantum

7) If a single-particle line closes to make a ring or both ends are connected with the same interaction line, affix to that particle line the convergence factor $\exp(\mathrm{i}\,\xi_n \Delta)$ $(\Delta \to +0)$ as the contribution from Δ or Δ' inserted in (5.8.7).

8) Sum up over m independent wave vectors $\boldsymbol{k}$ or $\boldsymbol{q}$ and m quasienergies ξ_n or ζ_n assigned to interior lines and interaction lines:

$$\sum_k kT \sum_{n=-\infty}^{\infty} \dots .$$

In Point (5) there are just m factors $1/V$, hence we can write the above sum over wave vectors as an integral in the thermodynamic limit.

As stated earlier, we need not take into account the factor $1/m!$ in the perturbational expansion (5.8.5) and the factor 1/2 relating to the interchange of $\boldsymbol{k}_1$ and $\boldsymbol{k}_2$ in the interaction (5.8.7), provided that we consider only topologically distinct diagrams. This makes our calculation very simple and easy to grasp.

Following to the above rules, let us look for an equation, called the *Dyson equation,* satisfied by the single-particle temperature Green's function for the model of a plasma repeatedly cited above as an example.

The diagram with $m=0$, i.e., of zeroth order, consists of a single-particle line and it is given by $\mathscr{K}_0(\boldsymbol{k}, \mathrm{i}\,\xi_n)$. The connected diagrams with $m=1$ are shown in Figs. 5.4c, d, but diagrams including parts of the "tadpole" form like 4c may be omitted because the interaction line concerned becomes that of $\boldsymbol{q}=0$ and has the factor $\Phi_0 = 0$. Consequently the first-order contribution comes from Fig. 5.6a, which is equivalent to

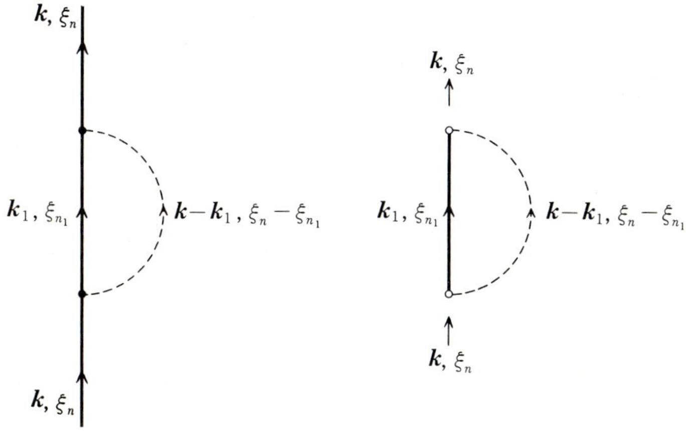

(a) $\mathscr{K}_1(\boldsymbol{k}, i\xi_n)$ (b) $\mathscr{M}_1(\boldsymbol{k}, i\xi_n)$

Fig. 5.6 a, b. The first-order diagram for $\mathscr{K}$. ○ denote exterior vertices

Fig. 5.4d, and is given by

$$\begin{aligned}\mathscr{K}_1(\boldsymbol{k}, \mathrm{i}\,\xi_n) &= (-1)^1\, \mathscr{K}_0(\boldsymbol{k}, \mathrm{i}\,\xi_n) \sum_{\boldsymbol{k}_1} kT \sum_{n_1=-\infty}^{\infty} \mathrm{e}^{\mathrm{i}\,\xi_{n_1}\Delta}\, \frac{\Phi_{\boldsymbol{k}-\boldsymbol{k}_1}}{V} \\ &\quad \times \mathscr{K}_0(\boldsymbol{k}_1, \mathrm{i}\,\xi_{n_1})\, \mathscr{K}_0(\boldsymbol{k}, \mathrm{i}\,\xi_n) \\ &= \mathscr{K}_0(\boldsymbol{k}, \mathrm{i}\,\xi_n)\, \mathscr{M}_1(\boldsymbol{k}, \mathrm{i}\,\xi_n)\, \mathscr{K}_0(\boldsymbol{k}, \mathrm{i}\,\xi_n)\,. \qquad (5.8.12)\end{aligned}$$

The factor $(-1)^1$ arises in accordance with Point (2). The convergence factor $\exp(\mathrm{i}\,\xi_n\Delta)$ is due to the prescription in Point (7).

The function $\mathscr{M}_1(\boldsymbol{k}, \mathrm{i}\,\xi_n)$ in the third member denotes the contribution from the part shown in Fig. 5.6b and is nothing but the first approximation of the mass operator defined below

$$\begin{aligned}\mathscr{M}_1(\boldsymbol{k}, \mathrm{i}\,\xi_n) &= (-1)^1 \sum_{\boldsymbol{k}_1} \frac{\Phi_{\boldsymbol{k}-\boldsymbol{k}_1}}{V}\, kT \sum_{n_1=-\infty}^{\infty} \exp(\mathrm{i}\,\xi_{n_1}\Delta)\, \mathscr{K}_0(\boldsymbol{k}_1, \mathrm{i}\,\xi_{n_1}) \\ &= -\int \frac{d\boldsymbol{k}_1}{(2\pi)^3}\, \Phi_{\boldsymbol{k}-\boldsymbol{k}_1}\, kT \sum_{n_1=-\infty}^{\infty} \frac{\exp(\mathrm{i}\,\xi_{n_1}\Delta)}{\mathrm{i}\,\xi_{n_1} + \mu - E(\boldsymbol{k})} \\ &= -\int \frac{d\boldsymbol{k}_1}{(2\pi)^3}\, \Phi_{\boldsymbol{k}-\boldsymbol{k}_1}\, f(E(\boldsymbol{k}_1))\,. \qquad (5.8.13)\end{aligned}$$

To get the last member we used

$$\lim_{\Delta\to+0} kT \sum_{n=-\infty}^{\infty} \frac{\exp(\mathrm{i}\,\xi_n\Delta)}{\mathrm{i}\,\xi_n - E} = \frac{-\eta}{\exp(\beta E) - \eta}\,. \qquad (5.8.14)$$

This formula is proved by the following technique.

To evaluate the sum over points of a series $\mathrm{i}\,\xi_{n'}$ like those appearing in the above equation, the common means is to change the sum into a contour integral by taking into account that all $\mathrm{i}\,\xi_{n'}$ are solutions of (5.6.13), i.e., the poles of $1/\exp(\beta z) \pm 1]$:

$$kT \sum_{n'=-\infty}^{\infty} g(\mathrm{i}\,\xi_{n'}) = \int_C \frac{dz}{2\pi\mathrm{i}}\, \frac{g(z)}{\exp(\beta z) \pm 1}\,, \qquad (5.8.15)$$

provided that the contour C encircles all poles $\mathrm{i}\,\xi_{n'}$ ($n' = 0, \pm 1, \pm 2, \ldots$) either in the negative direction for fermions $\eta = -1$, or in the reverse direction for bosons $\eta = +1$, and $g(z)$ is analytic inside C, i.e., poles of $g(z)$ should be outside the contour C. The contour C so chosen can be reformed into C' which encircles only the poles of $g(z)$ in the positive or negative direction (Fig. 5.7).

The diagram for $\mathscr{M}_1(\boldsymbol{k}, \mathrm{i}\,\xi_n)$ has two *exterior vertices* as shown in Fig. 5.6b by open circles. The directions of the particle lines to be connected outside with the exterior vertices are specified by short arrows.

Figure 5.8 illustrates second-order diagrams, from which we may omit diagram including "tadpole" parts as in Fig. 5.9. In general, if a

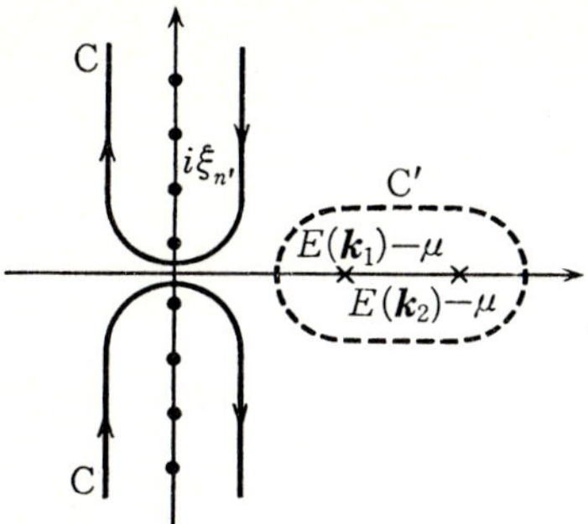

Fig. 5.7. The contours (C, C') for the integral (5.8.5). E denotes one of the poles of $g(z)$

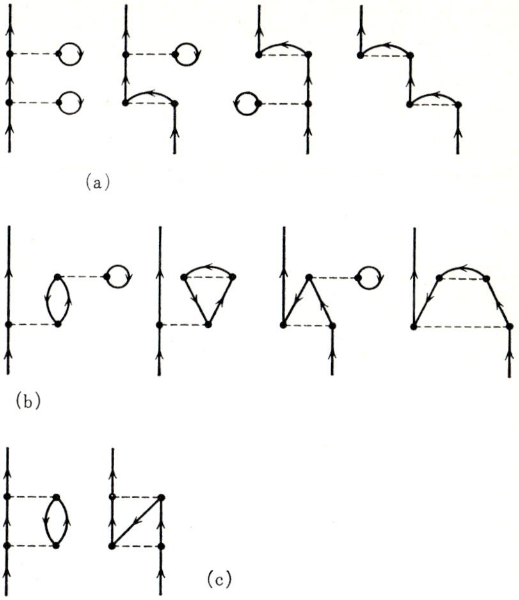

Fig. 5.8. (**a**) Repetition of first-order connected diagrams. (**b**) Examples obtained by insertion of first-order diagrams into first-order connected diagrams. (**c**) Properly second-order connected diagrams

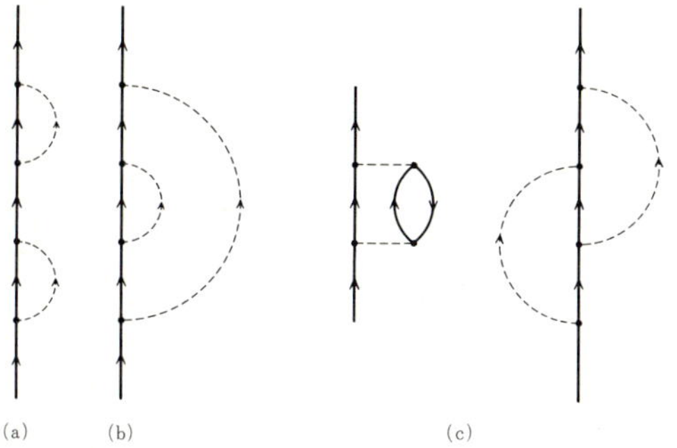

Fig. 5.9 a – c. Nonvanishing second-order diagrams. Different types of diagrams give different contributions (see text)

diagram is separated into two by cutting a single interaction line (as for the diagram second from the left in Fig. 5.8 b), that interaction line has the wave vector $\boldsymbol{q}=0$ and may be omitted. Figure 5.9 a has two $\mathscr{M}_1(\boldsymbol{k}, \mathrm{i}\,\xi_n)$'s connected in series through three $\mathscr{K}_0(\boldsymbol{k}, \mathrm{i}\,\xi_n)$'s and is found to give the following contribution

$$\mathscr{K}_0(\boldsymbol{k}, \mathrm{i}\,\xi_n)\,\mathscr{M}_1(\boldsymbol{k}, \mathrm{i}\,\xi_n)\,\mathscr{K}_1(\boldsymbol{k}, \mathrm{i}\,\xi_n) \tag{5.8.16}$$

through (5.8.12). Figures 5.9 b, c, if summed up, are obviously written in the form

$$\mathscr{K}_0(\boldsymbol{k}, \mathrm{i}\,\xi_n)\,\mathscr{M}_2(\boldsymbol{k}, \mathrm{i}\,\xi_n)\,\mathscr{K}_0(\boldsymbol{k}, \mathrm{i}\,\xi_n) \tag{5.8.17}$$

by introducing the second-order term $\mathscr{M}_2(\boldsymbol{k}, \mathrm{i}\,\xi_n)$ of the mass operator. Consequently, the second-order contribution is the sum of (5.8.16, 17) and becomes

$$\begin{aligned}\mathscr{K}_2(\boldsymbol{k}, \mathrm{i}\,\xi_n) = \mathscr{K}_0((\boldsymbol{k}, \mathrm{i}\,\xi_n)\,[\mathscr{M}_1(\boldsymbol{k}, \mathrm{i}\,\xi_n)\,\mathscr{K}_1(\boldsymbol{k}, \mathrm{i}\,\xi_n)\\ + \mathscr{M}_2(\boldsymbol{k}, \mathrm{i}\,\xi_n)\,\mathscr{K}_0(\boldsymbol{k}, \mathrm{i}\,\xi_n)]\,.\end{aligned} \tag{5.8.18}$$

Similarly, we can determine higher-order contributions and sum them up:

$$\begin{aligned}\mathscr{K}(\boldsymbol{k}, \mathrm{i}\,\xi_n) = \sum_{m=0}^{\infty} \mathscr{K}_m(\boldsymbol{k}, \mathrm{i}\,\xi_n) = \mathscr{K}_0(\boldsymbol{k}, \mathrm{i}\,\xi_n)\\ + \mathscr{K}_0(\boldsymbol{k}, \mathrm{i}\,\xi_n)\,\mathscr{M}_1(\boldsymbol{k}, \mathrm{i}\,\xi_n)\,[\mathscr{K}_0(\boldsymbol{k}, \mathrm{i}\,\xi_n) + \mathscr{K}_1(\boldsymbol{k}, \mathrm{i}\,\xi_n) + \dots]\\ + \mathscr{K}_0(\boldsymbol{k}, \mathrm{i}\,\xi_n)\,\mathscr{M}_2(\boldsymbol{k}, \mathrm{i}\,\xi_n)\,[\mathscr{K}_0(\boldsymbol{k}, \mathrm{i}\,\xi_n) + \dots]\\ + \dots\,.\end{aligned}$$

Since all diagrams of the perturbational expansion of $\mathscr{K}(\boldsymbol{k}, \mathrm{i}\,\xi_n)$ will appear below the lower exterior vertex of a mass operator, the contents of the square brackets in the above equation ought to be summarized in the form $\mathscr{K}(\boldsymbol{k}, \mathrm{i}\,\xi_n)$. Therefore, introducing the mass operator

$$\mathscr{M}(\boldsymbol{k}, \mathrm{i}\,\xi_n) = \sum_{m=-\infty}^{\infty} \mathscr{M}_m(\boldsymbol{k}, \mathrm{i}\,\xi_n)\,, \tag{5.8.19}$$

we obtain the *Dyson equation* [5.20]

$$\mathscr{K}(\boldsymbol{k}, \mathrm{i}\,\xi_n) = \mathscr{K}_0(\boldsymbol{k}, \mathrm{i}\,\xi_n) + \mathscr{K}_0(\boldsymbol{k}, \mathrm{i}\,\xi_n)\,\mathscr{M}(\boldsymbol{k}, \mathrm{i}\,\xi_n)\,\mathscr{K}(\boldsymbol{k}, \mathrm{i}\,\xi_n)\,. \tag{5.8.20}$$

When this equation is written in the following form, it elucidates the meaning of an approximation in which we retain only terms up to a certain order in the perturbational expansion of the mass operator (5.8.19):

$$[\mathscr{K}(\boldsymbol{k}, \mathrm{i}\,\xi_n)]^{-1} = [\mathscr{K}_0(\boldsymbol{k}, \mathrm{i}\,\xi_n)]^{-1} - \mathscr{M}(\boldsymbol{k}, \mathrm{i}\,\xi_n)\,. \tag{5.8.21}$$

Substituting the form (5.8.11) of $\mathscr{K}_0(\boldsymbol{k}, \mathrm{i}\,\xi_n)$, then

$$\mathscr{K}(\boldsymbol{k}, \mathrm{i}\,\xi_n) = \frac{1}{\mathrm{i}\,\xi_n + \mu - E(\boldsymbol{k}) - \mathscr{M}(\boldsymbol{k}, \mathrm{i}\,\xi_n)}\,. \tag{5.8.22}$$

Associating $\mathscr{K}(\boldsymbol{k}, \mathrm{i}\,\xi_n)$ with a thick line, the Dyson equation (5.8.20) may be represented by Fig. 5.10, in which diagrams are drawn horizontally.

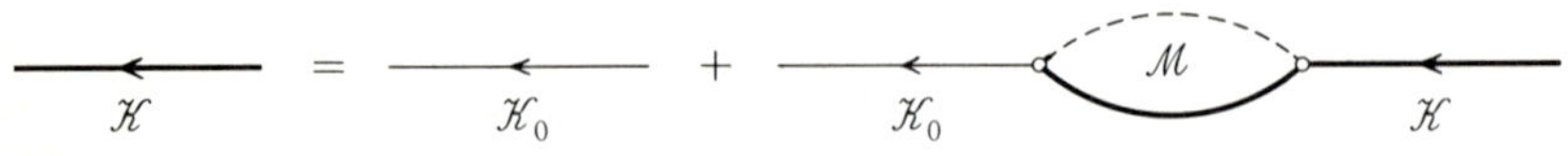

Fig. 5.10. The Dyson equation for $\mathscr{K}$

In all diagrams except the zeroth-order one, there is a part which is connected with the remainder of the diagram by two particle lines only. The simplest example is a part obtained by removing the two external lines. Such a part is called the *self-energy part,* borrowing the name used in quantum field theory. Among the self-energy parts there are those which are partitioned into two if we cut a certain particle line included in them. That obtained by removing the external lines from the diagrams shown in Fig. 5.9a is the simplest example. Such a self-energy part is said to be *reducible,* while that which is not reducible is called the *irreducible self-energy part.* We call the sum over all possible choices of irreducible self-energy parts the *mass operator.* This name also came from quantum field theory.

We can rewrite (5.8.20) in the following form

$$\mathcal{G}(\boldsymbol{k}, \tau; \boldsymbol{k}', \tau') = \mathcal{G}_0(\boldsymbol{k}, \tau; \boldsymbol{k}', \tau') + \int_0^\beta d\tau'' \int_0^\beta d\tau''' \sum_{\boldsymbol{k}'', \boldsymbol{k}'''} \mathcal{G}_0(\boldsymbol{k}, \tau; \boldsymbol{k}'', \tau'')$$
$$\times \mathcal{M}(\boldsymbol{k}'', \tau''; \boldsymbol{k}''', \tau''') \, \mathcal{G}(\boldsymbol{k}''', \tau'''; \boldsymbol{k}', \tau') \,. \tag{5.8.23}$$

Here $\mathcal{M}(\boldsymbol{k}'', \tau''; \boldsymbol{k}''', \tau''')$ is defined by the Fourier series like (5.6.12):

$$\mathcal{M}(\boldsymbol{k}'', \tau''; \boldsymbol{k}''', \tau''') = kT \sum_{n''} \exp[-\mathrm{i}\,\xi_{n''}(\tau'' - \tau''')] \, \mathcal{M}(\boldsymbol{k}'', \mathrm{i}\,\xi_{n''}) \, \delta_{\boldsymbol{k}'', \boldsymbol{k}'''} \,. \tag{5.8.24}$$

Remembering the unperturbed value (5.6.19), we may rewrite (5.8.20) as

$$\sum_{\boldsymbol{k}''} \delta_{\boldsymbol{k}\boldsymbol{k}''} \{[\mathrm{i}\,\xi_n + \mu - E(\boldsymbol{k})] - \mathcal{M}(\boldsymbol{k}, \mathrm{i}\,\xi_n)\} \, \mathcal{K}(\boldsymbol{k}'', \boldsymbol{k}'; \mathrm{i}\,\xi_n) = \delta_{\boldsymbol{k}\boldsymbol{k}'} \,. \tag{5.8.25}$$

5.9 Relationship Between the Thermodynamic Potential and the Temperature Green's Function

Various methods of obtaining the thermodynamic potential from the temperature Green's function have been devised: the following gives one of them. The reader is also referred to the work of *De Dominicis* and *Martin* [5.25].

Differentiating (5.5.11) for the thermodynamic potential with respect to the parameter g for perturbational expansion, by virtue of (5.5.9) then

$$\frac{\partial J}{\partial g} = -\frac{kT \langle \partial \hat{\mathcal{S}}(\beta)/\partial g \rangle_0}{\langle \hat{\mathcal{S}}(\beta) \rangle_0} = kT \int_0^\beta d\tau \, \frac{\langle T_\tau \{\mathcal{H}^{\mathrm{int}}(\tau) \, \hat{\mathcal{S}}(\beta)\} \rangle_0}{\langle \hat{\mathcal{S}}(\beta) \rangle_0} \,. \tag{5.9.1}$$

Assuming the form (5.7.1) of interaction, we can express this in terms of the two-particle temperature Green's function (5.6.26).

Since (5.8.25) suggests the Fourier transform of the equation of motion for the single-particle Green's function, let us follow the same procedure on the single-particle Green's function (5.8.1), i.e.,

$$\mathcal{G}(\boldsymbol{k}, \tau; \boldsymbol{k}', \tau') = -\theta(\tau - \tau') \langle \check{a}_{\boldsymbol{k}}(\tau)\, \check{a}^{\dagger}_{\boldsymbol{k}'}(\tau') \rangle - \eta \theta(\tau' - \tau) \langle \check{a}^{\dagger}_{\boldsymbol{k}'}(\tau')\, \check{a}_{\boldsymbol{k}}(\tau) \rangle ,$$

as that taken in Sect. 5.2 to set up the equation of motion for the double-time Green's function. Differentiating the above equation with τ gives

$$\frac{\partial}{\partial \tau} \mathcal{G}(\boldsymbol{k}, \tau; \boldsymbol{k}', \tau') = -\delta(\tau - \tau')\, \delta_{\boldsymbol{k}\boldsymbol{k}'} - \left\langle T_{\tau} \left\{ \frac{\partial \check{a}_{\boldsymbol{k}}(\tau')}{\partial \tau}\, \check{a}^{\dagger}_{\boldsymbol{k}'}(\tau') \right\} \right\rangle$$

via the commutation relations (5.7.2, 3). On the other hand, from (5.6.2)

$$\frac{\partial \check{a}_{\boldsymbol{k}}(\tau)}{\partial \tau} = [\mathsf{H}, \check{a}_{\boldsymbol{k}}(\tau)] . \tag{5.9.2}$$

Assuming the form of H as (5.5.1, 4.9, 7.1), then

$$\begin{aligned} \frac{\partial}{\partial \tau} \mathcal{G}(\boldsymbol{k}, \tau; \boldsymbol{k}', \tau') = & -\delta(\tau - \tau')\, \delta_{\boldsymbol{k}\boldsymbol{k}'} - \{E(\boldsymbol{k}) - \mu\}\, \mathcal{G}(\boldsymbol{k}, \tau; \boldsymbol{k}', \tau') \\ & + \frac{1}{V} \sum_{\boldsymbol{k}_2, \boldsymbol{k}_3, \boldsymbol{k}_4} \Phi_{\boldsymbol{k}-\boldsymbol{k}_4}\, \delta_{\boldsymbol{k}+\boldsymbol{k}_2, \boldsymbol{k}_3+\boldsymbol{k}_4}\, \mathcal{G}(\boldsymbol{k}_3, \tau, \boldsymbol{k}_4, \tau; \boldsymbol{k}_2, \tau + \Delta, \boldsymbol{k}', \tau') \end{aligned} \tag{5.9.3}$$

by calculating the commutator. As before, we should take the limit $\Delta \to +0$ at the end of the calculations. Here, too, is a characteristic of the many-body problem: in an equation for the single-particle Green's function, another function of a higher order, i.e., the two-particle temperature Green's function, appears. Changing to the Fourier coefficients, (5.9.3) becomes

$$\begin{aligned} & [\mathrm{i}\,\xi_n + \mu - E(\boldsymbol{k})]\, \mathcal{K}(\boldsymbol{k}, \boldsymbol{k}'; \mathrm{i}\,\xi_n) \\ & = \delta_{\boldsymbol{k}\boldsymbol{k}'} - \frac{1}{V} \sum_{\boldsymbol{k}_2, \boldsymbol{k}_3, \boldsymbol{k}_4} \Phi_{\boldsymbol{k}-\boldsymbol{k}_4}\, \delta_{\boldsymbol{k}+\boldsymbol{k}_2, \boldsymbol{k}_3+\boldsymbol{k}_4} \int_0^{\beta} d\tau\, \mathrm{e}^{\mathrm{i}\xi_n \tau}\, \mathcal{G}(\boldsymbol{k}_3, \tau, \boldsymbol{k}_4, \tau; \boldsymbol{k}_2, \tau + \Delta, \boldsymbol{k}', 0). \end{aligned}$$

Comparing this with (5.8.25) results in the relationship between the mass operator and the two-particle Green's function

$$\begin{aligned} & \int_0^{\beta} d\tau'' \sum_{\boldsymbol{k}''} \mathcal{M}(\boldsymbol{k}, \tau; \boldsymbol{k}'', \tau'')\, \mathcal{G}(\boldsymbol{k}'', \tau''; \boldsymbol{k}', \tau') \\ & \quad = -\frac{1}{V} \sum_{\boldsymbol{k}_2, \boldsymbol{k}_3, \boldsymbol{k}_4} \Phi_{\boldsymbol{k}-\boldsymbol{k}_4} \delta_{\boldsymbol{k}+\boldsymbol{k}_2, \boldsymbol{k}_3+\boldsymbol{k}_4}\, \mathcal{G}(\boldsymbol{k}_3, \tau, \boldsymbol{k}_4, \tau; \boldsymbol{k}_2, \tau + \Delta, \boldsymbol{k}', \tau') . \end{aligned} \tag{5.9.4}$$

From (5.9.1), by using (5.9.4), we arrive at

$$\frac{\partial J}{\partial g} = -\frac{kT}{2g} \int_0^{\beta} d\tau \sum_{\boldsymbol{k}} \int_0^{\beta} d\tau'' \sum_{\boldsymbol{k}''} \mathcal{M}(\boldsymbol{k}, \tau; \boldsymbol{k}'', \tau'')\, \mathcal{G}(\boldsymbol{k}'', \tau''; \boldsymbol{k}, \tau) . \tag{5.9.5}$$

Making the Fourier transformation and using (5.8.20), then

$$\begin{aligned}\frac{\partial J}{\partial g} &= -\frac{kT}{2g}\sum_{n=-\infty}^{\infty}\sum_{k}\mathscr{M}(\boldsymbol{k}, \mathrm{i}\,\xi_n)\,\mathscr{K}(\boldsymbol{k}, \mathrm{i}\,\xi_n)\\ &= -\frac{kT}{2g}\sum_{n=-\infty}^{\infty}\sum_{k}\frac{\mathscr{K}(\boldsymbol{k}, \mathrm{i}\,\xi_n)-\mathscr{K}_0(\boldsymbol{k}, \mathrm{i}\,\xi_n)}{\mathscr{K}_0(\boldsymbol{k}, \boldsymbol{k}; \mathrm{i}\,\xi_n)}\,. \end{aligned} \tag{5.9.6}$$

Noting that $J = J_0$ at $g = 0$, and integrating the above equation over g from 0 to g, we arrive at the relation connecting the thermodynamic potential to the temperature Green's function:

$$J = J_0 - \frac{kT}{2}\int_0^g \frac{dg}{g}\sum_{n=-\infty}^{\infty}\sum_{k}\frac{\mathscr{K}(\boldsymbol{k}, \mathrm{i}\,\xi_n)-\mathscr{K}_0(\boldsymbol{k}, \mathrm{i}\,\xi_n)}{\mathscr{K}_0(\boldsymbol{k}, \mathrm{i}\,\xi_n)}\,. \tag{5.9.7}$$

As already mentioned, for graphical summation the temperature Green's function is more convenient than the expansion formula (5.7.32). The troublesome weight $1/m$ for the mth-order diagrams appearing in the determination of the thermodynamic potential is taken care of by the integral over the parameter g: $\int g^m g^{-1}\,dg \propto 1/m$ if we expand in powers of g.

5.10 Special Case of Two-Particle Green's Function

According to the procedure described in Sect. 5.8.1, the two-particle temperature Green's function (5.6.26) can be written in the form

$$\begin{aligned}&\mathscr{G}(\boldsymbol{k}_1, \tau_1, \boldsymbol{k}_2, \tau_2; \boldsymbol{k}_3, \tau_3, \boldsymbol{k}_4, \tau_4)\\ &\quad = + \frac{\langle T_\tau\{\hat{a}_{\boldsymbol{k}_1}(\tau_1)\,\hat{a}_{\boldsymbol{k}_2}(\tau_2)\,\hat{a}^\dagger_{\boldsymbol{k}_3}(\tau_3)\,\hat{a}^\dagger_{\boldsymbol{k}_4}(\tau_4)\,\hat{\mathscr{S}}(\beta)\}\rangle_0}{\langle\hat{\mathscr{S}}(\beta)\rangle_0}\\ &\quad = + \langle T_\tau\{\hat{a}_{\boldsymbol{k}_1}(\tau_1)\,\hat{a}_{\boldsymbol{k}_2}(\tau_2)\,\hat{a}^\dagger_{\boldsymbol{k}_3}(\tau_3)\,\hat{a}^\dagger_{\boldsymbol{k}_4}(\tau_4)\,\hat{\mathscr{S}}(\beta)\}\rangle_0^{\mathrm{con}}\,. \end{aligned} \tag{5.10.1}$$

Therefore, we need only to sum up the terms associated with connected diagrams. In a diagram associated with the two-particle temperature Green's function, four lines correspond to two terminal points $(\boldsymbol{k}_1, \tau_1)$ and $(\boldsymbol{k}_2, \tau_2)$ and two initial points $(\boldsymbol{k}_3, \tau_3)$ and $\boldsymbol{k}_4, \tau_4)$. The zeroth-order term with respect to $g\,\mathscr{H}^{\mathrm{int}}$ is given, according to (5.7.10), by

$$\begin{aligned}&\mathscr{G}_0(\boldsymbol{k}_1, \tau_1, \boldsymbol{k}_2, \tau_2; \boldsymbol{k}_3, \tau_3, \boldsymbol{k}_4, \tau_4)\\ &\quad = \mathscr{G}_0(\boldsymbol{k}_1, \tau_1; \boldsymbol{k}_4, \tau_4)\,\mathscr{G}_0(\boldsymbol{k}_2, \tau_2; \boldsymbol{k}_3, \tau_3)\\ &\qquad + \eta\,\mathscr{G}_0(\boldsymbol{k}_1, \tau_1; \boldsymbol{k}_3, \tau_3)\,\mathscr{G}_0(\boldsymbol{k}_2, \tau_2; \boldsymbol{k}_4, \tau_4)\,. \end{aligned} \tag{5.10.2}$$

The associated Feynman diagrams are considered to be two lines with which four external lines are connected pairwise (Fig. 5.11). There are two

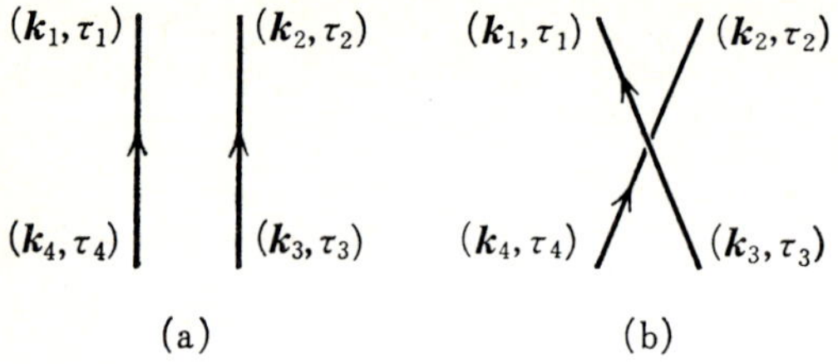

Fig. 5.11 a, b. The Feynman diagrams associated with two terms of the two-particle temperature Green's function of zeroth order

kinds of connected diagrams associated with expansion terms of first or higher order. Diagrams of the first kind are such that self-energy parts are attached to each of the two lines shown in Fig. 5.11, and those of the second kind are such that parts bridging the two lines are added. As discussed for connected diagrams for the single-particle temperature Green's function, to affix self-energy parts to particle lines and sum over terms corresponding to all possible types of self-energy parts is equivalent to replacing the single-particle temperature Green's function of zeroth order $\mathscr{G}_0$ with $\mathscr{G}$. Thus, the sum over all expansion terms associated with diagrams of the first kind yields a value obtained by using $\mathscr{G}$ in place of $\mathscr{G}_0$ on the right-hand side of (5.10.2). The *bridge parts* are the parts obtained by omitting from a diagram of the second kind four external lines or four parts, which are the four lines with self-energy parts attached to them. The *vertex part* is the sum of factors associated with all possible types of bridge parts, denoted by $\mathscr{V}(\boldsymbol{k}_1', \tau_1', \boldsymbol{k}_2', \tau_2'; \boldsymbol{k}_3', \tau_3', \boldsymbol{k}_4', \tau_4')$. Summation over expansion terms associated with diagrams of the second kind is carried out by partially summing up over bridge parts and over four parts connected with the four external lines. In this way, the perturbational expansion of (5.10.1) can be brought into the following form:

$$
\begin{aligned}
&\mathscr{G}(\boldsymbol{k}_1, \tau_1, \boldsymbol{k}_2, \tau_2; \boldsymbol{k}_3, \tau_3, \boldsymbol{k}_4, \tau_4) \\
&\quad = \mathscr{G}(\boldsymbol{k}_1, \tau_1; \boldsymbol{k}_4, \tau_4)\, \mathscr{G}(\boldsymbol{k}_2, \tau_2; \boldsymbol{k}_3, \tau_3) + \eta\, \mathscr{G}(\boldsymbol{k}_1, \tau_1; \boldsymbol{k}_3, \tau_3)\, \mathscr{G}(\boldsymbol{k}_2, \tau_2; \boldsymbol{k}_4, \tau_4) \\
&\quad\quad - \int_0^\beta d\tau_1' \int_0^\beta d\tau_2' \int_0^\beta d\tau_3' \int_0^\beta d\tau_4' \sum_{\boldsymbol{k}_1', \boldsymbol{k}_2', \boldsymbol{k}_3', \boldsymbol{k}_4'} \mathscr{G}(\boldsymbol{k}_1, \tau_1; \boldsymbol{k}_1', \tau_1') \\
&\quad\quad \times \mathscr{G}(\boldsymbol{k}_2, \tau_2; \boldsymbol{k}_2', \tau_2')\, \mathscr{V}(\boldsymbol{k}_1', \tau_1', \boldsymbol{k}_2', \tau_2'; \boldsymbol{k}_3', \tau_3', \boldsymbol{k}_4', \tau_4') \\
&\quad\quad \times \mathscr{G}(\boldsymbol{k}_3', \tau_3'; \boldsymbol{k}_3, \tau_3)\, \mathscr{G}(\boldsymbol{k}_4', \tau_4'; \boldsymbol{k}_4, \tau_4)\,. \qquad (5.10.3)
\end{aligned}
$$

Substituting (5.10.3) representing the two-particle Green's function by the vertex part in (5.9.4), the mass operator $\mathscr{M}$ is represented by the vertex part $\mathscr{V}$. Substituting $\mathscr{M}$ obtained in this way into (5.8.23 or 25), we obtain another form of equation which should be satisfied by the single-particle temperature Green's function. This equation is called the *Dyson equation*, which can be derived by direct substitution of the two-particle Green's function (5.10.3) into (5.9.3). The Dyson equation is usually solved by replacing the vertex part with its approximate value. Obviously this approximation is equivalent to summing up over terms associated with Feynman diagrams for which bridge parts are of a special type in the perturbational expansion of the single-particle Green's function.

Among the first-order terms in $g\,\mathscr{H}^{\text{int}}$ in the perturbation expansion (5.10.1) of two-particle Green's function, those associated with diagrams obtained by bridging the diagrams of Fig. 5.11 with one interaction line are

$$-\int_0^\beta d\tau' \frac{1}{V} \sum_{\boldsymbol{k}_1', \boldsymbol{k}_2', \boldsymbol{k}_3', \boldsymbol{k}_4'} \mathscr{G}_0(\boldsymbol{k}_1, \tau_1; \boldsymbol{k}_1', \tau_1')\, \mathscr{G}_0(\boldsymbol{k}_2, \tau_2; \boldsymbol{k}_2', \tau')\, (\Phi_{\boldsymbol{k}_1'-\boldsymbol{k}_4'} + \eta\, \Phi_{\boldsymbol{k}_1'-\boldsymbol{k}_3'})$$
$$\times\, \delta_{\boldsymbol{k}_1'+\boldsymbol{k}_2', \boldsymbol{k}_3'+\boldsymbol{k}_4'}\, \mathscr{G}_0(\boldsymbol{k}_3', \tau'; \boldsymbol{k}_3, \tau_3)\, \mathscr{G}_0(\boldsymbol{k}_4', \tau'; \boldsymbol{k}_4, \tau_4)\,. \tag{5.10.4}$$

Here we have assumed that $\Phi(-\boldsymbol{r}) = \Phi(\boldsymbol{r})$ and hence $\Phi_{-\boldsymbol{k}} = \Phi_{\boldsymbol{k}}$. Comparing the above expression with the integral term on the right-hand side of (5.10.3), the lowest approximation of the vertex part is given by

$$\mathscr{V}_0(\boldsymbol{k}_1', \tau_1', \boldsymbol{k}_2', \tau_2'; \boldsymbol{k}_3', \tau_3', \boldsymbol{k}_4', \tau_4')$$
$$= \frac{1}{V}(\Phi_{\boldsymbol{k}_1'-\boldsymbol{k}_4'} + \eta\, \Phi_{\boldsymbol{k}_1'-\boldsymbol{k}_3'})\, \delta_{\boldsymbol{k}_1'+\boldsymbol{k}_2', \boldsymbol{k}_3'+\boldsymbol{k}_4'}\, \delta(\tau_1' - \tau_4')\, \delta(\tau_1' - \tau_2')\, \delta(\tau_3' - \tau_4')\,. \tag{5.10.5}$$

5.10.1 Two-Particle Green's Function of Zeroth-Order for a Plasma

Let us look at the problem of a plasma discussed in Sects. 5.2, 3 from the standpoint of the diagram technique. As mentioned in Sect. 5.6.4, the temperature Green's function corresponding to the basic double-time Green's function (5.2.7) is a special form (5.6.28) of the double-temperature Green's function (5.6.26). We shall examine what form of Feynman diagrams in the perturbational expansion of the double-temperature Green's function should be summed up to obtain the decoupling approximation described in Sect. 5.2. The details are given in Sect. 5.10.3; here we examine the analytic continuation of the zeroth-order term in the perturbational expansion, given by

$$\mathscr{G}_0(\boldsymbol{k}_1, \boldsymbol{k}_2, \tau; \boldsymbol{k}_3, \boldsymbol{k}_4, \tau')$$
$$= \mathscr{G}_0(\boldsymbol{k}_2, \tau; \boldsymbol{k}_3, \tau')\, \mathscr{G}_0(\boldsymbol{k}_4, \tau'; \boldsymbol{k}_1, \tau)$$
$$- \mathscr{G}_0(\boldsymbol{k}_2, \tau; \boldsymbol{k}_1, \tau + \Delta)\, \mathscr{G}_0(\boldsymbol{k}_4, \tau'; \boldsymbol{k}_3, \tau' + \Delta')\,, \tag{5.10.6}$$

where we have assumed fermions. Fourier transforming according to (5.6.16) yields

$$\mathscr{K}_0(\boldsymbol{k}_1, \boldsymbol{k}_2; \boldsymbol{k}_3, \boldsymbol{k}_4; \mathrm{i}\,\xi_n)$$
$$= \int_0^\beta d\tau\, \mathrm{e}^{\mathrm{i}\,\xi_n \tau}\, \mathscr{G}_0(\boldsymbol{k}_1, \boldsymbol{k}_2, \tau; \boldsymbol{k}_3, \boldsymbol{k}_4, 0)$$
$$= kT \sum_{n'=-\infty}^{\infty} \mathscr{K}_0(\boldsymbol{k}_2, \boldsymbol{k}_3; \mathrm{i}\,\xi_{n'})\, \mathscr{K}_0(\boldsymbol{k}_4, \boldsymbol{k}_1; \mathrm{i}\,\xi_{n'-n})\,. \tag{5.10.7}$$

With the value (5.6.19) of the Fourier coefficient $\mathscr{K}_0$ of the single-particle temperature Green's function $\mathscr{G}_0$ of the zeroth-order,

$$\begin{aligned}
&\mathscr{K}_0(\boldsymbol{k}_1, \boldsymbol{k}_2; \boldsymbol{k}_3, \boldsymbol{k}_4; \mathrm{i}\,\xi_n) \\
&= kT \sum_{n'=-\infty}^{\infty} \frac{\delta_{\boldsymbol{k}_2\boldsymbol{k}_3}}{\mathrm{i}\,\xi_{n'} + \mu - E(\boldsymbol{k}_2)} \, \frac{\delta_{\boldsymbol{k}_1\boldsymbol{k}_4}}{\mathrm{i}\,\xi_{n'} - \mathrm{i}\,\xi_n + \mu - E(\boldsymbol{k}_1)} \\
&= \frac{\delta_{\boldsymbol{k}_1\boldsymbol{k}_4}\,\delta_{\boldsymbol{k}_2\boldsymbol{k}_3}}{\mathrm{i}\,\xi_n + E(\boldsymbol{k}_1) - E(\boldsymbol{k}_2)} \, kT \sum_{n'=-\infty}^{\infty} \left[\frac{1}{\mathrm{i}\,\xi_{n'} + \mu - E(\boldsymbol{k}_1)} - \frac{1}{\mathrm{i}\,\xi_{n'} + \mu - E(\boldsymbol{k}_2)} \right],
\end{aligned} \tag{5.10.8}$$

where n' replaces $n' - n$ in the first term in square brackets on the right-hand side. The summation over n' can be carried out via (5.8.15) and the procedure described below it:

$$\begin{aligned}
&\mathscr{K}_0(\boldsymbol{k}_1, \boldsymbol{k}_2; \boldsymbol{k}_3, \boldsymbol{k}_4; \mathrm{i}\,\xi_n) \\
&= \frac{\delta_{\boldsymbol{k}_1\boldsymbol{k}_4}\,\delta_{\boldsymbol{k}_2\boldsymbol{k}_3}}{\mathrm{i}\,\xi_n + E(\boldsymbol{k}_1) - E(\boldsymbol{k}_2)} \int_{C'} \frac{dz}{2\pi\mathrm{i}} \, \frac{1}{\exp(\beta z) + 1} \left[\frac{1}{z + \mu - E(\boldsymbol{k}_1)} - \frac{1}{z + \mu - E(\boldsymbol{k}_2)} \right] \\
&= \delta_{\boldsymbol{k}_1\boldsymbol{k}_4}\,\delta_{\boldsymbol{k}_2\boldsymbol{k}_3} \frac{f(E(\boldsymbol{k}_1)) - f(E(\boldsymbol{k}_2))}{\mathrm{i}\,\xi_n + E(\boldsymbol{k}_1) - E(\boldsymbol{k}_2)},
\end{aligned} \tag{5.10.9}$$

where $f(E)$ stands for the Fermi distribution function (5.4.12). Transferring this equation into the double-time Green's function by applying the Abrikosov-Gor'kov-Dzyaloshinskii-Fradkin theorem yields the result (5.2.19).

5.10.2 Polarization Operator

Since we have represented the interaction by quanta of a Bose field in Sect. 5.8.2, an interaction line is considered to be a free particle line for the quantum. The corresponding single-particle temperature Green's function is given by (5.8.8) so that its Fourier coefficient $\mathscr{U}_0(\boldsymbol{q}, \mathrm{i}\,\zeta_n)$ becomes $\Phi_{\boldsymbol{q}}$ by virtue of (5.8.9):

$$\begin{aligned}
\mathscr{D}_0(\boldsymbol{q}, \tau; \boldsymbol{q}', \tau') &= \delta_{\boldsymbol{q},\boldsymbol{q}'}\,\Phi_{\boldsymbol{q}}\,\delta(\tau - \tau') \\
&= kT \sum_{n=-\infty}^{\infty} \exp[-\mathrm{i}\,\zeta_n(\tau - \tau')]\,\delta_{\boldsymbol{q},\boldsymbol{q}'}\,\mathscr{U}_0(\boldsymbol{q}, \mathrm{i}\,\zeta_n),
\end{aligned} \tag{5.10.10}$$

$$\therefore \mathscr{U}_0(\boldsymbol{q}, \mathrm{i}\,\zeta_n) = \Phi_{\boldsymbol{q}}. \tag{5.10.11}$$

Fig. 5.12. The Dyson equation for $\mathscr{U}$

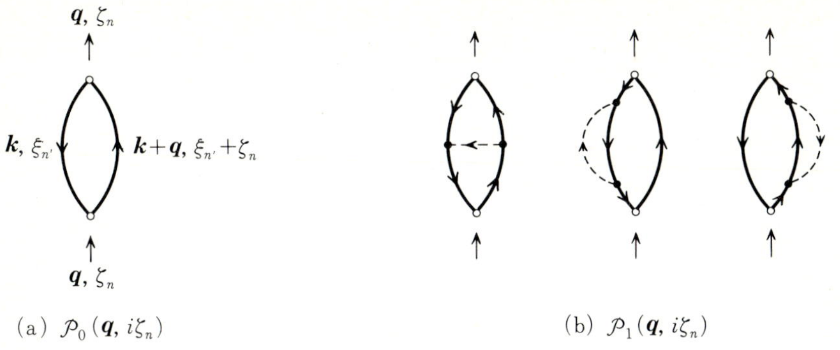

Fig. 5.13 a, b. Diagrams of zeroth and first-order for $\mathscr{P}$

This is the zeroth value of the Fourier coefficient $\mathscr{U}(\boldsymbol{q}, \mathrm{i}\,\zeta_n)$ of the single-particle temperature Green's function corresponding to the field quanta

$$\mathscr{D}(\boldsymbol{q}, \tau; \boldsymbol{q}', \tau') = kT \sum_{n=-\infty}^{\infty} \exp[-\mathrm{i}\,\zeta_n(\tau-\tau')]\,\delta_{\boldsymbol{q},\boldsymbol{q}'}\,\mathscr{U}(\boldsymbol{q}, \mathrm{i}\,\zeta_n), \tag{5.10.12}$$

which represents an interaction line associated with parts like the mass operator for the particle line. For the function (5.10.12) we can set up the Dyson equation like (5.8.20). The polarization operator $\mathscr{P}$ appears in place of the mass operator $\mathscr{M}$ (Fig. 5.12):

$$\mathscr{U}(\boldsymbol{q}, \mathrm{i}\,\zeta_n) = \mathscr{U}_0(\boldsymbol{q}, \mathrm{i}\,\zeta_n) + \mathscr{U}_0(\boldsymbol{q}, \mathrm{i}\,\zeta_n)\,\mathscr{P}(\boldsymbol{q}, \mathrm{i}\,\zeta_n)\,\mathscr{U}(\boldsymbol{q}, \mathrm{i}\,\zeta_n), \tag{5.10.13}$$

$$\therefore\ [\mathscr{U}(\boldsymbol{q}, \mathrm{i}\,\zeta_n)]^{-1} = [\mathscr{U}_0(\boldsymbol{q}, \mathrm{i}\,\zeta_n)]^{-1} - \mathscr{P}(\boldsymbol{q}, \mathrm{i}\,\zeta_n). \tag{5.10.14}$$

According to (5.10.11),

$$\mathscr{U}(\boldsymbol{q}, \mathrm{i}\,\zeta_n) = \frac{\Phi_{\boldsymbol{q}}}{\mathscr{E}(\boldsymbol{q}, \mathrm{i}\,\zeta_n)}, \tag{5.10.15}$$

where we have introduced

$$\mathscr{E}(\boldsymbol{q}, \mathrm{i}\,\zeta_n) = 1 - \Phi_{\boldsymbol{q}}\,\mathscr{P}(\boldsymbol{q}, \mathrm{i}\,\zeta_n). \tag{5.10.16}$$

Since a thick particle line corresponding to a field quantum yields, as it were, an *effective interaction*, $\mathscr{E}(\boldsymbol{q}, \mathrm{i}\,\zeta_n)$ is a *screening factor* for the Coulomb interaction $\Phi_{\boldsymbol{q}}$ according to (5.10.15), and is a function to be analytically continued with the complex dielectric function $\varepsilon[\boldsymbol{q}, \omega]$ described in Sect. 4.5.3. This is shown in Sect. 5.10.3.

The perturbational expansion of the polarization operator

$$\mathscr{P}(\boldsymbol{q}, \mathrm{i}\,\zeta_n) = \sum_{m=0}^{\infty} \mathscr{P}_m(\boldsymbol{q}, \mathrm{i}\,\zeta_n) \tag{5.10.17}$$

begins from the zeroth-order. Figure 5.13 depicts its diagrams of up to first order. Like the diagrams for $\mathscr{M}(\boldsymbol{k}, \mathrm{i}\,\xi_n)$, they have two exterior vertices. The contribution from the zeroth-order diagram is given by

$$\mathscr{P}_0(\boldsymbol{q}, \mathrm{i}\,\zeta_n) = \int \frac{d\boldsymbol{k}}{(2\pi)^3} kT \sum_{n'=-\infty}^{\infty} \mathrm{e}^{\mathrm{i}\,\xi_{n'}\Delta} \mathscr{K}_0(\boldsymbol{k}, \mathrm{i}\,\xi_{n'})\, \mathscr{K}_0(\boldsymbol{k}+\boldsymbol{q}, \mathrm{i}\,\xi_{n'+n})\,. \tag{5.10.18}$$

Calculations can be carried out by using (5.8.15) as in (5.10.8):

$$\mathscr{P}_0(\boldsymbol{q}, \mathrm{i}\,\zeta_n) = \int \frac{d\boldsymbol{k}}{(2\pi)^3} \frac{f(E(\boldsymbol{k})) - f(E(\boldsymbol{k}+\boldsymbol{q}))}{\mathrm{i}\,\zeta_n + E(\boldsymbol{k}) - E(\boldsymbol{k}+\boldsymbol{q})}\,. \tag{5.10.19}$$

5.10.3 Electric Charge Density Green's Function

In Sect. 5.2.2, where we wanted to make the relationship with the kinetic equation clear, we did not construct the double-time Green's function corresponding to the electric charge density $\hat{\varrho}_{\boldsymbol{k}}^{\mathrm{ind}}$. Now let us consider the corresponding double-temperature Green's function

$$\begin{aligned} e^2\, \mathscr{C}(\boldsymbol{k}, \tau; \boldsymbol{k}', \tau') &= -\langle T_\tau \{\check{\varrho}_{\boldsymbol{k}}^{\mathrm{ind}}(\tau)\, \check{\varrho}_{\boldsymbol{k}'}^{\mathrm{ind}\,\dagger}(\tau')\}\rangle \\ &= kT \sum_{n=-\infty}^{\infty} \exp[-\mathrm{i}\,\zeta_n(\tau-\tau')]\, \delta_{\boldsymbol{k},\boldsymbol{k}'}\, e^2 \mathscr{N}(\boldsymbol{k}, \mathrm{i}\,\zeta_n)\,, \end{aligned} \tag{5.10.20}$$

where we have separated the charge e^2 for later convenience. We may omit the case $\boldsymbol{k} = 0$ because of (5.2.6).

The perturbational expansion can be carried out by means of

$$\begin{aligned} \mathscr{C}(\boldsymbol{k}, \tau; \boldsymbol{k}', \tau') = -\sum_{\boldsymbol{q},\boldsymbol{q}'} \langle T_\tau \{&\hat{a}^\dagger_{\boldsymbol{q}-\boldsymbol{k}/2}(\tau+\Delta)\, \hat{a}_{\boldsymbol{q}+\boldsymbol{k}/2}(\tau) \\ &\times \hat{a}^\dagger_{\boldsymbol{q}'+\boldsymbol{k}'/2}(\tau'+\Delta')\, \hat{a}_{\boldsymbol{q}'-\boldsymbol{k}'/2}(\tau')\, \hat{\mathscr{S}}(\beta)\}\rangle_0^{\mathrm{con}} \end{aligned} \tag{5.10.21}$$

as for (5.10.1). The Feynman rules for the Fourier coefficient $\mathscr{N}(\boldsymbol{k}, \mathrm{i}\,\zeta_n)$ are obtained by modifying Points (1, 2 and 8) of the rules for $\mathscr{K}(\boldsymbol{k}, \mathrm{i}\,\xi_n)$ described in Sect. 5.8.3 as follows.

(1) Draw all topologically distinct connected diagrams composed of m interaction lines and $2m+2$ particle lines, which meet at two exterior vertices, and having vertices as shown in Fig. 5.5. Determine a mathematical expression specified by the following points for each diagram, and sum up those associated with all diagrams.

(2) For each diagram, attach the factor $(-1)^m$ resulting from the order $(-g)^m$ of the perturbational expansion, and the factor -1 due to the definition of (5.10.20). [This factor is not present in the two-particle Green's function (5.6.26)].

(8) Sum up over $m+1$ independent wave vectors $\boldsymbol{k}$ or $\boldsymbol{q}$ and $m+1$ independent quasienergies ξ_n or ζ_n affixed to particle lines or interaction

lines, respectively:

$$\sum_{k} kT \sum_{n=-\infty}^{\infty} \dots .$$

In Point (5) there are m factors $1/V$. If we rewrite the above $m+1$ summations over wave vectors into the integral forms in the thermodynamic limit, we thus obtain one remaining factor V, showing that $\mathscr{N}$ is an extensive quantity.

The definition of a connected diagram requires only that the diagram is connected with at least one of the two exterior vertices. Nevertheless, it should be noted that if two pieces each connected with a different exterior vertex are separated from each other as in the zeroth diagram Fig. 5.14, then $\boldsymbol{k}=0$. In that case the diagram may be omitted because of (5.2.6).

The zeroth-order contribution is, as evident from (5.10.21),

$$\begin{aligned} \mathscr{N}_0(\boldsymbol{k}, \mathrm{i}\,\zeta_n) &= (-1)^2 \sum_{\boldsymbol{k}_1} kT \sum_{n_1=-\infty}^{\infty} \exp(\mathrm{i}\,\xi_{n_1}\Delta)\, \mathscr{K}_0(\boldsymbol{k}_1, \mathrm{i}\,\xi_{n_1})\, \mathscr{K}_0(\boldsymbol{k}_1+\boldsymbol{k}, \mathrm{i}\,\xi_{n_1+n}) \\ &= V\mathscr{P}_0(\boldsymbol{k}, \mathrm{i}\,\zeta_n)\,. \end{aligned} \tag{5.10.22}$$

The factor $(-1)^2$ comes from -1 due to the definition of $\mathscr{C}$ and from -1 for a closed ring of a particle line. The corresponding diagram is shown in Fig. 5.13a. The first-order diagrams are shown in Figs. 5.13b, 15:

$$\mathscr{N}_1(\boldsymbol{k}, \mathrm{i}\,\zeta_n) = V\mathscr{P}_0(\boldsymbol{k}, \mathrm{i}\,\zeta_n)\frac{\Phi_{\boldsymbol{k}}}{V}\, V\mathscr{P}_0(\boldsymbol{k}, \mathrm{i}\,\zeta_n) + V\mathscr{P}_1(\boldsymbol{k}, \mathrm{i}\,\zeta_n)\,. \tag{5.10.23}$$

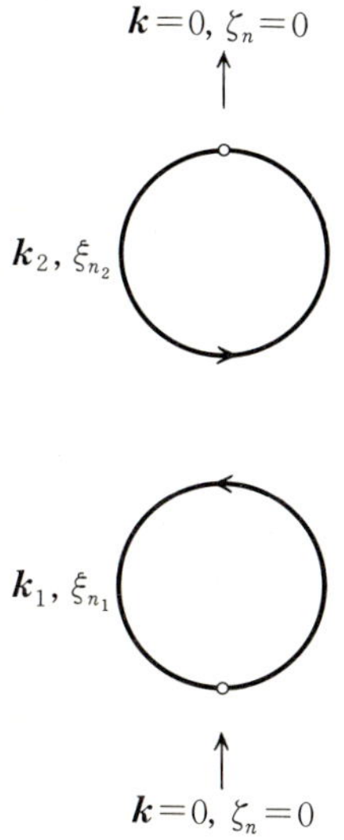

Fig. 5.14

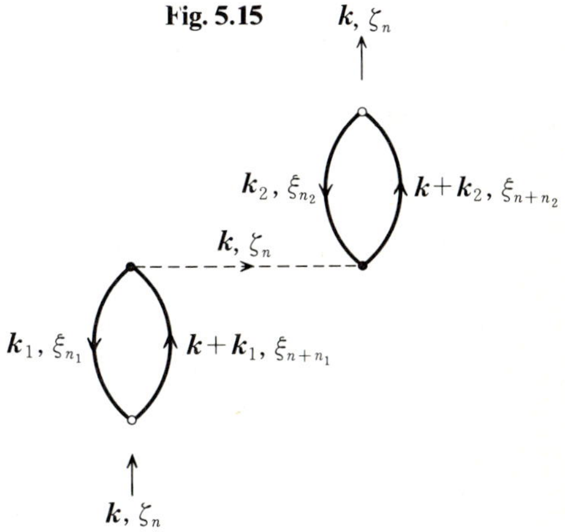

Fig. 5.15

Fig. 5.14. Zeroth-order connected diagrams to be omitted

Fig. 5.15. First-order connected diagrams corresponding to the first terms of (5.10.23)

As will be obvious on drawing higher-order diagrams, the Dyson equation is set up in the present case:

$$\mathcal{N}(\boldsymbol{k}, \mathrm{i}\,\zeta_n) = V\mathcal{P}(\boldsymbol{k}, \mathrm{i}\,\zeta_n) + V\mathcal{P}(\boldsymbol{k}, \mathrm{i}\,\zeta_n)\frac{\Phi_{\boldsymbol{k}}}{V}\mathcal{N}(\boldsymbol{k}, \mathrm{i}\,\zeta_n), \tag{5.10.24}$$

$$\therefore [\mathcal{N}(\boldsymbol{\varkappa}, \mathrm{i}\,\zeta_n)]^{-1} = [V\mathcal{P}(\boldsymbol{k}, \mathrm{e}\,\zeta_n)]^{-1} - \frac{\Phi_{\boldsymbol{k}}}{V}, \tag{5.10.25}$$

$$\therefore \mathcal{N}(\boldsymbol{k}, \mathrm{i}\,\zeta_n) = V\frac{\mathcal{P}(\boldsymbol{k}, \mathrm{i}\,\zeta_n)}{\mathcal{E}(\boldsymbol{k}, \mathrm{i}\,\zeta_n)}. \tag{5.10.26}$$

Let us return to the plasma problem discussed in Sects. 5.2, 3. As was stated in Sect. 5.6.4, the double-temperature Green's function (5.10.20) determines the double-time Green's function

$$\begin{aligned} e^2 C^{\mathrm{ret},-}(\boldsymbol{k}, t; \boldsymbol{k}', t') &= \theta(t-t')\left\langle \frac{1}{\mathrm{i}\hbar}[\hat{\varrho}_{\boldsymbol{k}}^{\mathrm{ind}}(t), \hat{\varrho}_{\boldsymbol{k}'}^{\mathrm{ind}\,\dagger}(t')]\right\rangle \\ &= \int_{-\infty}^{\infty}\frac{d\omega}{2\pi}\exp[-\mathrm{i}\,\omega(t-t')]\,\delta_{\boldsymbol{k},\boldsymbol{k}'}\,e^2 N^{\mathrm{ret}}[\boldsymbol{k}, \omega] \end{aligned} \tag{5.10.27}$$

by means of the analytical continuation. On the other hand, according to the theory of linear response described in Sect. 4.5.5, the Nozières-Pines formula (4.5.55) is written in the form

$$\frac{1}{\varepsilon[\boldsymbol{k}, \omega]} = 1 + \frac{\Phi_{\boldsymbol{k}}}{V}N^{\mathrm{ret}}[\boldsymbol{k}, \omega]. \tag{5.10.28}$$

Equation (5.10.26) can be written as

$$\frac{1}{\mathcal{E}(\boldsymbol{k}, \mathrm{i}\,\zeta_n)} = 1 + \frac{\Phi_{\boldsymbol{k}}}{V}\mathcal{N}(\boldsymbol{k}, \mathrm{i}\,\zeta_n) \tag{5.10.29}$$

by dividing both sides by $V\mathcal{P}(\boldsymbol{k}, \mathrm{i}\,\zeta_n)$ and making use of (5.10.24). This equation is of the same form as (5.10.28), and the complex dielectric function $\varepsilon[\boldsymbol{k}, \omega]$ is determined by the analytical continuation from $\mathcal{E}(\boldsymbol{k}, \mathrm{i}\,\zeta_n)$. Therefore, by virtue of (5.10.16) and by using a function $\Pi[\boldsymbol{k}, \omega]$ determined from $\mathcal{P}(\boldsymbol{k}, \mathrm{e}\,\zeta_n)$ by the analytical continuation, the complex dielectric function can be written as

$$\varepsilon[\boldsymbol{k}, \omega] = 1 - \Phi_{\boldsymbol{k}}\,\Pi[\boldsymbol{k}, \omega]. \tag{5.10.30}$$

This in an exact relation.

As is evident by comparison with (5.2.20) in which (5.2.19) is substituted, the decoupling approximation employed in Sect. 5.2 is nothing but the approximation of $\Pi[\boldsymbol{k}, \omega]$ in (5.10.30) by a function

$$\Pi_0[\boldsymbol{k}, \omega] = \lim_{\Delta\to+0}\int\frac{d\boldsymbol{k}_1}{(2\pi)^3}\,\frac{f(E(\boldsymbol{k}_1)) - f(E(\boldsymbol{k}_1+\boldsymbol{k}))}{\hbar(\omega+\mathrm{i}\,\Delta) + E(\boldsymbol{k}_1) - E(\boldsymbol{k}_1+\boldsymbol{k})} \tag{5.10.31}$$

analytically continued from the function (5.10.19)

$$\varepsilon[\boldsymbol{k}, \omega] = 1 - \Phi_k \Pi_0[\boldsymbol{k}, \omega] . \tag{5.10.32}$$

It is found that this approximation is, in the terminology of the diagram technique for the electric charge density Green's function (5.10.20), equivalent to summing up only over all special diagrams obtained by connecting the lowest-order polarization parts called "bubble", shown in Fig. 5.13a with interaction lines Φ_k having the same wave vector $\boldsymbol{k}$ as in Fig. 5.15. For $\boldsymbol{k} \approx 0$, $\Phi_k \propto \boldsymbol{k}^{-2}$ is very large, so that contributions from such diagrams are biggest for each order in the perturbational expansion. This is a reasonable method of partial summation, deserving consideration. These beautiful results are credited to *Larkin* [5.26].

Many questions remain to be discussed or solved: the relationship between the method of kinetic equations and the diagram technique, the relationship between the method of generalized Brownian motions and the diagram technique, and so on. As to the former question, we can, for example, easily generalize (5.3.16) into an exact expression including the exact dielectric function and obtain an exact kinetic equation within at least the linear approximation with respect to the external field. The latter question is suggested by comparing (5.10.28, 30) in the diagram technique with (4.5.54–57) in the theory of generalized Brownian motion. Hopefully such questions are a challenge to the reader.

General Bibliography of Textbooks

Abrikosov, A. A., Gor'kov, L. P., Dzyaloshinskii, I. E.: *Method of Quantum Field Theory in Statistical Physics* (Prentice-Hall, New York 1963)

Balescu, R.: *Equilibrium and Nonequilibrium Statistical Mechanics* (Wiley, New York 1975)

Bartlett, M. S.: *An Introduction to Stochastic Processes* (Cambridge U. Press, Cambridge 1956)

Bonch-Bruewich, V. L., Tjablikov, S. V.: *The Green Function Method in Statistical Mechanics* (North-Holland, Amsterdam 1962)

Economou, E. N.: *Green's Functions in Quantum Physics,* 2nd ed., Springer Ser. in Solid-State Sci. Vol. 7 (Springer, Berlin Heidelberg 1983)

Fetter, A. L., Walecke, J. D.: *Quantum Theory of Many-Particle Systems* (McGraw-Hill, New York 1971)

Groot, S. R. de, Mazur, P.: *Non-equilibrium Thermodynamics* (North-Holland, Amsterdam 1962)

Kadanoff, L. P., Baym, G.: *Quantum Statistical Mechanics* (Benjamin, New York 1962)

Kampen, N. G. van: *Stochastic Processes in Physics and Chemistry* (North-Holland, Amsterdam 1981)

Mahan, G. D.: *Many-Particle Physics* (Plenum, New York 1981)

Nelson, E.: *Dynamical Theories of Brownian Motion* (University Press, Princeton, NJ 1967)

Resibois, P., Leener, M.: *Classical Kinetic Theory of Fluids* (Wiley, New York 1977)

Rickayzen, G.: *Green's Functions and Condensed Matter* (Academic, New York 1980)

Thouless, D. J.: *The Quantum Mechanics of Many-Body Systems* (Academic, New York 1961)

Wax, N. (ed.): *Selected Papers on Noise and Stochastic Processes* (Dover, New York 1954)

Zubarev, D. N.: *Nonequilibrium Statistical Thermodynamics* (Consultants Bureau, New York 1974)

References

Chapter 1

1.1 R. Brown: Philos. Mag. **4,** 161 (1828); **6,** 161 (1829)
1.2 A. Einstein: Ann. Physik **17,** 549 (1905); **19,** 371 (1906);
Investigations on the Theory of the Brownian Motion, ed. by R. Fürth (Dover, New York 1956)
1.3 Min Chen Wang, G. E. Uhlenbeck: Rev. Mod. Phys. **17,** 323 (1945)
1.4 R. Kubo: *The Fluctuation-Dissipation Theorem,* Rep. Prog. Phys. **29** Part I, 255 (1966)
1.5 E. Kappler: Ann. Phys. **11,** 233 (1937)
1.6 W. Feller: *An Introduction to Probability Theory and Its Applications, Vol. 2* (Wiley, New York 1966);
H. Cramer: *The Elements of Probability Theory* (Wiley, New York 1955)
1.7 H. Cramer: *Mathematical Methods of Statistics* (University Press, Princeton, NJ 1945)
1.8 K. Toda, R. Kubo, N. Saito: *Statistical Physics I, Equilibrium Statistical Mechanics,* Springer Ser. Solid-State Sci. Vol. 30 (Springer, Berlin, Heidelberg 1982)
1.9 N. Wiener: Acta Math. **55,** 117 (1930);
A. I. Khintchine, Math. Ann. **109,** 604 (1934)
1.10 G. E. Uhlenbeck, L. S. Ornstein: Phys. Rev. **36,** 823 (1930)
1.11 L. L. Landau, E. M. Lifschitz: *Fluid Mechanics* (Addison-Wesley, Reading, MA 1959) p. 523;
A. Widom: Phys. Rev. A**3,** 1394 (1971);
K. M. Case: Phys. Fluid **14,** 2091 (1971)
1.12 B. J. Adler, T. E. Wainwright: Phys. Rev. A**1,** 18 (1970)
1.13 H. Nyquist: Phys. Rev. **32,** 110 (1928)
1.14 R. Kubo: J. Phys. Soc. Jpn. **12,** 570 (1957)
1.15 B. U. Felderhof: J. Phys. A**11,** 921 (1978)

Chapter 2

2.1 P. W. Anderson, P. R. Weiss: Rev. Mod. Phys. **25,** 269 (1954)
2.2 P. W. Anderson: J. Phys. Soc. Jpn. **9,** 316 (1954)
2.3 R. Kubo: J. Phys. Soc. Jpn. **9,** 935 (1954)
2.4 R. Kubo: In *Fluctuation, Relaxation and Resonance in Magnetic Systems,* ed. by D. ter Haar (Oliver and Boyd, London 1982);
R. Kubo: In *Stochastic Processes in Chemical Physics,* ed. by K. Shuler (Wiley, New York 1969)
2.5 N. Bloembergen, E. M. Purcell, R. V. Pound: Phys. Rev. **70,** 679 (1948);
J. H. Van Vleck: Phys. Rev. **72,** 1168 (1948)
2.6 R. Kubo: J. Math. Phys. **4,** 174 (1962)
2.7 H. Kramers: Physica **7,** 284 (1940)
2.8 W. Feller: *An Introduction to Probability Theory, Vol. 1* 2nd ed. (Wiley, New York 1962);
J. L. Doob: *Stochastic Processes* (Wiley, New York 1953)

2.9 R. Kubo: In *Perspectives in Statistical Physics*, ed. by J. Raveché (North-Holland, Amsterdam 1981)
2.10 A. Fokker: Dissertation, Leiden University (1913)
M. Planck: Berl. Ber. p. 324, 1927;
See references in G. E. Uhlenbeck, L. S. Ornstein: Phys. Rev. **36,** 823 (1930);
H. Risken: *The Fokker-Planck Equation*, Springer Syn., Vol. 18 (Springer, Berlin, Heidelberg 1984)
2.11 K. Ito: Proc. Imp. Acad. Tokyo **20,** 519 (1944);
K. Ito, H. McKean, Jr.: *Diffusion Processes and their Sample Paths* (Springer, Berlin, Heidelberg, New York 1965)
2.12 R. Stratonovich: J. SIAM Control **4,** 362 (1966)
2.13 R. E. Mortensen: J. Stat. Phys. **1,** 271 (1969);
M. Suzuki: Prog. Theor. Phys. Suppl. **69,** 160 (1980)
2.14 J. L. Doob: Ann. Am. Stat. **15,** 229 (1944); Ann. Math. **43,** 351 (1942);
M. C. Wang, G. E. Uhlenbeck: Rev. Mod. Phys. **17,** 323 (1945)
2.15 S. Nakajima: Prog. Theor. Phys. **20,** 948 (1958);
R. Zwanzig: J. Chem. Phys. **33,** 1338 (1960);
H. Mori: Prog. Theor. Phys. **33,** 424 (1965);
The projection operator method is essentially the same as the damping theory used in quantum mechanics: see W. Heitler: *The Quantum Theory of Radiation*, 2nd ed. (Oxford Univ. Press, Oxford 1944), and
A. Messiah: *Mechanique Quantique* Dunod, Paris 1959), *Quantum Mechanics* (North-Holland, Amsterdam 1961)
2.16 W. Pauli: In *Probleme der Modernen Physik, Festschrift zum 60. Geburtstag A. Sommerfelds*, ed. by P. Debye (Hirzel, Leipzig 1928)
2.17 L. Van Hove: Physica **21,** 517 (1955); **23,** 441 (1957);
R. Brout, I. Prigogine: Physica **22,** 621 (1956)
2.18 F. Englert: J. Phys. Chem. Solids **11,** 78 (1959). This work ist probably the first which formulated this problem in the form of (2.7.24). There have been a great number of papers which repeat essentially the same thing
2.19 W. Weidlich, F. Haake: Physik **185,** 30 (1965)
2.20 L. Boltzmann: *Vorlesungen über Gastheorie*, 2 Bde. (Barth, Leipzig 1912); *Lectures on Gas Theory* (Univ. of California Press, Berkeley 1964)
2.21 N. Bogolyubov: In *Studies in Statistical Mechanics, Vol. 1* ed. by G. E. Uhlenbeck, J. Boer (North-Holland, Amsterdam 1962)
2.22 J. R. Dorfman: In *Perspectives in Statistical Physics*, ed. by H. J. Raveché (North-Holland, Amsterdam 1981). This is a recent review of the problem
2.23 L. D. Landau: J. Exp. Theor. Phys. (USSR) **30,** 1058 (1956); Sov. Phys. **3,** 920 (1957);
D. Pines, P. Nozières: *The Theory of Quantum Liquids, Vol. 1* (Benjamin, New York 1966)
2.24 H. Mori: Prog. Theor. Phys. **33,** 424 (1965)
2.25 R. Kubo: Rep. Prog. Phys. **29,** Part I, 255 (1966)
2.26 J. Kirkwood: J. Chem. Phys. **14,** 180 (1946)
2.27 M. Huberman, G. V. Chester: Adv. Phys. **24,** 489 (1975) See also R. Kubo: In *Transport Phenomena*, ed. by G. Kirczenow and J. Maro, Lecture Notes Phys., Vol. 31 (Springer, Berlin, Heidelberg 1974)

Chapter 3

3.1 S. R. de Groot, P. Mazur: *Non-Equilibrium Thermodynamics* (North-Holland, Amsterdam 1963);
I. Prigogine: *Introduction to Thermodynamics of Irreversible Processes* 2nd ed. (Interscience, New York 1961)

3.2 H. Fröhlich: *Theory of Dielectric Constant and Dielectric Loss* (Oxford Univ. Press, Oxford 1949);
C. J. F. Böttcher: *Theory of Electric Polarization, Vol. I* (Elsevier, Amsterdam 1973);
C. J. Böttcher, P. Bordewijk: *Theory of Electric Polarization, Vol. II* (Elsevier, Amsterdam 1978);
L. Onsager: J. Am. Chem. Soc. **58,** 1486 (1936)
3.3 L. Onsager: Phys. Rev. **37,** 405 (1931); **38,** 2265 (1931);
H. B. G. Casimir: Rev. Mod. Phys. **17,** 343 (1945)
3.4 P. Debye: *Polare Molekeln* (Hirzel, Leipzig 1929)
3.5 J. H. Van Vleck, V. F. Weisskopf: Rev Mod. Phys. **17,** 227 (1945)
3.6 A. G. Redfield: IBM Journal **1,** 19 (1957)
3.7 F. Bloch: Phys. Rev. **70,** 460 (1946);
A. Abragam: *The Principle of Nuclear Magnetism* (Clarendon, Oxford 1961);
C. P. Slichter: *Principles of Magnetic Resonance,* Springer Ser. Solid-State Sci. Vol. 1, (Springer, Berlin, Heidelberg, New York 1978)
3.8 R. Kubo, N. Hashitsume: Prog. Theor. Phys. Suppl. **46,** 210 (1970)
3.9 J. M. Luttinger: Phys. Rev. **135**A, 1505 (1964)
3.10 R. Kubo, M. Ichimura: J. Math. Phys. **13,** 1454 (1972)
3.11 M. P. Vernotte: C. R. (Paris) **246,** 3154 (1958);
D. Forster: *Hydrodynamic Fluctuations, Broken Symmetry and Correlations* (Benjamin, London 1975)
3.12 M. Chester: Phys. Rev. **131,** 2013 (1963);
J. B. Brown, D. Y. Chung, P. W. Mathews: Phys. Lett. **21,** 241 (1963)

Chapter 4

4.1 M. Toda, R. Kubo, N. Saitô: *Statistical Physics I, Equilibrium Statistical Mechanics,* Springer Ser. Solid-State Sci. Vol. 30 (Springer, Berlin, Heidelberg 1983)
4.2 J. Kirkwood: J. Chem. Phys. **7,** 39 (1939);
H. Fröhlich: *Theory of Dielectrics, Dielectric Constant and Dielectric Loss.* (Oxford Univ. Press, Oxford 1949)
4.3 R. Kubo: J. Phys. Soc. Jpn. **12,** 570 (1957)
4.4 M. Suzuki: Physica **51,** 277 (1971);
P. Mazur: Physica **43,** 533 (1969);
H. Falk: Phys. Rev. **165,** 602 (1968);
R. M. Wilcox: Phys. Rev. **174,** 624 (1968)
4.5 L. Onsager: Phys. Rev. **37,** 405 (1931); **38,** 2265 (1931)
4.6 H. Nyquist: Phys. Rev. **32,** 110 (1928)
4.7 H. B. Callen, R. F. Welton: Phys. Rev. **86,** 702 (1952)
4.8 H. Takahasi: J. Phys. Soc. Jpn. **7,** 439 (1952);
R. Kubo: Int. J. Quantum Chem. Symp. **16,** 25 (1982)
4.9 R. Kubo: Rep. Prog. Phys. **29,** Part 1, 255 (1966)
4.10 J. M. Zimann: Adv. Phys. **13,** 89 (1964);
S. F. Edwards: Proc. Phys. Soc. **86,** 977 (1965);
M. Huberman, G. V. Chester: Adv. Phys. **24,** 489;
R. Kubo: in *Transport Phenomena,* J. Marro, ed. by Kirczenow, Lecture Notes Phys. Vol. 31 (Springer, Berlin, Heidelberg 1974)
4.11 P. Nozières, D. Pines: Nuovo Cimento X**9,** 470 (1958)
4.12 H. A. Kramers: tti congr. intern fis. Como **2,** 545 (1927)
R. de L. Kronig: J. Opt. Soc. Am. **12,** 547 (1926)
4.13 A. Einstein: Ann. Phys. (Leipzig) **19,** 371 (1906)
4.14 J. Kirkwood: J. Chem. Phys. **14,** 180 (1946); **15,** 72 (1947)
4.15 M. S. Green: J. Chem. Phys. **20,** 1281 (1952); **22,** 398 (1954)

4.16 R. Zwanzig: Phys. Rev. **124,** 983 (1961)
4.17 H. Mori, H. Fujisaka: Prog. Theor. Phys. **49,** 764 (1973);
H. Mori, H. Fujisaka, H. Shigematsu: Prog. Theor. Phys. **51,** 109 (1974)
4.18 M. Fixman: J. Chem. Phys. **36,** 310 (1962);
K. Kawasaki: Phys. Rev. **150,** 291 (1966); Ann. Phys. **61,** 1 (1970)
4.19 R. Kubo, M. Yokota, S. Nakajima: J. Phys. Soc. Jpn. **12,** 1203 (1957)
4.20 D. N. Zubarev: Sov. Phys. Dokl. **6,** 776 (1962); **10,** 452, 526, 850 (1965); *Nonequilibrium Statistical Thermodynamics* [transl. by P. J. Shepherd] (Consultants Bureau, New York 1974)
4.21 J. A. MacLennan, Jr.: Phys. Fluids **3,** 493 (1969); **4,** 1319 (1961); Phys. Rev. **115,** 1405 (1959)
4.22 B. Robertson: Phys. Rev. **144,** 151 (1966); **164,** 175 (1967); J. Math. Phys. **11,** 2482 (1970)
4.23 H. A. Kramers, W. Heisenberg: Z. Phys. **31,** 681 (1925)
4.24 N. G. Van Kampen: Phys. Norv. **5,** 279 (1971)
4.25 P. M. Bakshi, E. P. Gross: Ann. Phys. **49,** 513 (1968)
4.26 L. P. Kadanoff, P. C. Martin: Ann. Phys. **24,** 419 (1963)
4.27 B. J. Alder, T. E. Wainwright: Phys. Rev. A**1,** 18 (1968)
4.28 J. R. Dorfman: In *Fundamental Problems in Statistical Mechanics III*, ed. by E. G. D. Cohen (North-Holland, Amsterdam 1975)
4.29 B. J. Alder, W. E. Alley: Phys. Today **37,** Jan. 56 (1984);
W. E. Alley, B. J. Alder: Phys. Rev. A**27,** 3158 (1983)

Chapter 5

5.1 N. N. Bogolyubov, S. V. Tyablikov: Sov. Phys. Dokl. **4,** 589 (1959)
5.2 V. L. Bonch-Bruevich, S. V. Tyablikov: *The Green Function Method in Statistical Mechanics* (North-Holland, Amsterdam 1962)
5.3 D. N. Zubarev: Sov. Phys. Usp. **3,** 320 (1960)
5.4 A. L. Fetter, J. D. Walecka: *Quantum Theory of Many-Particle Systems* (McGraw-Hill, New York 1971), pp. 21 – 25
5.5 S. V. Tyablikov: *Methods in Quantum Theory of Magnetism* (Plenum, New York 1967)
5.6 D. Bohm, D. Pines: Phys. Rev. **92,** 609 (1953);
D. Pines: *The Many-Body Problems* (Benjamin, Reading MA 1962)
5.7 L. P. Kadanoff, G. Baym: *Quantum Statistical Mechanics* (Benjamin Reading MA 1962)
5.8 S. Ichmaru: Rev. Mod. Phys. **54,** 1017 (1982)
5.9 J. Lindhard: Dan. Mat.-Fys. Medd. **28,** No 28 (1954)
5.10 E. Wigner: Phys. Rev. **40,** 749 (1932)
5.11 Iu. L. Klimontovich: Sov. Phys. JETP **6,** 753 (1958)
5.12 A. A. Vlasov: Zh. Eksp. Teor. Fiz. **8,** 291 (1938)
5.13 L. D. Landau: J. Phys. USSR. **10,** 25 (1946)
5.14 P. Debye, E. Hückel: Phys. Z. **24,** 185 (1923)
5.15 A. A. Abrikosov, L. P. Gor'kov, I. E. Dzyaloshinskii: *Method of Quantum Field Theory in Statistical Mechanics* (Prentice-Hall, New York 1963)
5.16 J. R. Schrieffer: *Theory of Superconductivity* (Benjamin, New York 1964)
5.17 G. C. Wick: Phys. Rev. **80,** 268 (1950)
5.18 R. P. Feynman: Phys. Rev. **76,** 749, 769 (1949)
5.19 T. Matsubara: Prog. Theor. Phys. **14,** 351 (1955)
5.20 F. J. Dyson: Phys. Rev. **75,** 486, 1736 (1949)
5.21 A. A. Abrikosov, L. P. Gor'kov, I. E. Dzyaloshinskii: Sov. Phys. JETP **9,** 636 (1959); **10,** 186 (1960)

5.22 E. S. Fradkin: Sov. Phys. JETP **9,** 912 (1959)
5.23 G. Baym, N. D. Mermin: J. Math. Phys. **2,** 232 (1961)
5.24 C. Bloch, C. De Dominicis: Nucl. Phys. **7,** 459 (1956); **10,** 181, 509 (1959)
5.25 C. De Dominics, P. Martin: J. Math. Phys. **5,** 15, 31 (1964)
5.26 A. I. Larkin: Sov. Phys. JETP **10,** 186 (1960)

Subject Index

Abel theorem 140
Abrikosov-Gor'kov-Dzyaloshinskii-Fradkin theorem 233, 234, 258
Adiabatic condition 149
Adiabatic susceptibility 149, 150
Advanced Green's function *see* Green's function
Admittance *see* Complex admittance
  static 117, 135, 149, 163
Admittance function, singularity of 200
Aftereffect function 114
Analytical continuation 138, 207, 225, 233, 257
Anticausality 206
Antisymmetric part of tensor 162, 163
Asymptotic expansion 145

Bloch-De Dominicis theorem 235, 238, 244
Bloch equation 128
  generalized 128
Boltzmann-Bloch (type transport) equation 197, 198
Boltzmann distribution function 123
Boltzmann equation 61, 90, 94, 191
  linearized 97
Born approximation 94
Bose statistics 201, 223
Boson 225, 231, 233
Boson field 247
Bridge part 256
Brownian motion 2, 27, 40, 48, 69, 73
  Einstein-Ornstein-Uhlenbeck theory of 35
  generalized 263
  harmonic 21
  ideal 69, 185
  of quantal oscillator 84
  of quantal spin 84
  on unit circle 44
  rotational 124, 130
  statistical theory of 38
Brownian particle 2, 14, 19, 20
Bubble 263
Burnett coefficients 201
Burnett equation 201
  super 201

Canonical average 205
Canonical correlation 106, 148, 154, 155, 158, 170, 171, 188, 189
  of density fluctuation 174
  positivity of 148
  stationarity of 154
  symmetry of 148
Canonical distribution 62
Canonical ensemble 147
Chain of equations of motion 208, 209
Cauchy distribution 10, 45
Cauchy integral 207
Cauchy integral formula 136
Cauchy principal value 137
Causal Green's function *see* Green's function
Causality 111, 138, 206
Central limit theorem 9, 13, 28
Chapman-Kolmogorov equation 57
Characteristic function 10, 22, 24, 25
  for normal distribution 11
  thermodynamic 11
Characteristic functional 25, 26, 28, 30
  of Gaussian process 25
Characteristic parameter (Kubo number) of random frequency modulation 42, 49
Chemical potential 174
Coarse graining (procedure) 51, 69, 112, 196
  in space 50
  of time 50, 55
  various levels of 202
Cole-Cole diagram 122, 127
Collision
  cross section of 92
  duration of 106
  invariant of 96
  inverse 94
  operator 97
  term 219

Commutation relations 211, 235
Complex admittance 32, 36, 116, 131, 133, 136, 138, 143, 153, 164–166, 170, 200, 204, 207
 angular frequency dependence of 163, 164
 magnetic field dependence of 163, 164
 of diffusion phenomenon 134, 135, 144
 real and imaginary parts of 118
 relation of complex admittance to relaxation function 116
 relation of complex admittance to response function 116
 wave number dependent complex admittance 131, 134
Complex conductivity 32, 155
Complex dielectric function 121, 126, 140, 183, 210, 213, 220, 221, 259, 262
 by Lindhard 215
 Debye-type 122, 139
 Van Vleck-Weisskopf-Fröhlich type 127, 130, 131, 139
 zeros of dielectric function 221
Complex susceptibility 156
 complex magnetic susceptibility 130
Compressibility, isothermal 174
Concept of nonequilibrium 197
Conductivity 180
 complex (electrical) 32, 140
 electronic 181, 200
 external 180
 formula 179
 Kubo formula for complex conductivity 55
 local 180
Conservation laws in diagram technique 247, 248
Constants of motion 157
Contraction
 of density matrix 84
 of information 69, 90
Correlation *see* Canonical correlation, symmetrized correlation
Correlation function 17, 18, 20–22, 24, 33, 41, 87, 103, 159
 characteristic times of 105
 computer simulation of velocity correlation function 36
 condition of the use of classical correlation function 167
 decay law of 188
 higher order 198
 long-time tail of 36, 201
 negative long tail of 105
 of temperature fluctuation 184
 of velocity 49–51, 103
 relation between correlation functions of true and random forces 103, 105
 spectral function for 159
 true force correlation 104
 vanishment of time integral of current-current correlation function 191
Correlation time 13, 20, 42
 of velocity 51
Coulomb interaction 211, 247
Cross correlation 186
Cumulant 12, 23, 24
 average 52
 correlation 201
 fourth cumulant correlation of velocity 201
 function 23
Current 119, 154, 161, 168, 171
 generalized 110
 random 175, 180
 systematic 175
Current density 172
 particle 195

Damping theory 72, 97, 99
Debye (-type) relaxation phenomena 121, 123
Debye screening radius 221
Debye shielding length 129
Decay rate 186
Decoupling approximation 209, 212, 213, 257, 262
Density 171
 of conserved quantity 193
 expansion in powers of 94
Density fluctuation
 canonical correlation of 174
 Langevin equation for 175
Density matrix 77, 149, 183
 at thermal equilibrium 84
 canonical 88, 147, 151
 contraction of 84
 equation of motion for 77, 150
 local equilibrium 186, 190, 194, 195
 long-time behavior of diagonal part of 84
 nonequilibrium 190, 196
 of stationary state 190, 193
 perturbational expansion of equilibrium density matrix 227
 reduced 199
Description
 degree of crudeness of 109
 incompleteness of 61
Detailed balance 6, 59

Diagram
  connected 241–243, 245, 246
  disconnected 241, 243, 245
  Feynman 235, 240
Diagram technique 235, 246, 263
  at absolute zero temperature 226
Diamagnetic current 138, 153
Dielectric constant 179, 180
  static 121, 124, 221
Dielectric function 181
  complex 121, 126, 140
  complex Debye-type 122, 131
  complex Van Vleck-Weisskopf-Fröhlich type 127, 130, 131
  static 124
Dielectric loss 140
Dielectric relaxation 120
  Debye-type 121
Dielectric shielding 179
Diffusion 5, 143
  complex admittance for diffusion (phenomenon) 135, 143, 144
  in low-dimensional fluids 201
  non-Markovian law of 143
Diffusion coefficient 134
  generalized 183
  nonlocal retarded 175
  rotational 125
Diffusion constant 2, 8, 13, 14, 31, 76, 175, 177, 183
  longitudinal 131
  static 140
  transverse 131
Diffusion equation 5, 6, 50, 62, 76, 134
  extended 202
  fundamental solution of 135
  in velocity space 29
Diffusion process 13, 30, 50, 69, 74, 183
Disconnected diagram *see* Diagram
Disconnected part 242
Dispersion electron 182, 183
Dispersion relation 34, 138, 166, 182, 207
  proof of 136, 164, 166
Displacement 146, 155, 169
  generalized 110
Distribution function 48, 61, 69, 83, 183
  Boltzmann 123
  canonical 100
  contracted 69
  one-particle 90, 93, 216, 219
  orientational 124, 125
  reduced 93, 186, 199
  spatial 74
  two-particle 93
Driving force 5
Dynamic coupling 187
Dynamic susceptibility 156, 157
Dynamical coherence 44
Dyson equation 256
  for electric charge density Green's function 260
  for single-particle temperature Green's function 249, 252
  for single-particle temperature Green's function for boson field 259
Dyson time-ordering operator 228

Effective field 176, 177
Effective Hamiltonian 181
Einstein relation 2, 6, 31, 37, 146, 183
  generalized 36, 178
Eigenvalue problem 58, 76
Elastic constant 140
Elastic loss 140
Electric charge distribution 216
Electric charge density 210
Electric conductivity *see* Conductivity
Electric displacement 120, 138, 179, 182
Electric polarization 120, 123, 124, 131, 182
Elementary excitation 222, 225
Energy conservation law 118
Energy current density 195
Energy density 194
Energy dissipation 31, 117, 140, 170, 171
Energy loss 131
Enskog-Chapman method 191
Entropy 96, 187–189
  density 95
  density of entropy flow 95
  generalized 61
  irreversible generation of 119
  irreversible production of 96
  maximum formulation 189
Equation of continuity 134
Equilibrium distribution 185
Equipartition law 21
Ergodic process 16
  nonergodic 16
Ergodic system 61
Ergodic theory 196, 197
Expectation
  equilibrium 147
  statistical 152
External conductivity 180
External line 245
External mobility 174, 177, 179
External parameter 146
External potential 171, 172, 175, 177

Fermi distribution (function) 215, 258
Fermi liquid, Landau's Fermi-liquid theory 97
Fermi statistics 211, 213, 223
Fermion 225, 231, 233, 235
Feynman diagram 235, 240, 255
Feynman rules 248, 260
Fick law of diffusion 134, 143
  generalized Fick law 143
Filter 17
Fluctuating force 88
  spectra of 88
Fluctuation-dissipation theorem (FD-theorem) 2, 31, 36, 37, 119, 146, 170, 171
  of the first kind 37, 103, 155, 171
  of the second kind 37, 103, 171, 175
  wider view of 171
Fluctuation of macrovariables 185
Fokker-Planck equation 62, 63, 66, 67, 185, 186
  applied to nonlinear problems 64
Fourier analysis 116
  into superposition of running plane waves 133
Fourier coefficient 15, 16, 27
Fourier integral 19
Fourier series 15, 231
Free energy 11
  generalized 62
  Helmholtz 147
Friction constant 5
Friction retardation 31
Friction(al) coefficient 14, 31, 65
  Kirkwood's expression of friction coefficient 105
Frictional force 5, 14
  retarded friction 37

Gaussian distribution 8, 22, 27, 29
Gaussian law of error 8, 9
Gaussian modulation 47
Gaussian process 20, 22, 42, 53, 67
  Markovian 68
  white 64
  linear transformation of 27
Gaussian property 27, 28
  non-Gaussian nature of critical fluctuations 27
Gaussian white noise 185
Generalized hydrodynamics 200, 201
Gibbs-Duhem relation 174
Golden rule 89
Graphical summation 246, 255
Green-Kubo formula 187
Green's function 158, 203, 209
  advanced 206
  auxiliary 223, 224
  boundary condition for 208, 226, 231
  causal 225, 226, 229, 231
  density 180
  double-temperature 234, 235, 257
  double-time 207, 232–234, 262
  electric charge density 260
  equation of motion for (double-time) Green's function 208, 212, 214, 219, 226, 255
  four groups of 203
  Fourier analysis of temperature Green's function 230
  Fourier coefficient of temperature Green's function 232, 234, 246, 258
  Fourier transform of equation of motion for 214
  Fourier transformation of 232–234
  many-particle 203
  multiparticle 234
  of anticommutator or commutator type 205, 206, 223
  of product type 205
  quadruple-temperature 235
  quadruple-time 235
  retarded 204, 207, 210, 217
  single-particle 215, 222, 223, 233
  single-particle Green's function for boson field 247
  single-particle Green's function for free particle 224
  single-particle temperature Green's function 232, 244, 246, 249, 254, 258
  single-particle temperature Green's function for boson field 259
  temperature 229, 231, 255
  two-particle 234
  two-particle temperature Green's function 253–255
  unperturbed single-particle 240
Green's method 187
Grand canonical ensemble 223
Grand canonical partition function 228
Gross variable 187
Gyromagnetic ratio 128

H-function 59–61, 96
H-theorem 59, 95
Half-width 127
Hamiltonian equations of motion 153, 198
Harmonic analysis 15, 19, 67, 116

Harmonic oscillator 68
average energy of 168
Hartree approximation 213
Heat bath
comment on two systems in contact with 89
system in contact with 62, 84
Heat capacity 150, 157
Heat conduction 143
Heat conductivity tensor 184
Heat dissipation 119
positivity of 119
Heaviside unit step function 113
Heisenberg-Kramers dispersion formula 197
Heisenberg equation of motion 98, 106, 208
Heisenberg operator 87, 152
Hermitian polynomial 76
Hidden variable 70
Hierarchy of equations 93
Hilbert transformation 138
Hydrodynamic backflow effect 35, 201
Hydrodynamic equation, validity of 201
Hydrodynamics, generalized 200

Idealization, levels of 5
Identity
for exponential operators 148
Kubo 154
Impedance, complex 36
Indecomposable set of states 60
Induced current 183
Induced space charge 180
Inner product (of operators) 99
In-phase component 117
Integral of motion, local 193
Integrodifferential equation 63
Intensity of noise 17
Intensity spectrum 17
Interaction
direction of interaction line 247
effective 259
line 241
magnitude of 82
range of 82
representation 80
representation of interaction by a boson field 247, 258
Interference effect 111
Interior line 244
Internal field 171
Interpolation formula 142
Invariant of motion, quasi-invariant of motion 193
Invariant part (of operator) 193
Ionic polarization 128
Irreversible process, linear 111
Irreversible thermodynamics 110, 163
Isothermal susceptibility 128
Ito-McKean interpretation of stochastic differentials 65

Jellium model 172

Kinetic coefficient 111, 112, 133, 163, 171, 184, 186, 187
Kinetic equation 199, 200, 216, 263
linearized 219
Kinetic method 197
versus linear response theory 196
Kinetic stage 219
Kinetic viscosity 35
Kirkwood formula
generalization of 157, 186
Kirkwood's expression of friction coefficient 105
Kirkwood's expression of static admittance 149
Klimontovich operator 216
Kramers equation 55, 84, 89, 124, 130
Kramers-Kronig relations 138, 181
Kubo formula 154, 170
of conductivity 155
Kubo identity 154

Landau damping constant 221
Langevin-Debye equation 124
Langevin equation 15, 20, 28, 31, 37, 52, 64, 68, 103, 175, 185, 190
generalized 31, 33, 64, 66, 97, 107
nonlinear 68
Langmuir plasma frequency 215
Larmor frequency 48, 131
Larmor precession 48, 131
Laser physics 84
Legendre polynomial 125
Linearization 197, 199
Linear relation 111, 133, 204
Fourier transform of 116
of non-Markovian form 112
non-Markovian nonlocal 132
Linear response formula 199
Linear response theory 103, 119, 197, 199
kinetic method versus 196
statistical mechanics of 146

Line shape of spectrum 40, 45
  Gaussian shape 43
  Lorentzian shape 45, 47
Liouville equation 51, 77, 86, 93, 186, 198
  in velocity space 52
  quantal 84
  stochastic 51, 52
Liouville operator 70, 98, 151, 153
  exponential operator of quantal Liouville operator 151
  quantal 150, 190
Local conductivity 180
Local field 177, 179
Local field correction 177
Local field potential 176
Local integral of motion 193
Long-time approximation 45, 47, 49, 50
Lorentzian distribution 45

Macroscopic linearity 198
Macrovariable 185, 187
  fluctuation of macrovariables 185
Markovian condition 56
  time range of Markovian description 77
Markovian law 112
Markovian process 7, 22, 40, 47, 53, 55, 62
  ergodic 59
  induced by binary collisions 81
  nonstationary 64
  stationary 57, 58
  uniform 83
Mass flow 195
Mass operator 250, 252, 253
  relationship between mass operator and two-particle Green's function 254
Master equation 61, 69, 82, 83, 93
  derivation of 77
Matsubara formula 242, 244
Matsubara-Green's function *see* Green's function
Maxwell-Boltzmann distribution 215, 222
Maxwellian distribution of velocity 30, 55, 67, 73, 96
Mean field potential 219
Mean free path 14, 49, 50, 75, 132
Mean free time 14, 106
Mechanical force 110
Memory
  duration time of 72
  function 72
  time 72
Method of graphical summation 246
Microscopic linearity 198
Mobility 6, 103
  complex 32, 36
  external 174, 177, 179
  external nonlocal mobility tensor 173
  local mobility (tensor) 175, 177
Modeling, stages of 61
Molecular collisions 61
Molecular field approximation 212
Moment 10, 12, 23, 24
  formulas 46
  of jump 63
  sum rules 139, 140, 142
Mori's method 103, 175
Motional narrowing 44, 47, 129

Narrowing condition 84
Natural frequency 20
Natural motion 87
Near-equilibrium process 197
Nonconservative force 196
Nonequilibrium statistical mechanics of dense gases 94
Nonlinear response theory 198
Nonohmic conduction 197
Normal distribution 8, 9, 12, 22
Nozières-Pines formula 180, 210, 262
Nuclear magnetic resonance 128
Nyquist formula 36, 146
Nyquist theorem 37, 170

Onsager reciprocity law 171, 184
Onsager reciprocity theorem 163, 164
Onsager relation of reciprocity 111
Onsager's postulate 184, 186, 188
  general formulation of 187
Optical response 197
  second order 198
Ornstein-Uhlenbeck process 28
Oscillator strength
  generalized 182
  sum rule 141

Paramagnetic relaxation 130
Particle line 240
  closed ring of 248, 261
Partition function 10
  grand canonical 228
Path integral 4
  formulation of quantum mechanics 1
Pauri equation 82
Pauri's exclusion principle 97
Perturbation 147
Perturbation(al) calculation 77, 79
Perturbational expansion
  basic formula for 228
  formula for temperature Green's function 245, 257

of (β) 238
of equilibrium density matrix 227
of mass operator 252
of polarization operator 259
of thermodynamic potential 228
of unitary transformation 227
Phase space distribution function 93, 151, 186
Phenomenological relation 110
Phenomenological theory 109
Plasma frequency 182, 221
Plasma oscillation 220
Poisson bracket 151, 152
physical meaning of 153
quantal 150
Poisson distribution 47
Poisson modulation 47
Poisson process 47
Polarization
circular 130
electric 120, 123, 124, 131, 182
ionic 128
Polarization operator 259, 260
Polar molecule 123
solution of polar molecules 123
Power loss 119, 120, 131, 140
Power spectrum 17, 19, 20, 33, 41
nonnegativity of 34
of noise voltage 36
white 20, 31
Principle of equal weight of microscopic states 196
Probability amplitude 222
Probability density 4
Probability distribution, joint 4
Projected process 69
non-Markovian property of 69
Projection 69, 70, 105, 175
of distribution function 70, 186
of process 109
Projection operator 71, 78, 85, 99, 106
Propagation velocity 144

Quadrature (out of phase) component 117
Quantized wave function 222
Quantum field theoretical methods in statistical mechanics 203
Quasi-invariant of motion 193
Quasienergy 248
Quasimomentum 248
Quasiparticle 225

Random field 131
Random force 14, 15, 19, 20, 31, 33, 37–39, 54, 101, 103, 104, 107, 185, 186
Gaussian white 55, 67
noncorrelation with velocity 39
relationship between random and true forces 108
velocity dependent 65
Random frequency modulation 40, 46, 48, 49, 129
characteristic parameter of random frequency modulation (Kubo number) 42, 49
magnitude of 42
rate of 42
Random phase approximation 213
Random variable 3
bounded 104
sum of independent random variables 9, 26
Random walk 13
Randomization by linearization 198
Rate of change of perturbation 83
Relaxation (phenomenon) 114
after (sudden) removal of force 113, 114
Debye (-type) 121, 123
fundamental equation for treating relaxation and fluctuation of quantal systems 88
linear 69
longitudinal 129
macroscopic law of 186, 187
of density response 174
paramagnetic 130
transverse 128
when the initial condition is specified 115
Relaxation function 113, 114, 116, 133, 135, 143, 155, 156, 169, 174, 199
Fourier transform of 117, 166
Relaxation time 82, 87, 121, 125, 143
longitudinal 128, 130, 131
transverse 128, 131
Resistance, complex 36
Resistivity 179
formula 178, 179
formula in terms of correlation function of force 106
of conduction electrons 105
Resonance, nuclear magnetic 128
Resonance absorption 126
of the Van Vleck-Weisskopf-Fröhlich type 126
Response
current response to potential 173, 176
dynamic response to external force 150
in isothermal condition 149
instantaneous 112

Response (continued)
  intrinsic instantaneous 153
  of current 170
  of displacement 155, 169, 170
  static 115, 148
  thermodynamic 140
  to harmonic force 115
  to pulsed force 112, 113
  to thermal internal force 183
Response function 113, 114, 116, 121, 126, 132, 133, 143, 152, 154, 155, 168, 169, 200, 204
  asymptotic behaviour of 20
  Debye-type 124
  for negative time 162, 164
  Fourier transform of 117, 164
  initial values of response function and its derivatives 158
  reality of 162
  with exponential damping 121
Retardation function 32
Retardation kernel 190
Retardation of viscous resistance 35
Retarded friction 37, 39
Retarded Green's function *see* Green's function

Saturation phenomena 130
Screened Coulomb potential 222
Screening factor 259
Screening of external field 220
Second law of thermodynamics 124
Second sound 144
Sedimentation equilibrium 6
Self-consistent field approximation 217
Self-consistent field equation 220
  linearized 220
Self-energy of electrons 212
Self-energy part 253
  irreducible 253
  reducible 253
Shielding (of external potential) 176, 179
Shielding factor 176, 179
Shielding length, Debye 179
Short-time approximation 44, 45, 49, 50
Single-particle energy 225
Singularities at $k=0$, $\omega=0$ 200
Source term
  added to Liouville equation 200
Spatial translation 132
Spatial translational symmetry 246
Specific heat 184
Spectral function 139, 158, 163, 167
  analytical continuation of 207, 225, 233
  convenience of 168
  importance of 207
  of commutator or anticommutator 168, 225
  of correlation function 159, 162
  of Green's function 204, 205, 213, 214, 224, 226, 233, 234
  poles of 225
Spectral intensity 65
Spectral line shape 40
Spin diffusion 143
Spin relaxation 84
Spin resonance 84
Static admittance 117, 135, 149, 163
Static susceptivity 174, 177
Stationary process 15, 16
  stationary stochastic process 16, 19
Statistical physics (mechanics)
  aim of 50
  foundation of nonequilibrium statistical mechanics 197
  fundamental logical structure of 69
  fundamental problem of statistical mechanics 196
Stochastic equation 15
  linear 19
  stochastic differential equation 19
  stochastic Liouville equation 51, 52
Stochastic operator 51, 57
Stochastic process 3, 22, 40
Stochastic variable (random operator) 3
Stochastization 196–199
Stokes law 14
Stosszahl Ansatz (assumption for the collision frequency) 92
Strong collision model 199, 200
Sum rule 131, 142, 143, 145, 181, 182
  meaning of 142, 181
  moment 139, 140, 142
  oscillator strength 141
Superposition principle 111, 115, 130
Susceptibility
  adiabatic 149, 150, 156, 157
  complex 156
  complex magnetic 130
  dynamic 156, 157
  isolated 156
  isothermal 128, 148, 156, 157
  nonlocal density 173, 174
Symmetric part of tensor 162, 163
Symmetrized correlation 167, 168, 170
  of current 168
  of displacement 169
Symmetry
  in current response 162
  in displacement response 163

of correlation function 158
of kinetic coefficients 184
time reversal 161, 164

Tadpole (part) 249, 250
Temperature Green's function (Matsubara-Green's function) *see* Green's function
Thermal conduction 143
Thermal conductivity tensor 194
Thermal diffusivity 184
Thermal (internal) force 110, 146, 147, 183, 187, 189
Thermal noise 2
Thermal fluctuation
average regression of 184
in equilibrium 171
measure of 171
Thermodynamic(al) limit 190, 191, 193, 199, 209, 211, 213, 215, 216, 249
Thermodynamic potential 229, 242–244, 246, 253
relationship between thermodynamic potential and temperature Green's function 253, 255
Thermodynamic susceptibility 157
Time average 16
long-time average 193
Time reversal 94
time-reversal symmetry 161, 164, 184
Time scale of velocity change 76
Transient phenomena 129
Transition probability 4, 6, 7, 28, 29, 50, 56
Transition rate 62
for collision 82
Transport equation, conventional equation of transport 89
Transport phenomena
in highly disordered system 198
near critical points 187
True force 103–105, 107, 108

Uhlenbeck-Ornstein process 28

Van Hove limit 84
Van Kampen's objection 198
Variance matrix 24
Variance of displacement 30
Velocity correlation function 49–51
Vertex (part) 241, 247, 256
exterior vertex 250
lowest approximation of vertex part 257
Viscosity 14
Viscosity coefficient 196
Vlasov equation 220
linearized 213

White noise 106
Wick time-ordering operator 225, 228
Wiener-Khintchine theorem 17, 20, 33, 41, 158
Wiener process 1
Wigner (phase space) distribution (function) 216, 219

Zero-frequency part of Heisenberg operator 156